The Demography of Disasters

Dávid Karácsonyi · Andrew Taylor · Deanne Bird
Editors

The Demography of Disasters

Impacts for Population and Place

CSFK

CHARLES DARWIN UNIVERSITY
NORTHERN INSTITUTE
People. Policy. Place.

Springer

Editors
Dávid Karácsonyi
Geographical Institute, Research Centre for Astronomy and Earth Sciences
Hungarian Academy of Sciences
Budapest, Hungary

Andrew Taylor
Northern Institute
Charles Darwin University
Darwin, NT, Australia

Deanne Bird
Faculty of Life and Environmental Sciences, Askja
University of Iceland
Reykjavik, Iceland

ISBN 978-3-030-49919-8 ISBN 978-3-030-49920-4 (eBook)
https://doi.org/10.1007/978-3-030-49920-4

This Springer imprint is published by the registered company Springer Nature Switzerland AG
The registered company address is: Gewerbestrasse 11, 6330 Cham, Switzerland

Foreword

It is my honor to introduce the present book titled *The Demography of Disasters, Impacts for Population and Place* to the reader. While the title suggests a social science inquiry on disasters from the angle of demography, the subtitle of the book clearly reflects that the geographer's perspective is inevitable when analyzing disasters. Geography plays an important role in understanding complex interactions of human–earth systems. Geography also bears the potential to help and coordinate policy-making, addressing the challenges of growing entropy in our technological society. These advantages of geography come from the holistic understanding of the world which enables geographers to see the disaster in a spatial perspective. That perspective is particularly helpful in addressing issues of global climate change.

The climate emergency impacts Australia immensely, as it is facing longer and more intensive bushfire seasons. Nevertheless, climate change recently affects less disaster-prone nations, such as Hungary as well. Landslides, floods, extreme rainfalls along with desertification in certain areas of Hungary constitute a major hazard risk affecting the livelihood of several thousand people. We have experienced over a long period of time the floods of the Tisza river, the second major waterway of Hungary after the Danube. The floods damaged entire villages or even sometimes cities located on the floodplains, altering the regional prospects of Alföld (The Great Hungarian Plain). Engineering design failures and poor technical maintenance often interplay with extreme rainfall, making the casualties even more serious. This was the case in 2010, when the Ajka alumina plant reservoir accident happened, which was the most significant technological disaster in the recent history of Hungary. The disaster caused a red mud spill, killed ten people, injured hundreds, and contaminated significant area with the caustic substance. The Hungarian Academy of Sciences and in particular the Geographical Institute played an important role in investigating these disasters and providing mitigation strategies, which explains our interest in a broader international cooperation of disaster studies.

The volume you have in your hands is the first result deriving from that cooperation between the Geographical Institute of the Research Centre for Astronomy and Earth Sciences (Hungary) and the Northern Institute of Charles Darwin

University. A large international team came together to present this book including scholars from Asia-Pacific, the USA, and the post-Soviet countries providing a broad spectrum of understandings on disasters. The thirteen chapters of the book range from theoretical studies on demography of disasters to discussions of technological accidents such as Chernobyl. The present volume includes two chapters on the 2011 Tohoku earthquake considered the most significant disaster impacting Japan since World War II. Hurricane Katrina in 2005, bush and forest fires in California and in Russia, heat waves in Australia are also discussed in the book, each of which directly can be linked to climate change.

Each author has demonstrated their commitment to this unique and valuable book project for better understanding the disaster–demography interaction. I would like to acknowledge in particular the role Andrew Taylor played in this project. From the very beginning, he invested tremendous energy arranging the academic cooperation between our institutions, securing financial resources and managing the preparation of the manuscripts. I sincerely look forward to the future prospects in the disaster–demography study field in which the present volume will probably be considered as a significant step forward.

Dr. Károly Kocsis
Academician
Director of the Geographical Institute
Research Centre for Astronomy
and Earth Sciences
Hungarian Academy of Sciences
Budapest, Hungary

Preface and Acknowledgements

The present book is the result of a collaboration between the Geographical Institute of the Research Centre for Astronomy and Earth Sciences, Hungary, the Northern Institute of Charles Darwin University, Australia, and the Faculty of Life and Environmental Sciences, University of Iceland. The book was first proposed some time ago by fellow editor Andrew Taylor when we discussed demographic issues around disasters following a seminar talk I gave during my first visit to Darwin. Since then Andrew has put a lot of energy and enthusiasm into making this book happen. Along with his contributions as editor and author, he also generously supported the cooperation between our institutions, particularly my visit to Darwin when we reviewed and discussed the chapter manuscripts provided by authors from various backgrounds. In the meantime, Deanne Bird, who is currently commuting between Reykjavík and Melbourne with responsibilities in disaster mitigation, also joined our editorial team and helped us with her experience in the disaster study field and in negotiating with Springer, the publisher of this volume. She also helped to get additional authors on board and with reviewing their submissions.

This book discusses a new and broadening field which links demography and disaster studies. Traditionally, people have viewed disasters as shocks to everyday routines, the result of an unanticipated natural hazard event or as engineering failures and therefore the domain of technological rather than social sciences. Under these paradigms, the links between disasters and demography seem relatively simplistic and unidirectional, with the focus on estimating post-disaster populations, measuring mortality or understanding out-migration impacts. Nevertheless, disasters have the capacity to fundamentally alter population profiles at local and regional levels. Impacts vary according to the type, rapidity, and magnitude of the disaster, but also according to the pre-existing population profile and its relationships to the economy and society. In all cases, the key to understanding impacts and avoiding them in the future is to understand the relationships between disasters and population change, both prior to and after a disaster.

The aim of this book is to provide a comprehensive discussion on the demography of vulnerability and resilience in the face of disasters, the demography of risk from disaster impacts for vulnerable groups. Demographic methods can help in post-disaster population estimation and in managing and mapping people. In many cases, human migration is a common response to disaster, and hence, there is a link to spatial population dynamics as well. The demography of policy and practice around disaster mitigation is also an important part of the present volume along with profiling future risks and opportunities. The demography of climate change highlights the disaster–demography link in a specific way. In line with the multifarious nature of the disaster–demography nexus, our book takes an interdisciplinary approach, with chapters ranging from geography to gendered understanding of disaster resilience.

The Demography of Disasters seeks to advance both practical and theoretical insights into our understanding of the role of demography in planning for and mitigating impacts from disasters in developed nations. We hope that the book will provide policy-makers, disaster recovery experts, planners, and academics in the field with a wide range of examples demonstrating the importance of the interplay between demography and disasters in regions and spatially. This book will be of interest to social scientists across a range of fields. The book is also helpful for policy-makers, planners, practitioners of disaster management, and environmental agencies who are interested in a demographer's perspective on disaster. Academics and students in areas such as demography, disaster management, climate change, social policy, and human studies are also the targeted audience.

Most of the contributors to this book have come from academia, spanning a broad range of countries, including but not limited to Japan, Germany, USA, Sweden, Ukraine, and Russia. It has at times been a difficult task to coordinate such a broad and diverse team with different perspectives and scientific backgrounds. It is the editors' hope that through this book project, the team of authors will become part of an academic community which will in the future further contribute to our understanding of the disaster–demography nexus.

The editors have taken the view that human geography is closely related to the disaster–demography nexus, and hence, we have made extensive use of maps, the most common communication language of geographical science. Special attention has been paid to presenting a uniform cartographic design and layout throughout the book which I hope will make this volume more visually appealing to the reader.

I would like to acknowledge here the tremendous help given by Shelly (Shell) Worthington who assisted us tirelessly in proofreading and editing all the chapter manuscripts. Special thanks also go to Ferenc Probáld, Professor Emeritus of Eötvös Loránd University (Budapest, Hungary), my former Ph.D. supervisor, who was always ready to check and comment on my work. Also many thanks to Kang-tsung (Karl) Chang for his rigorous but helpful review comments, along with all the other reviewers of each chapter including Tony Barnes, Sigurd Dyrting, Kat Haynes, Richie Howitt, Anita Maertens, Jane Mullet, and Jan Salmon. This project would not have been possible without the support of Károly Kocsis, Director of the Geographical Institute in Budapest, and László Kiss, General Director of the Research Centre for Astronomy and Earth Sciences who also arranged funding from the Hungarian Academy of Sciences to cover some of the costs related to the present volume. Thanks also go to the Japan Society for the Promotion of Science

(JSPS) for the one year I spent at the International Research Institute of Disaster Science at Tohoku University in Japan which helped me greatly in gaining a better understanding of disaster science as a distinguished academic field.

I hope you enjoy reading this book on the various perspectives of the disaster–demography nexus. There are inevitably more approaches than what we were able to include in the following chapters some of which raise more questions than answers. However, I believe the collaborations featured in this project will push further the discussion on the demography of disasters.

Darwin, Australia
January 2020

Dávid Karácsonyi

Contents

Editors and Contributors

About the Editors

Dávid Karácsonyi has been working at the Geographical Institute of Hungarian Academy of Sciences (Budapest) since 2009 where he is currently a research fellow. He is a geographer, specialized on human and regional geography. Dávid is interested in the geography of peripheral and rural areas and has conducted his research in Ukraine and Belarus, studying rural spaces and their populations. Recently, he has spent two and half years in East Asia first at the International Research Institute of Disaster Science, Tohoku University, Japan, where he compared the impacts of disasters like Chernobyl and Fukushima on regional population trends, than at the Department of Geography, National Taiwan University. Currently, he is also a research associate at the Northern Institute of Charles Darwin University.

Andrew Taylor Northern Institute, Charles Darwin University, Darwin, Australia. Andrew researches the causes and consequences of population change for the Northern Territory of Australia and northern regions more broadly. He undertakes both quantitative and qualitative researches to understand impacts from policy, economic, and structural changes for communities. In his Ph.D., he investigated policy and theoretical implications from changing migration practices for Indigenous Territorians. Prior to academia, Andrew worked for a decade with the Australian Bureau of Statistics.

Deanne Bird is Research Specialist at the Faculty of Life and Environmental Sciences, University of Iceland. She specializes in evaluating societal resilience and vulnerability in relation to emergency preparedness, response, and recovery. Deanne has collaborated on a variety of projects with community groups and, government and non-government organizations and is Associate of Monash University Disaster Resilience Initiative, Monash University. Deanne also works as Senior Advisor—Engagement for the Victorian Department of Health and Human

Services within the Latrobe Health Innovation Zone. As Senior Advisor, Deanne is responsible for local project leadership, coordination, monitoring, and reporting relating to specific actions arising from the 2014 Hazelwood Mine Fire. Through her position at the University of Iceland, and as an editor of Demography of Disasters, Deanne is supported by the Nordic Centre of Excellence for Resilience and Societal Security—NORDESS, which is funded under the Nordic Societal Security Programme.

Contributors

Per Axelsson is Associate Professor at the Department of Historical, Philosophical and Religious Studies and affiliated researcher at the Centre for Demographic and Ageing Research and the Centre for Sami Research (Vaartoe) at Umeå University. Per publishes widely on the demography and health of Sami people in the eighteenth and nineteenth centuries. He has a particular interest in improving sources of data about Sami health and demography, and has undertaken comparative studies in Australia, New Zealand, and Canada among other places.

Jessica L. Barnes PLA, ASLA, graduated from the Disaster Resilience Leadership Academy at Tulane University in New Orleans in 2018. She is now the founder and principal at Third Coast Collaborative, a US-based landscape architecture firm dedicated to promoting community resilience and inclusive access to public spaces through design.

Doris A. Carson is an economic and human geographer interested in the socioeconomic development of communities in sparsely populated, rural, and remote areas. She completed her Ph.D. in 2011 at James Cook University in Australia, looking at the dynamics of regional tourism innovation systems in remote resource peripheries. Doris also holds a Master degree in tourism and leisure management from the IMC Krems University of Applied Sciences in Austria. Doris moved to Umeå in northern Sweden in 2014 to work on tourism and lifestyle-related mobilities in the Arctic and their impacts on innovation capacity in small rural communities. She is now a researcher and lecturer at the Department of Geography and Economic History at Umeå University, specializing in tourism, lifestyle mobilities and migration, and small community development in sparsely populated areas.

Dean B. Carson has spent the past 20 years researching who lives in, works in, and visits sparsely populated areas, and how and why these patterns change over time. Dean has worked in Australia, Canada, Scotland, Australia, and Sweden. He is currently the inaugural Visiting Professor at the Arctic Research Centre at Umeå University (Arcum) and Guest Professor with Sweden's Centre for Rural Medicine. His primary research interest is demographic change in small towns and villages.

Serhii Cholii is an associate professor (docent) at the chair of history, Igor Sikorsky Kyiv Polytechnic Institute. He is a historian, investigating the processes of implementation of different military strategies in Europe during the period of European Modernity and evolution of civil–military relations. Serhii is also interested in processes of forced migration during the nineteenth and twentieth centuries. During last years, he conducted field trips and interviews in different locations of European Polesia, studying the case of population relocation as an aftermath of Chernobyl power plant disaster in 1986.

Stephen T. Garnett is Professor of Conservation and Sustainable Livelihoods at the Research Institute for the Environment and Livelihoods, Charles Darwin University. Much of his research has been on threatened species, particularly birds, but he has studied and written on a wide variety of environmental issues. He is particularly interested in the synergies arising from combining biodiversity conservation and natural resource-based livelihoods, particularly Indigenous livelihoods.

Yetta Gurtner is a human geographer, lecturer, and researcher at James Cook University, Townsville. With a multidisciplinary background in Southeast Asian Studies, her core research interests include integrated disaster management, tourism crisis management, and planning for sustainable development/disaster risk reduction. Related research interests include social impact assessment and an appreciation of dynamic contexts of vulnerability, capacity, adaptation, mitigation, and resilience, particularly within developing nations.

Kazumasa Hanaoka is an associate professor at Department of Geography, Ritsumeikan University, Japan. He has been a deputy director of the Institute of Disaster Mitigation for Urban Cultural Heritage, Ritsumeikan University since 2018. From 2012 to 2016, he was an assistant professor at the International Research Institute of Disaster Science, Tohoku University, Japan. His main research areas are human geography and geographic information science, in particular spatial analysis using big data. His research activities cover a number of topics related to spatial microsimulation modeling, spatio-temporal analysis using mobile phone data, census data analysis on foreign population in Japan, and population decline and recovery after a large-scale natural disaster. Also, he has numerous experiences in fieldwork research in Japan, Philippines, Bangladesh, Belarus, and Ukraine.

David King is an Associate Professor of Geography in the College of Science and Engineering at James Cook University, and is director of the Centre for Disaster Studies, and the Centre for Tropical Urban and Regional Planning. He is a Fellow of the Planning Institute of Australia. He has taught and researched at James Cook University for 28 years and was formerly at the University of Papua New Guinea for ten years. He began his career as a school teacher on the diamond fields of Sierra Leone, turning that experience into a doctorate on the social impact of the diamond rush. His research specializes in social impact and evaluation in such areas as planning, natural hazard vulnerability, climate change adaptation and resilience, disaster mitigation and recovery, as well as the social impact of mining projects, development projects, and census analysis.

Michimasa Matsumoto International Research Institute of Disaster Science (IRIDeS), Tohoku University, Sendai, Japan

Osamu Murao is a professor at the International Research Institute of Disaster Science (IRIDeS) at Tohoku University, which was established in order to disseminate learning from the 2011 East Japan Earthquake and Tsunami Disaster, and the founder of the International Strategy for Disaster Mitigation Laboratory (ISDM). His current researches focus on post-disaster recovery process and urban design, and the relationship between physical environment (architecture and urban design) and disaster. To date, with research grants, he has investigated the post-disaster recovery process for damaged areas in Taiwan, Turkey, Sri Lanka, Thailand, Indonesia, Peru, Philippines, and World Trade Center in New York. Particularly he kept tracking the recovery process of Chi-Chi Township since the 1999 Chi-Chi earthquake in Taiwan as a visiting researcher of National Taiwan University in 2005. Dr. Murao has been involved in some research projects about post-disaster urban recovery and disaster risk reduction in the world.

Tatiana Nefedova is a chief researcher, Institute of Geography, Russian Academy of Science. Tatiana is interested in spatial organization and problems of agriculture and rural areas, interaction between the cities and countryside, Russia's suburbs and dacha suburbanization, social and economic problems of Russia's regions.

Tetsuya Okada is a Ph.D. graduate, who studied human geography at Macquarie University, Australia. His Ph.D. research project investigated community development through disaster recovery in four case study areas in Australia and Japan to better understand the sociocultural, economic, and political factors of and impacts on the local sociality. He has worked on various research projects including reports on disaster recovery for the National Climate Change Adaptation Research Facility (NCCARF), the impacts of flash flooding for the New South Wales State Emergency Service in Australia, along with other academic publications. In addition, Tetsuya is also a certified professional translator in Australia and has translated a number of technical writings on a wide range of natural hazards and associated disasters between English and Japanese.

Carmen Richerzhagen is a Senior Researcher at the German Development Institute in Bonn, Germany. She is an agricultural and environmental economist with the research focus on climate and biodiversity policies and ecosystem-based adaptation.

Ethan Sharygin Demographic Research Unit, California Department of Finance, USA. Ethan is research manager with the Demographic Research Unit at the State of California Department of Finance. His research focuses on modeling fertility, mortality, and migration rates, and other topics related to housing and population estimates and projections. Prior to joining the State of California, Ethan was a postdoctoral fellow with the all-cause mortality team at the University of

Washington Institute for Health Metrics and Evaluation (IHME). His previous studies involved marriage markets, imbalanced sex ratios at birth, and demographic consequences of famine. Ethan holds a Ph.D. in demography from the University of Pennsylvania.

Emma A. Singh Risk Frontiers, Department of Environmental Sciences, Macquarie University, New South Wales, Australia

Gabriella Sköld was Coordinator of the Arctic Research Centre at Umeå University (Arcum). She completed her Ph.D. at Umeå University in 2010, investigating fertility patterns among Sami populations in northern Sweden in the nineteenth century. Since then she has worked as a postdoctoral research fellow at the Centre for Sami Research (Vaartoe) at Umeå University, and has examined patterns of demographic change linked to hydro-power developments in the north of Sweden in the mid-twentieth century.

Peter Sköld is Director of the Arctic Research Centre at Umeå University (Arcum). Peter is an historian who has focused on researching the history of the Sami people of northern Sweden. His Ph.D. looked at the impact of smallpox on Sami populations during the Swedish colonization period. Peter has previously been the Director of the Centre for Sami Research (Vaartoe) at Umeå University and has served in a number of academic and political leadership positions in the Arctic and the Barents Region, including as Chair of the International Association of Arctic Social Scientists.

Yelizaveta Skryzhevska is an associate professor of Geography at Miami University of Ohio, USA. Her areas of specialization are human geography and Geographic Information Systems. Her research interests are in the fields of regional development of the post-Soviet countries such as Ukraine and Russia, and geography education. She has published numerous articles on inequalities of regional development and disparities in distribution of rural population in Ukraine, adaptation to climate change as a tool for regional development in Russia, and geography education in the US universities. She currently serves as Associate Dean for Academic Affairs at the College of Liberal Arts and Applied Science at Miami University Regionals.

Kerstin K. Zander is an Associate Professor at the Northern Institute at Charles Darwin University. She is interested in the economics of climate change impacts and adaptation, including migration. She is a Humboldt Fellow and an Adjunct Researcher at the German Development Institute in Bonn.

Chapter 1
Introduction: Conceptualising the Demography of Disasters

Dávid Karácsonyi and Andrew Taylor

Abstract Understanding and documenting intersects between disasters and human demographic change is an emerging academic field. Both the study of disasters and demographic issues are broad constructs in their own rights. While it may seem obvious to link the two, as we have in this book, disasters can impact on populations and population change in multifarious, obtuse and complex ways. Our aim in this book is to extend the nascent work to improve disaster policy and planning processes through enhancing knowledge about the demography-disaster nexus. In this chapter, we overview contemporary debates and paradigm shifts within the field of disaster studies to provide conceptual links between these and the field of demography. To conclude, we outline the topics and case studies which form the basis for individual chapters in this book.

Keywords Disaster-demography nexus · Hazard · Vulnerability · Social embeddedness · Non-routine event

1.1 Introduction

Understanding and documenting intersects between disasters and human demographic change is an emerging scientific field. This demography-disaster nexus is growing in importance as efforts to build structured and cohesive planning to prevent and mitigate disasters grow in line with the anticipated and realised increases in extreme events (Coleman 2006; Eshghi and Larson 2008; Okuyama and Sahin 2009). According to Donner and Rodríguez (2008, p. 1090), the contemporary growth in

---

D. Karácsonyi (✉)
Geographical Institute, CSFK Hungarian Academy of Sciences, Budaörsi út 45, Budapest 1112, Hungary
e-mail: karacsonyi.david@csfk.mta.hu

A. Taylor
Northern Institute, Charles Darwin University, Ellengowan Dr., Darwin, Casuarina NT 0810, Australia
e-mail: andrew.taylor@cdu.edu.au

D. Karácsonyi et al. (eds.), *The Demography of Disasters*,
https://doi.org/10.1007/978-3-030-49920-4_1

population densities in vulnerable areas and regions, and especially mass urbanisation in developing nations, is the most important factor that has increased population-related hazard exposure. In addition to this, the increased frequency and intensity of extreme weather events due to climate change (Katz and Brown 1992; Easterling et al. 2000; Frich et al. 2002; Coumou and Rahmstorf 2012; Jakab et al. 2019) have already contributed and soon will more significantly contribute to increased population vulnerabilities. Furthermore, global entropy has increased through technological change (Ellul 1964) and as a result of complex and systemic issues where nuclear, chemical, biotech industries and artificial intelligence along with 'traditional' threats such as pandemic, violence, war and famine bear the potential to trigger catastrophic events (Quarantelli et al. 2007). Each of these factors have the potential to alter demographic profiles and negatively affect people's lives through disasters.

The demographic make-up of populations at the time of disaster determines who is impacted and the extent of impacts on residents and others. Improving technologies, improved warning systems, investments in disaster mitigating infrastructure and improving community preparedness and response in the face of disasters are examples of attempts to reduce disaster impacts. While a range of studies have looked at demographic impacts from individual disasters, these are generally short-term focused and concentrated on post-event analysis and evaluation. Our aim in this chapter and generally in the book is to extend the nascent work to improve disaster policy and planning process through the growing knowledge about the demography-disaster nexus.

Both the study of disasters and demographic issues themselves are broad constructs in their own rights. While it may seem obvious to link the two, as we have in this book, disasters can impact on populations and change populations in multifarious, obtuse and complex ways. Populations may be the root cause of a localised disaster or indeed be the main 'victims'. Disasters can speed up pre-existing demographic changes or create new population profiles through immediate impacts and human reactions to such events (as we learn in Chap. 2 by Karácsonyi et al., in Chap. 5 by Carson et al. and in Chap. 7 by Bird and Taylor). Their impacts on demographic change can occur on a complex time–space continuum which may involve feedback loops. Most obvious are deaths and injuries, out-migration and the temporary relocation of residents and others from affected areas which rapidly and noticeably alter the pre-existing demographic profile of a town or region. More vulnerable cohorts (like the elderly) are often disproportionately impacted. Disasters may also affect the gendered demography of places, as discussed in Chap. 9 by Barnes. Disasters may encourage populations not immediately affected by the disaster at hand to adjust their demographic behaviours, such as around fertility or migration. Such feedback loops may be complex, unknown and unpredictable, altering not only the population profile but the economies and social fabric of towns and regions well into the future (see for example, Chap. 5 by Carson et al., Chap. 6 by King and Gurtner, or Chap. 8 by Zander et al.).

This introductory chapter discusses paradigm shifts in the field of disaster studies. This provides important theoretical context to posit our examination of the disaster-demography nexus which is the core theoretical contribution of the book. In the latter part of this chapter, we outline the structure of the book to facilitate a digestible overview for the reader.

1.2 Disaster Studies: From 'Acts of God' to a Distinguishable Scientific Field

In the face of the conventional perception of disasters as 'acts of God' (Robinson 2003) that are chaotic and random situations with anarchy and panic (Quarantelli and Dynes 1972; Webb 2007), the majority of disaster experts now perceive disasters are never 'accidental' (Quarantelli et al. 2007; Lavell and Ginnetti 2013). Thus, there is a need for systematic research and analysis to enhance the potential for the prevention of disasters. Samuel Prince's doctoral thesis (Prince 1920) on the 1917 Halifax explosion[1] is recognised by many as the first systematic disaster study (Drabek and McEntire 2003; Perry 2007). However, the real flourishing in disaster studies started during the early Cold War era with studies on the impact of bombings on civilian's morale in World War II (see United States Strategic Bombing Survey 1947). Many of these studies attempted to address how American communities might react to a possible nuclear war. Consequently, they had strong strategic-military origins and aims along with consummate funding provided by the armed forces (Bolin 2007; Perry 2007; Rodríguez et al. 2007). A notable step forward in the institutionalisation of civilian disaster studies was the establishment of the Disaster Research Centre at Ohio State University, the first of such schools, in 1963. The School published several synthesising works on disaster studies and played a pioneering role in developing theoretical frameworks rooted in the field of social sciences (see for example, Dynes 1970; Mileti 1975; Quarantelli 1978; Oliver-Smith 1979; Drabek 1986). This nascent adjunction between disasters and social sciences more than half a century ago was arguably the beginnings of the field we recognise today.

Although early links were evident between disaster studies and social sciences, disaster studies are multifarious and are strongly multidisciplinary by nature (Quarantelli 2006). Because of this, disaster studies are not ontologically limited to a clearly defined field (see for example the debate summarised in Perry and Quarantelli 2005). Alexander (2005, p. 24) proposed, for example, that disaster studies encompass the fields of geography, anthropology, sociology, social psychology, development studies, health sciences, geophysical sciences and engineering. Consequently, members of risk, hazard and disaster study communities with various backgrounds have often taken very different approaches which Cutter (2005) proposed made it

[1] A cargo ship loaded with munition exploded during the World War killing approximately 2000 people.

difficult to develop a recognisable and distinctive academic field with a universally understood inner logic and framework.

As such, the field of disaster studies is constantly emerging and reshaping. This brings many benefits. Not least, the knowledge and understanding about the risks of and impacts from disasters can be sourced and considered from a wide range of disciplines; including those reflected on here. The theoretical and applied 'stock' of knowledge is, therefore, not constrained to ontological paradigms, and its evolving nature protects disaster studies from the dangers of received wisdom which may be pervasive in singular and more focused fields. Disaster studies are, therefore, plural in a conceptual, theoretical and applied sense. While the benefits are many, this also brings challenges, as we will now discuss in relation to the debate on defining precisely what constitutes a disaster and therefore disaster studies.

1.3 Disaster: A Non-routine Phenomenon or Embedded in Society?

To meaningfully consider the disaster-demography nexus, it is necessary to plot the debate and literal discussion on what constitutes a disaster. As an illustration, Dynes (1970), Rodríguez et al. (2007) and Webb (2007) stressed that disaster is understood by many as an agent for bad luck to be bought to bear on the physical environment, human system and society in general. However, disasters are now widely recognised as complex events with society and populations intricately linked to the causes and consequences of individual disasters, rather than simply as victims or passive actors.

Two seminal books have sought to address the definition of a disaster; *What is a disaster?* by Perry and Quarantelli (2005) and *At risk*, by Blaikie et al. 1994 (its newer edition is Wisner et al. 2004). In the former, most contributors emphasised that a disaster can oftentimes be understood as 'departure from normal', a 'non-routine event', as echoed by Kreps (1989) and Drabek (1989). In Wisner's work, authors understood disasters as being embedded in the 'normal' functioning of the society and in particular rooted in social inequalities. The main criticism on this 'non-routineness' position is that, if disaster occurs when society cannot function normally under severe environmental change, then members of society are usually able to cope with everyday problems and needs when there is no external hazard stress on them (Donner and Rodríguez 2008, p. 1092).

McEntire (2013) distinguished two main schools in disaster studies, the social vulnerability school (emphasising social, political and economic structures) and the holistic school (in which disasters are seen as non-routine events, related to non-routine social problems). In the holistic school, for example, disaster definitions can be related to the scale and speed of disasters. The scale is emphasised by Quarantelli (1998, 2006) who distinguished a catastrophe from disaster and by Bissel (2013) who distinguished an emergency, hazard, disaster and catastrophe from each other based on the size of the impacted area and population. A further element in the definitional quandary is based on the different speeds of disaster impacts (Alexander 2001; Robinson 2003; Quarantelli 2006; Bissel 2013); those slowly emerging (such as sea level rise, industrial pollution, landscape degradation—see Kertész and Křeček 2019) and rapid onset disasters (e.g. hurricanes, earthquakes and floods).

Furthermore, in the holistic school (non-routineness perspective), the definitional debate has focused on the root causes of disasters including that they may be sudden, unexpected and unwanted events (Gencer 2013) and a result of natural or human made catastrophes (Robinson 2003). Alexander (2001) categorised disasters based on the geophysical agent, such as earthquake, volcanic eruption, desertification and soil erosion. Alexander (2001) and others, such as Kapuchu and Özerdem (2013), also stressed that natural and man-made disasters are usually very different. They highlighted that a disaster is a situation where people are confronted with a sudden and uncontrollable catastrophic change caused by natural phenomena or external human action. Hence, Kapuchu and Özerdem (2013, p. 12) pointed out that disasters induced by natural hazards can be predicted to some extent, but that it is impossible to prevent them.

In contrast to the stance on disasters by Alexander, Kapuchu and Özerdem, the majority of disaster experts (e.g. Quarantelli et al. 2007) agree that each disaster is influenced by humans and can therefore be prevented. This viewpoint stresses that human actions have the potential to reduce disaster impacts. Conversely, poor or wrong decisions can worsen the situation. In support of this, Perry (2007) bought to the fore definitional differences between the terms hazard and disaster. In his view, a disaster occurs when a hazard (an extreme physical event, such as extreme rainfall, heat wave or earthquake) intersects with the social system (vulnerable human population). In a practical sense, therefore, a landslide in a remote, uninhabited area is just a geological phenomenon if there is no injury, loss of life or property damage. It is important to emphasise that only a very small fraction of hazards lead to a disaster (Quarantelli et al. 2007) because usually there is no impact on human population (see Alexander 2001 on the 1964 Sherman debris avalanche in sparsely populated Alaska).

Robinson (2003), Wisner et al (2004) and Cutter (2005) argued that a disaster is an interplay between human–social root causes and hazard event and hence can lead to a combination of multiple events (e.g. conflict causing hunger, disease and displacement). Blaikie et al. (1994) and Wisner et al. (2004) used the ‘pressure and release model’ to show how disasters can occur when natural hazards affect vulnerable populations. In the model, vulnerability and natural hazards are seen as intersecting forces generating systemic pressure which are released in the form of a disaster. The pressure and release model suggests that, even if the triggering event is natural, the root causes are always society-based. This view helped shift attention to the vulnerability of human–social systems. As Cutter (2005, p. 39) stated, the question is not about a disaster but about vulnerability to environmental threats and extreme events. Cutter underlines that vulnerability is embedded in human, natural and technological systems, which are interconnected, and this interaction is of utmost importance in grasping vulnerability.

Alexander (2001) surmised that human vulnerability has often been a result of underestimating the limited degree to which technology and economic systems are able to mitigate *natural disasters*. In other words, natural hazard events have the power to exert a substantial and consistent influence on modern society (ibid. p. 3). Hence, in Alexander's perception, disasters induced by these events are in fact 'natural' disasters. Alexander's view is further supported by the recent cultural turn in disaster studies (see Webb 2007, Ekström and Kverndokk 2015). Consistent with that paradigm, there are some studies on how communities living in hazard-prone areas have been influenced through their continuous disaster experiences, including in Japan (Bajek et al. 2008; Kitagawa 2016) and the Philippines (Bankoff 2003). A cross-cultural study by Paton et al. (2010) comparing Japan and New Zealand (two disaster-prone nations) revealed some universal similarities on how hazard beliefs and social characteristics interact despite the fundamental cultural differences between the two countries.

In spite of the earth science-based criticisms by Alexander and the findings by recent cultural studies, Wisner's 'social embeddedness' view is widely accepted by disaster study experts with a sociology background. Hence, these scholars tend to avoid the use of the term 'natural disaster' (see O'Keefe et al. 1976; Cannon 1994; Cohen and Werker 2008). Despite this, the term 'natural disaster' is still common particularly in the economic, demographic and geographic literature (see for instance, UN General Assembly 1989; Cavallo and Noy 2011; Cavallo et al. 2013). Some authors have concluded that the 'natural disaster' events they studied have different impacts for developing nations (Kahn 2005; Toya and Skidmore 2007; Loayza et al. 2012; Chen et al. 2013), for women (Enarson 2000; Neumayer and Plümper 2007) or for low-income people (Masozera et al. 2007). These all emphasise that the severity of a disaster is related to social, rather than natural characteristics, in support of Wisner and his colleague's (2004) view that disasters are rooted in 'social' rather than in 'natural' causes.

Wisner et al. (2004, p. 92) went further to suggest a more 'radical' social embeddedness view, stating even nature itself can be considered as a part of the resources that are allocated by social processes, such that, under routine social functioning, people become more or less vulnerable to hazard impacts. Hence, vulnerability to natural hazards is a social construct (Lavell et al. 2012). According to Wisner, vulnerability describes a set of conditions that people derive from their historical, cultural, social, environmental, political and economic contexts, as well as socio-economic status. In Wisner's definition (ibid. p. 11), vulnerability is 'the characteristics of a person or group and their situation that influence their capacity to anticipate, cope with, resist and recover from the impact of natural hazard events'.

Wisner's assumption has been further elaborated on by Oliver-Smith (2009) who argued that "social systems generate the conditions that place people differentiated along axes of class, race, ethnic, gender or age, at different levels of risk from the same hazard and subject to different forms of suffering from the same event" (p. 120). Such perspectives were the first to mark direct links to population and their characteristics according to demographic traits such as age, race, gender and socio-economic status. Hence, vulnerable groups are not only at risk because they are exposed to a hazard but as a result of their marginality, of everyday patterns of social interaction and organisation and their constrained access to resources (Cardona et al. 2012, p. 71). Further, Wisner et al. (2004) accentuated that levels of vulnerability change during the life-course (e.g. partner formation, child bearing and later years) such that vulnerability for individuals may be temporary or in flux based on a range of factors in their lives including changes in occupation, immigration status or residence and again, the links to the field of demography become evident.

Of course, in the academic discourse, disasters are not only approached through root causes and triggering events, but also in terms of consequences. While the 'social embeddedness' approach accentuates social conditions as both the root cause and consequence from disasters, the 'non-routineness' perspective has a strong social science emphasis regarding consequences, which are understood as 'non-routine' social conditions generated by the disaster. These consequences may include significant loss of resources and threats to life (Frankenberg et al. 2014, p. 3), causing severe physical injuries, emotional distress and substantial property damage (Flanagan et al. 2011). According to Robinson (2003) and Smith (2005, p. 233), disasters change social and human life and livelihood in a negative way for certain groups hit directly or indirectly by the event, leading to serious disturbances and disruptions to the functioning of a society. In addition, Lindell (2013) summarised disaster consequences into two main groups, physical impacts (e.g. casualties, damage) and social impacts (psychological, demographic, economic and political impacts).

In summary, analysing literature in which disasters are defined, debated and understood highlights some successive paradigm shifts in disaster science (Fig. 1.1), the move away from seeing disasters as non-routine cyclic events, meaning simply a departure from, then a return to 'normal'. Even the 'normal' is often a different, deteriorated or improved situation compared to the initial stage. Hence, the disaster life cycle, a fundamental framework for disaster studies (Coetzee and Niekerk 2012; Lavell et al. 2012), which was proposed by Mileti (1975), Baird et al. (1975) and Drabek (1986, 1999), was broadly used by key institutions such as UN Office for Disaster Risk Reduction (UNISDR 2009) and Federal Emergency Management Agency (FEMA) of the USA has been criticised as being overly simplistic (see Lewis 1999). Disasters are complex, multidimensional events where root causes and consequences are embedded in the functioning of society, and they are related to policy and political management failures, social injustice and exclusion. Ultimately, this conception of disasters focuses attention on the relevance of the disaster-demography nexus in understanding, managing and recovering from disasters.

1.4 Method and Structure of the Book

The book consists of eleven, mostly multiple-authored chapters which aim to compare jurisdictions and provide insights from different cases. The cases range from technological to natural hazard events and population vulnerabilities. It is important

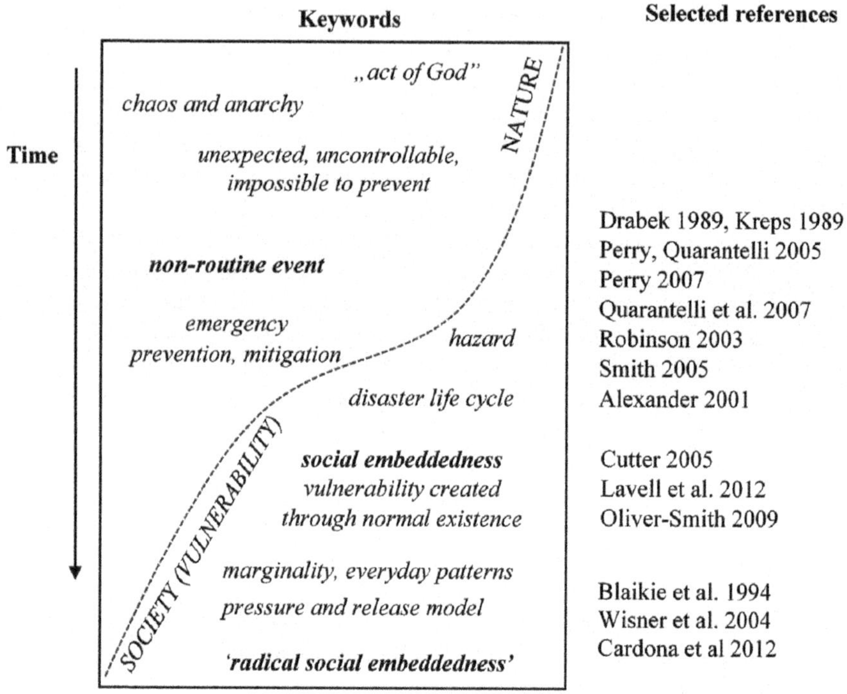

Fig. 1.1 Paradigm shifts in disaster studies

to highlight that several other shock events, other than natural hazards, technological accidents, infectious diseases or climate change could also unfold vulnerabilities and hence lead to disastrous demographic consequences for populations. Unrest, war and ethnic cleansing could be also understood as part of demography of disasters (see Drabek 1999; Wisner et al. 2004) along with pandemics or economic crises because they change the demographic profile of the impacted populations and lead to mass displacements. Global disasters (nuclear war, climate change) are part of the disaster literature dialect as well (see Giddings 1973), and hence, the concepts of 'risk society' or 'global risk society' have been introduced by Wisner et al. (2004) and Cutter (2005). In addition, Wisner et al. (2004) suggested disasters caused by natural hazards are not the greatest threat to humanity in comparison with violent conflicts or famines which interact with each other, making disaster impacts worse. Of course, the cases presented in this volume are limited in scope. That is why we summarised other ways of disaster-demography nexus in Chap. 13. Hence, the final chapter is not only summarising the learnings from the present volume but also serves as an extended literature review.

This volume contains significant spatial-geographical analysis, and this is the basis for its subtitle 'impacts for population and place'. For example, Chap. 2 has a strong emphasis on spatiality through permanent mass displacements. Chap. 3 and

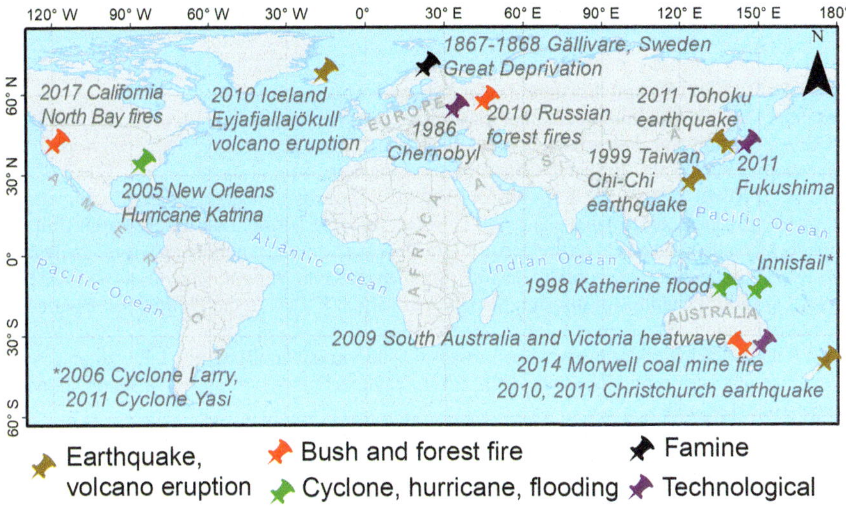

Fig. 1.2 Locations and triggering events of the disasters represented in this book

the comparative example in Chap. 4 draws attention to how bushfires impacted population migration in California while in Chap. 4 we learn of the contributing effects of a lack of forest maintenance to catastrophic forest fires in Russia. Geography is also represented by the spatial variation of case study areas in this book (Fig. 1.2) covering Europe, Asia, North America, Australia and New Zealand. Altogether 13 disaster cases are discussed in this volume with the majority occurring in the twenty-first century.

The following chapters aim to broaden links between demography and disasters. Chapters 2–8 focus on holistic population-disaster linkages where hazard events alter population trends (disaster-induced mass displacements, migration as heat adaptation strategy, long-term impact of disasters in sparsely populated areas) or population change generates disaster risks. Chapters 9–11 focus on issues related to demography more broadly, such as the disruption of lifeline networks as key infrastructure elements, gender aspects of resilient urban design and the role of communities in disaster recovery. Chapter 12 is about international collaboration in the field of disaster studies and highlights, through the author's personal experiences, how scientific ideas and approaches can be exchanged and new connections can be built through case studies across nations.

References

Alexander D (2001) Natural disasters. Taylor & Francis Group, New York.

Alexander D (2005) An interpretation of disaster in terms of changes in culture, society and international relations. in: Perry RW, Quarantelli EL eds. What is a disaster? New answers to old questions. International Research Committee on Disasters.

Baird A, O'Keefe P, Westgate KN, Wisner B (1975) Towards an explanation and reduction of disaster proneness. Occasional paper no.11, University of Bradford, Disaster Research Unit.

Bajek R, Matsuda Y, Okada N (2008) Japan's Jishu-bosai-soshiki community activities: analysis of its role in participatory community disaster risk management. *Natural Hazards*, 44:281–292.

Bankoff G (2003) Cultures of disaster. Society and natural hazard in the Philippines. Routledge-Curzon, London.

Bissel R (2013) What is a catastrophe and why is this important? in: Bissel R ed. Preparadness and response for catastrophic disasters. CRC Press, Taylor&Francis Group. Boca Raton, London, New York. 185–224.

Blaikie PT, Cannon T, Davis I, Wisner B (1994) At risk: Natural hazards, people, vulnerability, and disasters. Routledge, London, UK.

Bolin B (2007) Race, class, ethnicity, and disaster vulnerability. in: Rodríguez H, Quarantelli EL, Dynes RR eds. Handbook of disaster research. Springer, pp. 113–129.

Cannon T (1994) Vulnerability analysis and the explanation of 'natural' disasters. in: Varley A ed. Disasters, development and environment. John Wiley&Sons.

Cardona OD, van Aalst MK, Birkmann J, Fordham M, McGregor G, Perez R, Pulwarty RS, Schipper ELF, Sinh BT (2012) Determinants of risk: exposure and vulnerability. in: Field CB, Barros V, Stocker TF, Qin D, Dokken DJ, Ebi KL, Mastrandrea MD, Mach KJ, Plattner GK, Allen SK, Tignor M, Midgley PM eds. Managing the Risks of Extreme Events and Disasters to Advance Climate Change Adaptation A Special Report of Working Groups I and II of the Intergovernmental Panel on Climate Change (IPCC). Cambridge University Press, Cambridge, UK, and New York, NY, USA, pp. 65–108.

Cavallo E, Noy I (2011) Natural disasters and the economy - A survey. *International Review of Environmental and Resource Economics*, (5):63–102.

Cavallo E, Galiani S, Noy I, Pantano J (2013) Catastrophic natural disasters and economic growth. *Review of Economics and Statistics,* 95(5):1549–1561.

Chen S, Luo Z, Pan X (2013) Natural disasters in China: 1900–2011. *Natural Hazards*, 69:1597–1605.

Coetzee C, Van Niekerk D (2012) Tracking the evolution of the disaster management cycle: A general system theory approach. *Jàmbá: Journal of Disaster Risk Studies* 4(1):9

Cohen C, Werker ED (2008) The political economy of "natural" disasters. *Journal of Conflict Resolution*, 52(6):795–819.

Coleman L (2006) Frequency of Man-Made Disasters in the 20th Century. *Journal of Contingencies and Crisis Management* 14(1):3–11.

Coumou D, Rahmstorf S (2012) A decade of weather extremes. *Nature Climate Change* (2):491–496.

Cutter SL (2005) Are we asking the right question? In Perry RW, Quarantelli EL eds. What is a disaster? New answers to old questions. International Committee on Disasters. pp. 39–48.

Donner W, Rodríguez H (2008) Population composition, migration and inequality: The influence of demographic changes on disaster risk and vulnerability. *Social Forces*, 87(2):1089–1114.

Drabek TE (1986) Human systems responses to disaster: an inventory of sociological findings. Springer, New York.

Drabek TE (1989) Disasters as nonroutine social problems. *International Journal of Mass Emergencies and Disasters*, 7(3):253–264.

Drabek TE (1999) Revisiting the disaster encyclopedia. *International Journal of Mass Emergencies and Disasters*, 17(2):237–257.

Drabek TE, McEntire DA (2003) Emergent phenomena and the sociology of disaster: lessons, trends and opportunities from the research literature. *Disaster Prevention and Management: An International Journal*, 12(2)97–112.

Dynes R (1970) Organized behaviour in disaster. Lexington Books, Lexington, MA.

Easterling DR, Evans JL, Groisman PYa, Karl TR, Kunkel KE, Ambenje P (2000) Observed variability and trends in extreme climate events: A brief review. *Bulletin of the American Meteorological Society* 81(3):417–426.

Ekström A, Kverndokk K (2015) Introduction: Cultures of disaster. *Culture Unbound*, 7:356–362.

Ellul J (1964) The technological society. Vintage Books, New York.

Enarson E (2000) Gender and natural disasters. Working paper. Recovery and Reconstruction Department, Geneva. Accessed on the 6 December 2019: https://www.ilo.int/wcmsp5/groups/public/---ed_emp/---emp_ent/---ifp_crisis/documents/publication/wcms_116391.pdf

Eshghi K, Larson RC (2008) Disasters: lessons from the past 105 years. *Disaster Prevention and Management* 17(1):62–82.

Flanagan BE, Gregory EW, Hallisey EJ, Heitgerd JL, Lewis B (2011) A social vulnerability index for disaster management. *Journal of Homeland Security and Emergency Management*, 8(1).

Frankenberg E, Laurito M, Thomas D (2014) The demography of disasters. in: Smelser NJ, Baltes PB eds. International encyclopedia of the social and behavioral sciences, 2nd edition (Area 3). North Holland, Amsterdam.

Frich P, Alexander LV, Della-Marta P, Gleason B, Haylock M, Klein Tank AMG, Peterson T (2002) Observed coherent changes in climatic extremes during the second half of the twentieth century. *Climate research* 19:193–212.

Gencer EA (2013) Chapter 2: Natural disasters, urban vulnerability, and risk management: A theoretical overview. in: The Interplay between Urban Development, Vulnerability, and Risk Management. *SpringerBriefs in Environment, Security, Development and Peace,* 7:7–43.

Giddings JC (1973) World population, human disaster and nuclear holocaust. *Bulletin of the Atomic Scientists*, 29(7):21–50.

Jakab G, Bíró T, Kovács Z, Papp Á, Sarawut N, Szalai Z, Madarász B, Szabó Sz (2019) Spatial analysis of changes and anomalies of intense rainfalls in Hungary. *Hungarian Geographical Bulletin*, 68(3):241–253.

Kahn ME (2005) The death toll from natural disasters: The role of income, geography, and institutions. *Review of Economics and Statistics*, 87(2):271–284.

Kapuchu N, Özerdem A (2013) Managing emergencies and crises. Jones&Bartlett Learning, Burlington, USA.

Katz RW, Brown BG (1992) Extreme events in a changing climate: Variability is more important than averages. *Climate Change* 21(3):289–302.

Kertész Á, Křeček J (2019) Landscape degradation in the world and in Hungary. *Hungarian Geographical Bulletin*, 68(3):201–221.

Kitagawa K (2016) Disaster preparedness, adaptive politics and lifelong learning: a case of Japan. International Journal of Lifelong Education, 35(6):629–647.

Kreps GA (1989) Future directions in disaster research: the role of taxonomy. *International Journal of Mass Emergencies and Disasters*, 7(3):215–241.

Lavell A, Oppenheimer M, Diop C, Hess J, Lempert R, Li J, Muir-Wood R, Myeong S (2012) Climate change: new dimensions in disaster risk, exposure, vulnerability, and resilience. in: Field CB, Barros V, Stocker TF, Qin D, Dokken DJ, Ebi KL, Mastrandrea MD, Mach KJ, Plattner GK, Allen SK, Tignor M, Midgley PM eds. Managing the Risks of Extreme Events and Disasters to Advance Climate Change Adaptation A Special Report of Working Groups I and II of the Intergovernmental Panel on Climate Change (IPCC). Cambridge University Press, Cambridge, UK, and New York, NY, USA, pp. 25–64.

Lavell C, Ginnetti J (2013) Technical paper: the risk of disaster-inducted displacement. Central America and the Caribbean. Internal displacement monitoring centre (IDMC). Geneva. Norwegian Refugee Council, p 54.

Lewis J (1999) Development in disaster-prone places: Studies of vulnerability. Intermediate Technology Publications, London.
Lindell MK (2013) Disaster studies. *Current Sociology,* 61(5–6):797–825.
Loayza NV, Olaberría E, Rigolini J, Christiaensen L (2012) Natural disasters and growth: Going beyond the averages. *World Development*, 40(7):1317–1336.
Masozera M, Bailey M, Kerchner C (2007) Distribution of impacts of natural disasters across income groups: A case study of New Orleans. *Ecological Economics*, 63(2–3):299–306.
McEntire DA (2013) Understanding catastrophes. A discussion of casuation, impacts, policy approaches, and organizational structures. in: Bissel, R, 2013. Preparadness and response for catastrophic disasters. CRC Press, Taylor&Francis Group. Boca Raton, London, New York.
Mileti DS (1975) Human systems in extreme environments: A sociological perspective. Colorado, Institute of Behavioural Sciences, The University of Colorado.
Neumayer E, Plümper T (2007) The gendered nature of natural disasters: The impact of catastrophic events on the gender gap in life expectancy, 1981–2002. Annals of the Association of American Geographers, 97(3):551–566.
O'Keefe P, Westgate K, Wisner B (1976) Taking the naturalness out of natural disasters. *Nature*, 260(April):566–567.
Okuyama Y, Sahin S (2009) Impact estimation of disasters: A global aggregate for 1960 to 2007. World Bank Policy Research Working Papers June 2009.
Oliver-Smith A (1979) Post disaster consensus and conflict in a traditional society: The 1970 avalanche of Yungay, Peru. *Mass Emergencies*, 4:39–52.
Oliver-Smith A (2009) Climate change and population displacement: Disasters and diasporas in the twenty-first century. in: Crate, S. A., Nuttal, M. eds. Anthropology and climate change: from encounters to actions. Left Coast Press Walnut Creek, CA.
Paton D, Bajek R, Okada N, McIvor D (2010) Predicting community earthquake preparedness: a cross-cultural comparison of Japan and New Zealand. *Natural Hazards*, 54:765–781.
Perry RW (2007) What is a disaster? in: Rodríguez H, Qarantelli EL, Dynes RR, eds. Handbook of disaster research. Springer.
Perry RW, Quarantelli EL eds. (2005) What is a disaster? New answers to old questions. International Research Committee on Disasters.
Prince SH (1920) Catastrophe and social change. Columbia University, New York.
Quarantelli EL (1978) Disasters: Theory and research. Sage, Beverly Hills, CA.
Quarantelli EL (1998) What is a disaster? Perspectives on the question. Routledge, London.
Quarantelli EL (2006) Catastrophes are different from disasters: Some implications for crisis planning and managing drawn from Katrina. Accessed on-line on 16/12/2019 at https://understandingkatrina.ssrc.org/Quarantelli/
Quarantelli EL, Dynes RR (1972) When disaster strikes. *Psychology Today*, 5(Feb.):66–71.
Quarantelli EL, Lagadec P, Boin A (2007) A heuristic approach to future disasters and crises: New, old, and in-between types. In: Rodríguez H, Quarantelli EL, Dynes RR eds. Handbook of disaster Research. Springer, pp 16–41.
Robinson C (2003) Overview of disaster. in: Robinson C, Hill K eds. Demographic methods in emergency assessment A Guide For Practitioners. Center for International Emergency, Disaster and Refugee Studies (CIEDRS) and the Hopkins Population Center; Johns Hopkins University Bloomberg School of Public Health, Baltimore, Maryland.
Rodríguez H, Quarantelli EL, Dynes RR (2007) Editors' introduction, in: Rodríguez, H, Quarantelli, EL, Dynes, RR eds. Handbook of disaster Research. Springer.
Smith D (2005) In the exes of the beholder? Making sense of the system(s) of disaster(s). in: Perry RW, Quarantelli EL eds. What is a disaster? New answers to old questions. International Research Committee on Disasters.
Toya H, Skidmore M (2007) Economic development and the impacts of natural disasters. Economics Letters, 94:20–25.

UNISDR (2009) United Nations International Strategy for Disaster Reduction (UNISDR), UNISDR Terminology on disaster risk reduction. Accessed on line on the 15/05/2019 at https://www.unisdr.org/files/7817_UNISDRTerminologyEnglish.pdf
UN General Assembly (1989) International decade for natural disaster reduction, Session 44 Resolution 236, 22 December 1989. Accessed on-line 10.12.2019 at https://www.un.org/ga/search/view_doc.asp?symbol=A/RES/44/236
United States Strategic Bombing Survey (1947) The effects of bombing on health and medical care in Germany. US Government Printing Office, Washington.
Webb GR (2007) The popular culture of disaster: Exploring a new dimension of disaster research. In: Rodríguez H, Quarantelli EL, Dynes RR eds. Handbook of disaster research. Springer. pp 430–440.
Wisner B, Blaikie P, Cannon T, Davis I (2004) At Risk. Natural hazards, people's vulnerability and disasters. Routlege, London-New York.

Chapter 2
Long-Term Mass Displacements—The Main Demographic Consequence of Nuclear Disasters?

Dávid Karácsonyi, Kazumasa Hanaoka, and Yelizaveta Skryzhevska

Abstract Human history has witnessed several major disasters that have affected the economic, social and environmental conditions of their respective regions. The nuclear disaster of Chernobyl (1986, Ukraine, that time the Soviet Union) and Fukushima (2011, Japan) appears to be the most significant disasters in terms of negative outcomes produced for their population over a long time. Despite this, the analysis of the socio-economic outcomes of these disasters has attracted much less scientific attention than health or radiation-related issues (UNDP 2002a; Lehman and Wadsworth 2009, 2011). Although nuclear accidents are deemed to be rare events, the Fukushima disaster occurred only 25 years after Chernobyl. These disasters highlighted the need for a detailed long-term socio-economic analysis of these accidents to acquire sufficient knowledge to be applied when considering new construction sites for nuclear power facilities (Lehman and Wadsworth 2011). This chapter focuses on the problem of permanent resettlement resulting from nuclear disasters and its effects on regional demographic trajectories and spatial shifts. Based on the results of this study we argue that mass displacement after a nuclear disaster rather than the radiation itself has a much more significant impact on deteriorating health, natural reproduction and economic performance of the affected population. Furthermore, given the differences in radio-ecological conditions, reconstruction policy and the time framework, Fukushima may demonstrate demographic consequences that are different from the Chernobyl case.

D. Karácsonyi (✉)
Geographical Institute, CSFK Hungarian Academy of Sciences, Budaörsi út 45, Budapest 1112, Hungary
e-mail: karacsonyi.david@csfk.mta.hu

K. Hanaoka
Department of Geography, Ritsumeikan University, 56-1 Tojiin-kita-machi, Kita-ku, Kyoto 603-8577, Japan
e-mail: kht27176@fc.ritsumei.ac.jp

Y. Skryzhevska
Miami University, 202 Mosler Hall, 1601 University Boulevard, Hamilton, OH 45011-3316, USA
e-mail: skryzhy@miamioh.edu

D. Karácsonyi et al. (eds.), *The Demography of Disasters*,
https://doi.org/10.1007/978-3-030-49920-4_2

Keywords Chernobyl · Fukushima · Radiation exposure · Mass displacement · Spatial-demographic consequences · Urbanisation

2.1 Introduction

Human history has witnessed several major disasters that have affected the economic, social and environmental conditions of their respective regions. The nuclear disaster of Chernobyl (1986, Ukraine, that time the Soviet Union) and Fukushima (2011, Japan) appears to be the most significant disasters in terms of negative outcomes produced for their population over a long time. Despite this, the analysis of the socio-economic outcomes of these disasters has attracted much less scientific attention than health or radiation-related issues (UNDP 2002a; Lehman and Wadsworth 2009, 2011). Although nuclear accidents are deemed to be rare events, the Fukushima disaster occurred only 25 years after Chernobyl. These disasters highlighted the need for a detailed long-term socio-economic analysis of these accidents to acquire sufficient knowledge to be applied when considering new construction sites for nuclear power facilities (Lehman and Wadsworth 2011). This chapter focuses on the problem of permanent resettlement resulting from nuclear disasters and its effects on regional demographic trajectories and spatial shifts, while Chap. 11 studies the community involvement during and after evacuation and resettlement process in two nuclear disaster cases.

At first glance, Chernobyl and Fukushima may not be comparable, given the tremendous difference in their socio-economic conditions. The geographic, cultural and socio-economic distances can hardly be greater between any two points on Earth than between Japan during the 2010s and the Soviet Union during the 1980s. However, according to Oliver-Smith (2013), geographically and culturally distanced societies present analogous issues during similar disaster events. For instance, Chernobyl and Fukushima both deployed similar policies to manage mass population displacements. An initial emergency evacuation was followed by organised or spontaneous resettlement of the vulnerable population to avoid further radiation risks. The radiation threat should be considered as both a rapid-onset disaster because of the urgent need of evacuation and a slow-onset disaster because of its long-lasting effect, making a return difficult or even impossible over time.

The Chernobyl and Fukushima disasters illustrate how long-term emergency mass displacement triggers demographic shifts. They are not the largest evacuations in human history; however, they are among the largest permanent resettlements in peacetime caused by a previously unforeseen, unplanned situation.

Geographically detailed data is necessary to understand spatial turbulences caused by mass displacement after the Fukushima and Chernobyl disasters. In this study, data series spanning over three decades was used to provide insights into long-term population shifts after Chernobyl based on census populations. In the Fukushima case, mobile phone location data was used alongside the census data in spatially fine-scale to interpret population changes. This provides short-term but spatially and structurally detailed data about the effect of mass displacement. This detail is used to explore the difference between de facto and de jure populations which could be significant during the post-disaster evacuation phases. Thus, the findings derived by analysing the two data sets can complement each other.

The core analysis used Geographic Information Systems (GIS) based on detailed spatial units that included the territories of Belarus, Ukraine, nine western provinces of Russia and the Tohoku area located on the northern part of Honshu the main island of Japan.

2.2 Spatial Demography Impact of Mass Displacements

Relocation or resettlement of disaster-stricken populations is a common strategy applied within disaster mitigation policy (Oliver-Smith 1996). Generally speaking, displacement is the impact of a disaster which results from the vulnerability of people to shocks or stresses, compelling them to relocate just for survival (Lavell and Ginnetti 2013). Migration has always been one of the most important survival strategies adopted by people facing natural hazards or human-caused disasters (Hugo 2008).

Along with the demographic loss (death tolls and injuries) caused directly during the disaster event, it could be argued that population displacement can be viewed as the other demographic consequence of a disaster. Mass displacement also has direct effects such as death, injury, disease (Robinson 2003) and could cause social insecurity and disrupted life prospects. Displacement, the demographic and social impacts of which are oftentimes underestimated, can be an even more significant consequence of a disaster than the direct death toll.

A large number of studies on the health and natural reproduction consequences of Chernobyl mostly explained the demographic losses resulting exclusively from radiation. However, they often neglected the impact of mass displacement itself. In the 30-year period after the Chernobyl disaster, the number of indirect victims (deaths caused by cancer, cardiovascular diseases, etc.) is still widely debated (TORCH 2006, 2016; Peplow 2006 or see Greenpeace 2006) because the linkage between radiation and cancer cannot be proven due to its stochastic occurrence within the population (see WHO 2006, 2016; IAEA-WHO-UNDP 2005; IAEA 2006). The only two exceptions are the increase of thyroid cancer cases among those who were young and adolescent during the disaster (4000 cases by 2002 according to IAEA 2006; OECD 2002) and leukaemia that occurred after 1990 among former on-site emergency clean-up workers (*likvidators*) (Hatch et al. 2005; Balonov et al. 2010; European Commission 2011).

Despite there being a large number of health studies, science still lacks full, final and objective information about the medical and biological consequences of Chernobyl (UNDP 2002b; Hatch et al. 2005; Baverstock and Wiliams 2006 and others). Studies investigating health issues resulting from nuclear disasters are limited as a result of the lapse between initial exposure and the presentation of symptoms. This lapse can be as long as 10–20 years after the accident (Lehmann and Wadsworth 2011). Extreme variations in individual exposure causing tragic outcomes are impossible to follow up over long periods of time when analysing large populations. Considering also that the screening effect inflates health statistics when populations in the affected area experience intense health control that identifies illnesses which would otherwise never be explored (UNDP 2002a). It is estimated that the indirect death toll varies from 4,000 (Peplow 2006; IAEA 2006) to 60,000 (TORCH 2006).

Studies on natural population reproduction trends often did not prove a mathematical correlation with radiation exposure after Chernobyl (IAEA 2006). According to Linge et al. (1997), the birth and death rates during 1986–1996 were similar to those in non-affected areas. However, their investigations were based on the populations of larger regions, potentially cloaking extreme local variations. Other studies (see Omelianets et al. 1988, 2016; Lakiza-Sachuk et al. 1994; Voloshin et al. 1996; Rolevich et al. 1996) claim radiation exposure temporarily decreased birth rates immediately after the accident in the affected areas. However, this decrease can be explained by the disinclination to bear children during the uncertain life prospects that eventuated post-Chernobyl resettlement (Abbott et al. 2006). It can also be explained by the fear of the effect of radiation exposure during pregnancy instead of the presence of radiation effect itself (Jaworowski 2010). This view is supported by Lehmann and Wadsworth (2011), who states that contamination levels have little or no influence on fertility, marriage behaviour and education performance. This is further illustrated by a lack of statistical correlation between radiation exposure and chromosome aberrations or birth defects (OECD 2002; Baverstock and Williams 2006). According to the research by Rolevich et al. (1996), Libanova (2007), and Mesle and Poniakina (2012), there is higher mortality among people living in the affected areas. However, the increase in mortality cannot be explained by the radiation alone (Shestopalov et al. 1996). The increase of psychological problems caused by the social disruption during the resettlement presented significant health consequences (Brenot et al. 2000; Balonov et al. 2010). Furthermore, as younger generations migrate out of the disaster area (Voloshin et al. 1996; Omelianets et al. 2016), a statistical consequence is an increase in mortality simply because of the shift in age structures towards aging.

Although approaches cited in the previous paragraph often appear opposing, it is very important to stress that even if the results are biased that the majority of scientists agree that the Chernobyl disaster as a whole played a significant role in the deteriorating natural reproduction during the 1990s. Natural reproduction decline was also related to general socio-economic decay during and after the collapse of the Soviet Union, such as growing poverty and unemployment, increasing alcoholism and poorer medical services (Ioffe 2007; Baranovski 2010; Marples 1993, 1996 and others). As such, it is hard to distinguish the two separate effects.

Out of the health and natural reproduction studies, only a few research focuses on wider demographic consequences. Lehmann and Wadsworth (2011) underline the lower market performance of those who were exposed to higher radiation. This effect however is based on the self-assessment of their own health condition as poor, not directly from the radiation. These people also have lower mobility. Abbott et al. (2006) approached the socio-economic effect of Chernobyl through the view of risk and uncertainty when analysing poor economic circumstances. The most significant documentation on Chernobyl-related social and economic issues has been launched by UNDP (2002a as well as 2002b, c), calling for the need of a new developmental approach. According to this document, a holistic approach integrating health, radio-ecology and economic aspects is needed to fully understand Chernobyl.

Based on the literature, we argue that the main direct negative demographic impact of Chernobyl was not the number of deaths or illnesses, not even the psychological consequences (see Rumyantseva et al. 1996; Lochard 1996; Brenot et al. 2000 or Jaworowski 2010) but the urgent need for the resettlement of ten-hundred thousands of people because of the long-term radiation threat. This resettlement resulted in the distortion of everyday life and changed natural reproduction due to post-disaster uncertainty and social insecurity. Post-Chernobyl and post-Fukushima displacement caused much more significant demographic shifts than the radiation itself. Thus, the effect of permanent mass displacement should be the focus when explaining demographic outcomes.

Permanent resettlements resulting from disasters are relatively rare events hence less discussed in the general disaster literature (Oliver-Smith 2013) despite having a long-term demographic impact. Flooding, earthquakes and volcano eruption can cause large evacuations but rarely long-term displacements. However, long-term demographic shifts can be caused by temporary resettlement during or following a natural hazard event as well as it discussed in Chaps. 5 and 6. If a displacement caused by natural hazards becomes permanent, this indicates failed remediation policies (Oliver-Smith 2013). Yet there is no clear distinction between temporary (short-term) and permanent (long-term) displacement in the literature. As previously identified, such a distinction can often be policy induced. In Fukushima, policy documents refer to evacuees as temporarily displaced people to maintain hope for a return and to keep communities together (see Chap. 11). A displacement can last for years, even for life and still, it is described as "temporary".

Furthermore, mass displacement in practice often does not solve the problem caused by the disaster itself but generates new challenges (Robinson 2003). In many cases, the resettlement results in a secondary disaster (Oliver-Smith 2013), which will further produce serious consequences in a badly planned or unplanned resettlement (Cernea 2004). At a new location, an appropriate settlement design, housing, services and an economic base need to be built to enable people to revitalise itself and achieve adequate levels of resilience (Oliver-Smith 2009). These challenges caused by mass displacement should also be considered as integral parts of the Chernobyl and Fukushima disasters. We argue that the lack of a holistic view led to an overemphasis of health risks by radiation which neglected the effect of the main consequence of nuclear disasters: permanent mass displacement and the uncertainties and disturbances caused by it.

2.3 Changing Region and Shifting People by Nuclear Disaster

Nuclear radiation levels significantly determine evacuation and resettlement policies rather than an understanding of the key findings from the literature summarised in the previous section whereby radiation exposure itself has a minor effect directly on general demographic trends. The radiation is not a homogenous phenomenon, it has changing levels, composition and characteristics over time, and thus, the evacuation and resettlement measures follow this change to cope with the changing radiation threat. Based on this, certain radiation phases can be distinguished over time which results in an adjustment of resettlement policies following the disaster. Thus, the level of radiation and change in population trends are strongly interrelated through the resettlement policies rather than through health consequences. There are important differences between Fukushima and Chernobyl in terms of the composition of emitted isotopes resulting in the slightly different evacuation and resettlement policies. In the following section, these different and changing characteristics of radiation will be presented to better understand evacuation and resettlement measures in the two cases.

2.3.1 *The Chernobyl Case*

In 1986, at the Chernobyl nuclear power station, the operation failure and poor engineering design that led to an explosion in Unit 4 within the reactor core[1] which damaged the shielding, released 3–4% (5–6 tonnes) of fragmented nuclear fuel[2] into the surrounding environment and was followed by a ten-day graphite fire. The accident caused a release of fission products into the environment, which was the

[1] A runaway nuclear chain reaction caused a steam blast within the reactor vessel that was followed by a larger chemical explosion seconds later either by hydrogen–air or carbon-monoxide–air ignition. The magnitude of the reactor core explosion and resulting release of fuel was unprecedented in the history of nuclear accidents, e.g. in Fukushima, in Three Mile Island or Windscale.

[2] The reactor that exploded in Chernobyl used low-enriched uranium-oxide fuel with 2% ^{235}U and 98% ^{238}U. The spent fuel still consists of ~96% ^{238}U as well as <1% ^{235}U, ^{236}U and ~1% transmuted isotopes such as ^{239}Pu, ^{240}Pu and other trans-uranium elements, which emit alpha radiation (high energy ^{4}He nuclei) when decaying. These isotopes are harmful when inhaled (causing lung cancer) or ingested in small quantities, but human skin stops alpha radiation.

On the other hand, during normal reactor operation, most of the fissile isotopes (82% of ^{235}U, 62–73% of ^{239}Pu and 72% of ^{241}Pu), representing ~2 to 3% of the total fuel mass, will undergo fission when capturing a neutron instead of transmutation, producing short and medium half-life products such as ^{131}I (beta emitter), ^{90}Sr (beta emitter) and ^{137}Cs (beta and gamma emitter, a decay product of very short-lived ^{137}Xe). Although the total amount of such material released was less than 0.5 kg for ^{131}I and less than 25 kg for ^{137}Cs, it represents a significant danger because beta radiation (emission of high energy electrons or positrons) and especially gamma radiation (high energy photons) can penetrate human tissue, making it hazardous even without actual intake of the isotope.

principal source of high initial effective doses[3] in large areas that extended outside of the Soviet Union. However, the isotopes with a short half-life[4] rapidly decay (UNDP 2002b), such as ^{131}I (Iodine-131) which was found to be the root cause of increased thyroid cancer cases after the disaster. One year after the accident, the total effective dose levels dropped to 2% of what it had been at the time of the accident, and after two years, it had fallen to 1% (IAEA 2006). Among those isotopes that remained over long time,[5] the largest area (Table 2.1) was polluted by ^{137}Cs (Caesium-137), which is the predominant source of the remaining dose levels causing a health risk since the third year after the disaster.[6]

After the Chernobyl disaster, two solutions—or their combination—were employed to mitigate the effects on the local population: radiological decontamination and the resettlement of people to non-contaminated areas. Establishing new homes for resettlement seemed clearly more expensive but was also the much safer solution (Tykhyi 1998).

The most severely contaminated area was the surroundings of the power plant with a radius of 30 km, the so-called "Exclusion Zone" including the plant itself, as well Pripyat city with 50 thousand people (1986), Chernobyl town and several villages. The total population of this area (116 thousand people) (Table 2.1) was evacuated during 1986–1987 (UNSCEAR 2000; IAEA 2006) mostly to large cities such as Kiev, Minsk, Chernihiv, Zhytomyr (Lehmann and Wadsworth 2011), was their communities quickly dissolved (Voloshin et al. 1996; IOM 1997). Many of them were settled later in Slavutich, a town established in 1988 for evacuees from Pripyat (Voloshin et al. 1996; Mesle and Poniakina 2012). The "Exclusion Zone" remained closed even until today. It is only opened in special cases such as for a very small number elderly and voluntary repatriates (*samosyoli*) (Lochard 1996).

The concentric zoning was adjusted in 1988 based on the survey results of ^{137}Cs surface activity levels[7] (Fig. 2.1). At the same time, the definition of "Contaminated Area" (^{137}Cs activity is above 37 kBq/m^2),[8] as well as an additional "Resettlement

[3]Effective dose is measured in Sieverts (Sv) defined as the total amount of energy from ionizing radiation absorbed by the human body, measured in J/kg.

[4]Half-life of these released isotopes: ^{132}Te 78 h, ^{133}Xe 5 days, ^{131}I 8 days. This means that half of the initial amount of each isotope present will have decayed over this time. E.g. ^{131}I 100% on day 1, 50% on day 8, 25% on day 16, 12.5% on day 24 and so on.

[5]^{137}Cs has a half-life of 30 years, ^{90}Sr 29 years, ^{239}Pu 24 thousand years and ^{240}Pu 6.5 thousand years.

[6]^{137}Cs is highly soluble in water. Thus, its salts are more easily integrated into parts of the food chain and easily adsorbed in human soft tissues (particularly the cardiovascular system). Its biological half-life is only 2 months; thus, it is rapidly excreted on intake. Unlike ^{137}Cs, ^{90}Sr has a biological half-life of 20 years and absorbed into the bones. This prolonged exposure caused leukaemia amongst clean-up workers in the evacuation zone (Balonov et al. 2010).

[7]Activity, given in becquerels (Bq) or curies (Ci), is a measure of the total number of nuclear decays per second occurring in a certain quantity of radioactive substance. Activity by weight of radioactive material can be expressed as Bq/g, and average surface activity level can be expressed as kBq/m^2 or Ci/km^2. 1 Ci = 3.7×10^{10} Bq.

[8]Staying in an area where ^{137}Cs activity is 37 kBq/m^2 is equivalent to receiving an effective dose of 0.25 μSv per hour (the human body absorbs ~0.25×10^{-6} J energy per kg every hour from beta

Table 2.1 Zoning in Chernobyl, compared to Fukushima (Based on IAEA 2006; UNDP 2002a)

Level of contamination (1986/2011)	Zoning in Chernobyl	Size of affected territory by Chernobyl (1986)	Population living in Chernobyl affected area (thousand people) (1986/2010)	Size of affected territory by Fukushima (2011)
Cs^{137} > 37 kBq/m^2 Cs^{137} > 1 Ci/km^2 or >0.25 μSV/h	Contaminated Area/zone of radiation monitoring	191,560 km^2 total, 57,900 km^2 in Russia 46,500 km^2 in Belarus 41,900 km^2 in Ukraine 45,260 km^2 in other countries	6000/5000	
Cs^{137} > 185 kBq/m^2 Cs^{137} > 5 Ci/km^2 or >1 μSV/h	Zone of voluntary/guaranteed resettlement	29,000 km^2 total, 16,000 km^2 in Belarus 8,000 km^2 in Russia 5,000 km^2 in Ukraine		1,700 km^2
Cs^{137} > 555 kBq/m^2 Cs^{137} > 15 Ci/km^2 or >4 μSV/h	Zone of (mandatory/obligatory) resettlement	6,400 km^2 in Belarus 2,440 km^2 in Russia 1,500 km^2 in Ukraine	400/200	
Various levels	Exclusion zone	2,230 km^2 in Ukraine 2,162 km^2 in Belarus	116/0	Initially 600 km^2, reduced to 207 km^2 (2018)

Zone" (^{137}Cs activity above 555 kBq/m^2) was established (UNSCEAR 2000; IAEA 2006). The latter was subject to further obligatory evacuation and resettlement. The general position was that in the mainly rural areas where healthy foodstuffs could no longer be produced in agriculture, it was futile to compel the local population to stay. Decontamination efforts were suspended in those areas where the local population

and gamma-decay). This effective dose over one day is equivalent to one dental X-ray check (which is ~5 μSv), or eating 2.5 bananas every hour (0.1 μSv) (Banana contains naturally occurring ^{40} K). During a standard air flight, the average dose is 2–4 μSv/h, while during a flight by Concorde, it was 9–10 μSv/h.

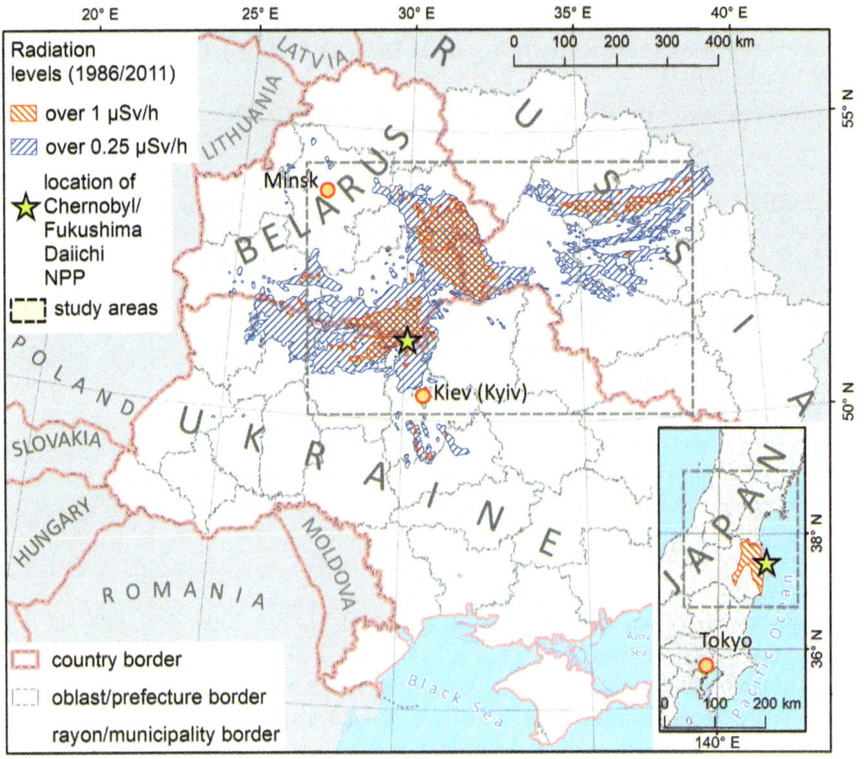

Fig. 2.1 Affected areas by Chernobyl and Fukushima disasters (Author and cartography by Karácsonyi, ^{137}Cs contamination data by USCEAR 2000)

would receive a 70-year ("lifetime") extra effective dose[9] of at least 350 mSv (milli-Sievert) (Malko 1998), and residents of these areas, 220 thousand people, were also resettled to non-contaminated areas at the very beginning of the 1990s (UNSCEAR 2000; IAEA 2006). The 350 mSv concept became the subject of sharp criticism (Malko 1998) given the impossibility of determining the dosage for each person leading to questioning the calculations made for the resident population as a whole. This inevitably caused mistrust among people.

Majority of subsequent resettlements happened in Belarus, until 1996, when these measures were finalised there. This second wave was followed by a more moderate one during the late 1990s only in Ukraine by the resettlement of additional 50 thousand people until 2005 when the process was officially ended there as well (IAEA 2006).

There are no exact records available on the total number of people that moved in the three affected countries, the numbers vary between 300 and 500 thousand which

[9]The average daily global background dose on the Earth's surface is 100 μSv which means, people who exposed 350 mSv extra dose throughout their lifetime, they exposed to at least ~10% higher background radiation caused by Chernobyl.

Table 2.2 Estimated number of people affected by post-Chernobyl resettlement/evacuation by country and by time frame (Based on UNDP 2002a; UN 2002; IOM 1997; UNSCEAR 2000; IAEA 2006; Lehmann 2011)

Resettled/evacuated	The entire time period (1986–2005)	1986–1987	1990–1996	1996–2005
By country	163,000 Ukraine (UNDP) 135,000 Belarus (UNDP) 52,000 Russia (UNDP)	24,000 Belarus (IOM) (the rest in Ukraine)	131,000 Belarus (IOM) 52,000 Russia (UNDP) (the rest in Ukraine)	50,000 Ukraine (IAEA)
Ukraine, Belarus, Russia total	326,000 (IAEA) 350,000 (UNDP) 492,000 (UN)	115,000 (UNSCEAR/IEA) 120,000 (Lehmann)	220,000 (UNSCEAR/IEA)	

includes voluntary resettlement as well (326,000—IAEA 2006; 350,000—UNDP 2002a; IAEA-WHO-UNDP 2005; 492,000—UN 2002) (Table 2.2). Outside the "Resettlement Zone", in areas with ^{137}Cs activity levels between 185 and 555 kBq/m^2, people were free to decide to stay or to leave. A legal act provided dwellings for those displaced. In these areas, economic restrictions, such as in agriculture, were coupled with increased health and food control.

It was reported that around 5–6 million people (IAEA-WHO-UNDP 2005; Balonov et al. 2010) lived in "Contaminated Areas" as of 2005, and still 200 thousand (IAEA 2006; Balonov et al. 2010) resided in the "Resettlement Zone". We estimate that by 2010, approximately 5 million people still lived in areas where radiation exposure exceeded 0.25 μSv/h (^{137}Cs activity over 37 kBq/m^2) in 1986 (Table 2.1). As a result of the natural isotope decay and purification processes, the exposure significantly decreased from 1986 which allowed the zoning to be readjusted (IOM 1997; UNDP 2002b). The only exception is the closer area surrounding the accident site in the "Exclusion Zone". It was polluted by ^{241}Pu which has a relative short half-life and decaying into ^{241}Am, a much more radio-toxic isotope.[10]

[10] ^{241}Pu (a beta emitter) has a relatively short half-life (14 years), but its decay product ^{241}Am (an alpha and gamma emitter, half-life: 400 years) is much more radio-toxic. In contrast to other isotopes, there will be a natural increase in ^{241}Am activity over time. By 2058, the ^{241}Am activity will surpass the cumulative activity of all trans-uranium isotopes (UNDP 2002b) and reach its maximum concentration a hundred years after the accident (IAEA 2006). Because of its longer half-life, it will surpass the activity of ^{137}Cs 300 years after the accident, significantly slowing the natural purification within the "Exclusion Zone".

2.3.2 Fukushima—The Accident, Zoning, Regulation, Consequences

On 11 March 2011, the Great East Japan Earthquake (moment magnitude of 9.0) and subsequent huge tsunami hit the two nuclear power plants (Fukushima I and II or Fukushima Daiichi and Daini) in Fukushima Prefecture. According to the official government report (NAIIC 2012), all nuclear reactors stopped safely after the earthquake. However, soon after the earthquake water from the tsunami wave, which was higher than 14 m, flowed into the nuclear power plants over the seawall. At the Fukushima I nuclear power plant, four out of six nuclear reactors lost their cooling functions because all the emergency backup generators were destroyed or drained by the huge tsunami. Three nuclear reactors then experienced a nuclear meltdown, and hydrogen–air chemical explosions occurred outside the reactor vessel.

Due to the accident, a large amount of fission products, mainly ^{131}I, ^{134}Cs and ^{137}Cs, were released into the ocean and the atmosphere. The total amount of radioactive materials released was estimated to be 770,000 TBq by the Nuclear and Industrial Safety Agency and 570,000 TBq by the Nuclear Safety Commission (TEPCO 2011). Steinhauser et al. (2014) reported around 5,300,000 TBq in case of Chernobyl and 520,000 TBq in case of Fukushima. These numbers mean that the long-lasting effect of Fukushima is significantly lower than Chernobyl; however, the threat during the accident was almost on a similar scale. In particular, radioactive materials spreading in a north-western direction from the nuclear power plant made a huge geographic area highly contaminated. During the evening of 11th March, the Japanese government declared a nuclear emergency. On the same day, an evacuation order was initially issued to the area within a 3 km radius of the power plant, but this was increased to a 30 km radius until 15th March. Group evacuations of residents were carried out by municipality offices on the morning of 12th March. At the same time, many residents who could drive cars evacuated individually.

After the nuclear accident, government organisations started to measure accurate aerial radiation levels with aeroplanes, helicopters and cars. Also, radiation levels were monitored continuously by at least 1600 monitoring posts installed across Fukushima Prefecture. The geographical extent of the highly contaminated area in Fukushima is limited compared to the Chernobyl case (Imanaka 2016; Steinhauser et al. 2014). However, ^{137}Cs is expected to remain for a long time (Fig. 2.1).

In April 2012, since the nuclear reactors were confirmed to be cooled down, the evacuation areas were reorganised into three based on their annual radiation dose as of March 2012: (1) areas where it is expected that the residents will have difficulty returning for a long time (above 50 mSv/year), (2) areas in which the residents are not permitted to live (20 mSv-50 mSv/year) and (3) areas for which the evacuation order is ready to be lifted (below 20 mSv/year). The Japanese government accepted the 20 mSv/year rule during rehabilitation period according to recommendations by ICRP Publication 111 (ICRP 2009) in comparison with the 5 mSv/year (or 350 mSv during 70 years) rule introduced after Chernobyl. Inside the evacuation areas, the total population was 81,291 people over an area of 1150 km^2. Among the population,

24,814 people resided within the highest contaminated area (above 50 mSv/year), whose size is 337 km^2 (Team in Charge of Assisting the Lives of Disaster Victims, Cabinet Office 2013). In 2015, the evacuation order was lifted for some parts of the evacuation areas, but five municipalities remain completely within evacuation areas and another three are partially included. In all the evacuation areas, previous residents are not allowed to stay overnight without special permission. In the highest contaminated area, all entrance roads to the area are blocked, and previous residents are currently not permitted to enter.

In the area affected by Fukushima disaster, large-scale decontamination is now underway (Ministry of Environment 2017). In fact, owing to natural degradation and decontamination work since the disaster, some parts of the evacuation areas meet the criteria for the order being lifted; the annual air dose has dropped to a level below 20 mSv per year, infrastructure and basic amenities such as supermarkets, hospitals and post offices can be reconstructed, and close consultation with municipalities achieved (Nuclear Emergency Response Headquarters 2015). In areas where the evacuation order was lifted, previous residents are permitted to return. On the other hand, in areas with the highest contamination levels, it was expected that residents would not be permitted to enter for a long time. The Japanese government, however, revised their legislation in an attempt to encourage residents to return there in five years or so by selecting prioritised sites for intensive decontamination and reconstruction which is named "reconstruction base".

2.4 Data and Methods

This research analysed and mapped population census data related to the Chernobyl (1979–2010) and Fukushima (2005–2015) affected regions. Demographic trends in the Chernobyl affected area at district (*raion*) level were derived from population censuses extending three decades following the disaster. In conjunction with the final two censuses under the Soviet Union (1979, 1989),[11] the censuses from the successor states were also used noting their time and methodological deviation. These included Belarus in 1999 and 2009, Ukraine in 2001 and Russia in 2002 and 2012. To further Ukraine's 2001 census data, the 2010 registered resident population number and composition data were included because there was no further census held in Ukraine after 2001 and until the publication of this chapter.

This investigation used a consolidated spatial system that included 846 units based on district-level (*rayon*) data. This data was free of administrative boundary changes that covered the entire territory of Ukraine and Belarus, as well as nine western regions (*oblasts*) of Russia (Fig. 2.1). This wider territory included the evacuation areas as well the evacuee receiving sites, where the post-Chernobyl evacuation and resettlement could have a fundamental impact upon the demographic processes. For an exact determination of the impact of the disaster, the share of ^{137}Cs contaminated

[11]The Soviet Union dissolved in 1991.

area within the area of each spatial units (districts) was taken into account and ArcGIS was utilised for calculations. The radiation level data was provided by UNSCEAR (2000).

As evacuation is still underway in the Fukushima-affected area, understanding displacement after the accident is a difficult task. Resident registration and questionnaire surveys of evacuees are important data sources. However, at the time of writing this chapter, the former provides information on registered residents only. Response rates to the latter are now around 50–60%, and the tabulated data published is inadequate for demographic analysis. Recently, the results of the Japanese population census 2015 were released. The census form, which asked about demographic characteristics as well as locations of residence five years ago, provides information on migration from highly contaminated areas. The geographic unit for this study is the municipality level (*Shi-Cho-Son*).

As a shortcoming of the census data, the data can only capture night-time populations who have resided in their current place continuously for at least three months. Although evacuation areas are lifted every year and temporal visitors increase after that, we cannot understand such changes in the ambient population by utilising census data. In Japan, as a new form of dataset, ambient population datasets based on 70 million mobile phone users' locations, named "mobile spatial statistics", are released at the 500 m grid cell level. The technical details of the dataset are explained in Terada et al. (2013) and Oyabu et al. (2013). The ambient population datasets we have shown the average of the hourly population in June 2015 and June 2016. Since the ambient population before the accident is not available, the census population in 2010 at the same 500 m grid cell level was used in combination to understand the geographical distributions of population change from 2010 to 2015.

2.5 Demographic Impacts on Regional Scale

2.5.1 Chernobyl Disaster—Shifts of Three Decades

The population of Polesye region, where Chernobyl disaster occurred, has been decreasing since 1970 (Khomra 1989), first to a moderate extent, then at a higher rate since 1986. The data series spanning three decades since the disaster reveals that the consequences of the Chernobyl accident are reflected most characteristically by the demographic trends of the 1990s even though evacuation measures can be reached back to the 1980s (Fig. 2.2). This is partly because during the census in 1989, three years after the disaster, many people still stayed in the evacuation zone and received permanent housing in the following years after that.

The annual population growth of Ukraine amounted barely to 200,000 in the 1980s. Such a wave of resettlement mobilising 100,000 in 1986–87 and additional 100,000 during the following ten years thoroughly reshaped the total population pattern in large part of the country. In Belarus, the population increased in the 1980s by

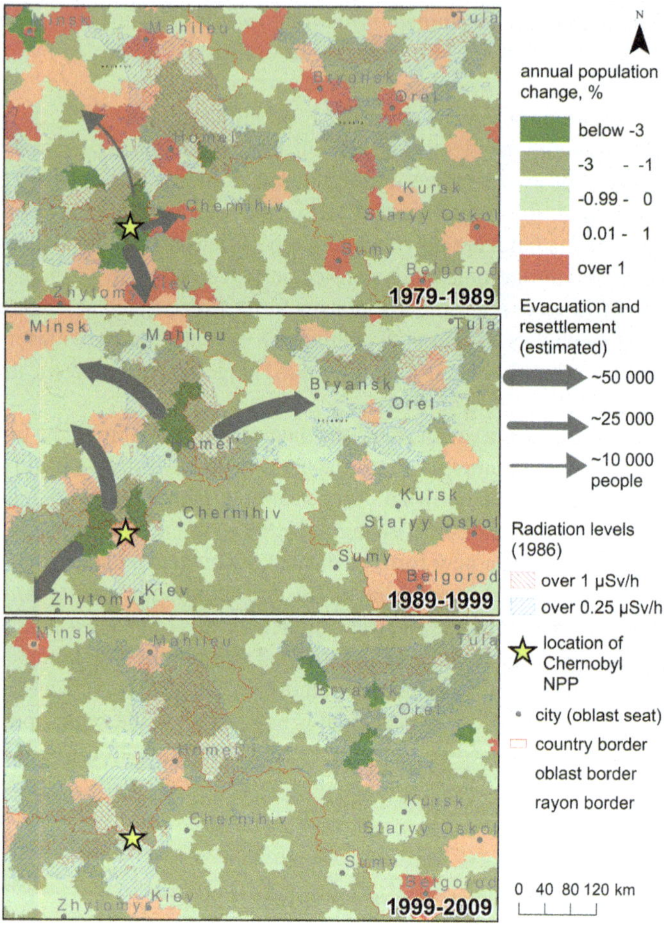

Fig. 2.2 Average annual population change by rayons (Author and cartography by Karácsonyi, calculation based on 1979, 1989 Soviet, 2001 Ukrainian, 1999, 2009 Belarusian and 2002, 2010 Russian census populations as well as de jure population in Ukraine by 2010)

around 30,000 people each year, while in consequence of the Chernobyl disaster, 25–30,000 people were resettled in 1986–87, followed by additional 100,000–130,000 during the 1990s in a country with a population of barely 10 million. The effect was even more dramatic because the evacuation affected around 1–1.6% of the country's total population. The corresponding figure was 0.4% in Ukraine and 0.04% in Russia. Even in 2010, Belarus had the highest share of population living in contaminated areas (Tables 2.3and 2.4). No other country has experienced the impact of a nuclear accident to the same degree. Given the large number of people who were resettled, the recipient regions—in particular the major towns and their environs—saw a relatively more favourable demographic trend in the 1990s.

Table 2.3 People living in administrative units in 2010, where radiation dose was higher than 0.25 μSv/h during Chernobyl (in 1986) and Fukushima (in 2011) disasters

Country	Population of affected administrative units (2010)	Area of affected administrative units (km^2)
Belarus	2,388,700 (34.4%)	60,463 (46.3%)
Japan[a]	1,875,210 (27%)	9,812 (7.5%)
Ukraine	1,407,811 (20.2%)	33,860 (25.9%)
Russia	1,281,781 (18.4%)	26,400 (20.2%)
Total	6,973,502 (100%)	130,535 (100%)

[a]Before the disaster, they became affected one year later

Table 2.4 Share of population and area of administrative divisions in selected countries in 2010, where radioactive contamination was higher than 0.25 μSv/h after Chernobyl (in 1986) and Fukushima (in 2011)

	Country	Administrative division level	From total population of the country (%)	From total area of the country (%)
1	Belarus	District	25.1	29.1
2	Ukraine	District	3.1	5.6
3	Japan[a]	Municipality	1.5	2.6
4	Russia	District	0.9	0.2

[a]Before the disaster, they became affected one year later

When reviewing the entire affected area within the three countries, a general dependence of the population change on the proportion of areas of radioactive contamination was only apparent in the 1990s (Table 2.5). The internal population change of entire Ukraine and Belarus was strongly under the influence of the resettlement measures rather than by natural change or other types of internal migration between 1989 and 2001. In Russia, the depopulation of affected areas was less significant. Because of dissolution of the Soviet Union, a large number of ethnic Russian political refugees from other republics arrived and resettled into these areas counterbalancing the out migration caused by the disaster (Veselkova et al. 1994).

From the 2000s, a significant correlation between population change and the share of contaminated areas could not be found. Demographic "waves" of resettlement calmed down by the 2000s, and even returning migration to the former places could be detected. The population of several small towns that lay in the contaminated areas in Belarus but had been cleaned–up (Naroŭlia, Brahin and Chojniki) began to grow once more (Table 2.6). In these towns that have undergone complex rehabilitation, people receive significant state assistance as well as apartments built with governmental funding. In such small towns, the presence of young families with small children is striking. For this reason, in the contaminated areas, the population is becoming urbanised more rapidly than elsewhere. These areas have become

Table 2.5 Correlation between share of contaminated surface (>1 μSv/h) of admin. districts and selected demographic indicators

Area of investigation	Number of administrative raions	Population change 1989–2000	Population change 2000–2010	Urban population change 2000–2010	Rural population change 2000–2010	Change of urbanisation level 2000–2010	Population density 2010	Rural population density 2010
In the total area	846	−0.23	−0.15	0.19	−0.12	0.28	−0.30	−0.41
In Belarus	119	−0.88	−0.13	0.31	−0.27	0.52	−0.37	−0.49
In Russia[a]	177	−0.21	−0.08	0.06	−0.09	0.05	−0.24	−0.05
In Ukraine[a]	239	−0.74	−0.35	0.15	−0.46	0.64	−0.31	−0.67

[a]Only those districts are considered where contaminated spot over 1 μSv/h is present or located closer than 200 km to the closest such contaminated spot

Table 2.6 Change of total population of some towns in contaminated area

Country	City	Total population, 1989 (persons)	Total population, 2000 (persons)	Total population, 2010 (persons)	Change between 1989 and 2000 (%)	Change between 2000 and 2010 (%)
Ukraine	Ovruch	19,121	17,031	16,792	−11	−1
	Ivankiv	10,282	10,563	9,768	3	−8
	Poliske	13,786	0	0	−100	0
Belarus	Lelchitsi	8,600	9,700	8,900	13	−8
	Hoyniki	17,100	15,000	13,100	−12	−13
	Brahin	5,900	3,400	3,954	−42	16
	Naroulya	11,000	7,200	8,400	−35	17
	Vetka	11,000	7,700	8,200	−30	6
	Elsk	9,600	10,400	9,600	8	−8
	Chechersk	9,700	7,400	7,700	−24	4
	Petrikov[a]	11,800	11,200	10,200	−5	−9
	Turov[a]	15,300	17,100	16,700	12	−2
Russia	Novozubkov	44,845	43,038	41,745	−4	−3
	Starodub	18,906	18,643	18,445	−1	−1

[a]Outside the contaminated area

Belarus's "most rapidly urbanising" regions (Table 2.5). The ratio of urban population is on the increase in Ukraine as well, but it is also a result of its close location to the Kyiv agglomeration.

Chernobyl did not change the direction of regional population dynamics. The decline in population would be significant even without Chernobyl; however, it did accelerate the process. The population density was low even before the disaster, and the evacuations merely accentuated this state of affairs. Natural population reproduction data (crude birth and crude death ratio) around 2010 does not reflect any correlation with higher radiation levels any more, suggesting the decline in the birth rate was only temporary after the disaster and connected with the uncertainties because of resettlement. The disaster did, however, fundamentally alter the urbanisation processes and the network of villages. Smaller villages in remote areas disappeared in significant numbers, whereas small towns and minor urban centres became relatively more "stable".

The negative demographic processes of the Polesye combined with the disaster-caused outmigration and resettlement poked a huge hole in the demographic space of the region, which is especially spectacular in the changing population density within rural areas. Even the districts outside the evacuated zone became the most sparsely populated areas of Belarus and Ukraine (Fig. 2.3).

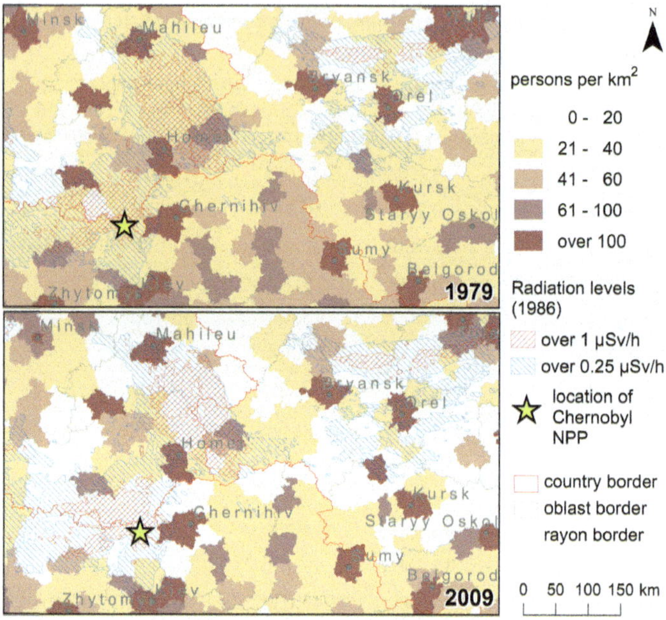

Fig. 2.3 Population density by rayons (Author and cartography by Karácsonyi, calculation based on 1979 Soviet, 2009 Belarusian, 2010 Russian census populations as well as de jure population in Ukraine by 2009)

2.5.2 *Fukushima—Recent Demographic Processes*

Regional population structure changed fundamentally after the construction of the nuclear power plants in the coastal area of Fukushima Prefecture. According to a case study conducted in Tomioka Town by Kajita (2014), residents consisted of three groups: (1) people who lived in the area originally, before the construction of the nuclear power plant, (2) "newcomers" who migrated there to work for construction and electric industries and had already settled for a long term and (3) short-term stayers who were sent by TEPCO and other related companies. Overall, the total population increased in the 1970s when people migrated for work. However, during the 1990–2000s, it gradually decreased again or levelled off. Along with such population decline, the elderly ratio (people aged 65 years and over) went up to around 20–30% in 2010.

The nuclear accident had almost irreversible impacts on regional population structure. As we explained above, group evacuation was organised by municipality offices, and evacuees temporally stayed at city halls, schools and hotels in nearby major cities such as Iwaki, Fukushima, Koriyama and Nihonmatsu. An exception is Futaba, which chose Saitama Prefecture, located around 200 kms away to the south. Understandably, some people outside the evacuation areas escaped farther, stayed at their friends' or relatives' houses temporarily and after that found new houses by themselves.

According to official statistics, the total number of evacuees including both ordered and voluntary evacuations reached the maximum number (164,000 people) in May 2012 and 79,000 people remained evacuated as of February 2017 (Asahi 2017a).

In Japan, there are two types of temporal houses provided for evacuees by municipality: prefabricated houses specially constructed after the accident and existing rented houses that municipalities leased. The maximum number of house units provided was 16,800 in 2013 for the former type and 25,554 in 2012 for the latter type (Fukushima Prefecture 2017). Evacuees from the same municipality were arranged to stay in the same prefabricated house complex to maintain the original community and human network (Fig. 2.4). As the map suggests, most of them are located in major cities where infrastructures are provided and daily necessities are easily purchased. In contrast, the detail spatial distribution of households staying in rented houses is not publicly reported, but it can be assumed to be more dispersed and close to major cities based on the locations of housing supply before the accident.

The Japanese population census 2015 provides insights about where people were displaced after the accident (Fig. 2.5). Table 2.7 provides the proportion of evacuees who live outside or inside Fukushima Prefecture by five municipalities which lost their population almost completely after the accident. For those who stay in the prefecture, the proportions of current residences in four major cities are presented. Analysing the proportions disaggregated by age group, we found that the younger they are, the more they are likely to leave Fukushima Prefecture. Around 30–40% of people aged below 40 years old choose to find new residence outside Fukushima Prefecture, for example, Tokyo Metropolitan Area and Sendai in Miyagi Prefecture. Young people can choose a new place which is distant from their home municipality because they have less economic and social capital which keeps them staying in an area closer to their previous residence (Isoda 2015).

As far as those who remain in Fukushima Prefecture are concerned, people are likely to select the nearest major city. For example, in the case of Iitate Village, 60–70% of residents among those who remain in Fukushima Prefecture chose Fukushima city located 35 kms away. Around 40% of residents from Tomioka selected Iwaki. These results partly reflect the fact that about 55,239 people still lived in temporary houses at the end of September 2015 (Fukushima Prefecture 2017), when the Japanese population census was carried out. The figures in Table 2.7 partially include people already resettled. According to questionnaire surveys conducted by the Reconstruction Agency (2016) in 2015–2016, around 30–40% of affected families bought a new house and settled in a new community. Furthermore, based on the number of recipients of special provisions for housing acquisition, 85% of them seem to have found a newly owned house in Fukushima Prefecture (Asahi 2016). The locations of new houses are likely to be in major cities because land prices in places such as Fukushima City, Iwaki City have increased or at least levelled off.

The sudden population increase caused by mass displacement unintentionally led to several complaints among residents in receiving municipalities. For example, local newspapers reported that traffic jams became more frequent, the queues in hospitals were longer, and housing rent increased. It should also be noted that junior and high school students evacuated from Fukushima-affected areas experienced bullying

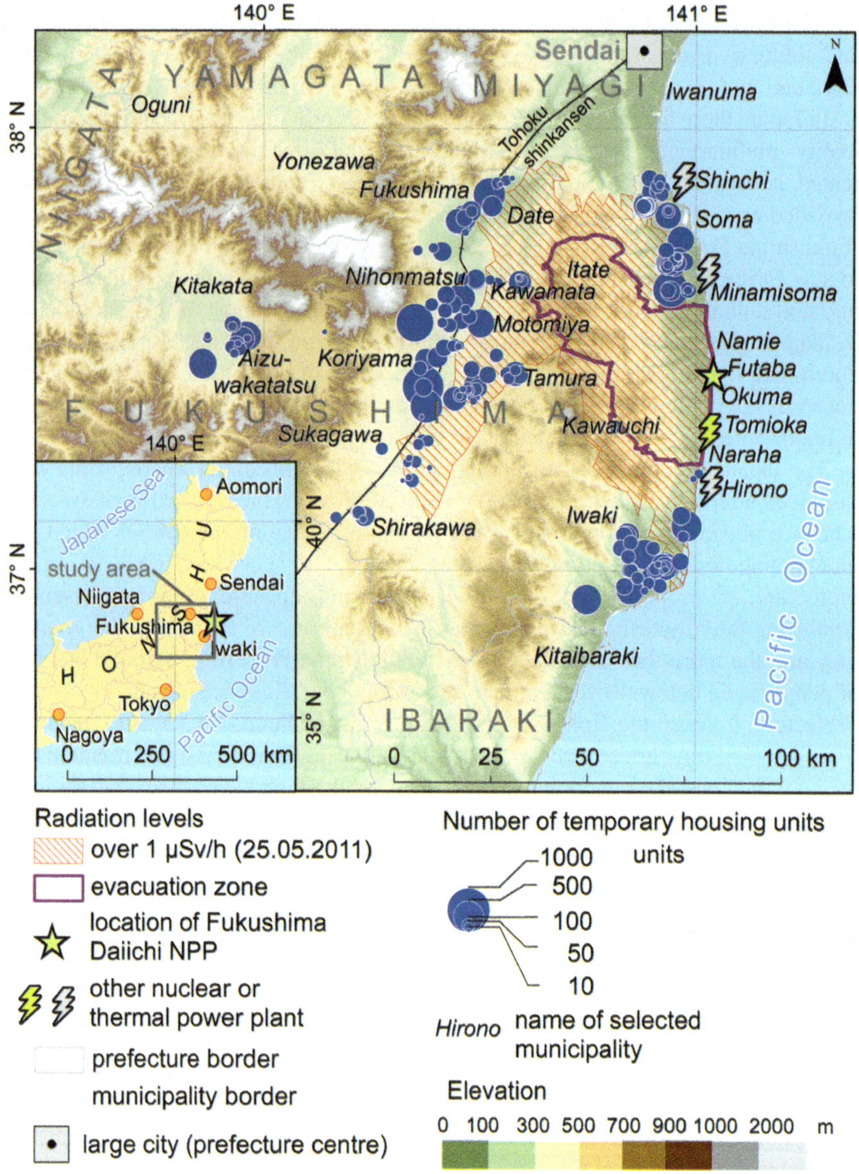

Fig. 2.4 Locations of temporary housing complexes in Fukushima Prefecture (Author Hanaoka, cartography by Hanaoka and Karácsonyi, data from www.pref.fukushima.lg.jp/sec/41065d/juutakutaisaku001.html)

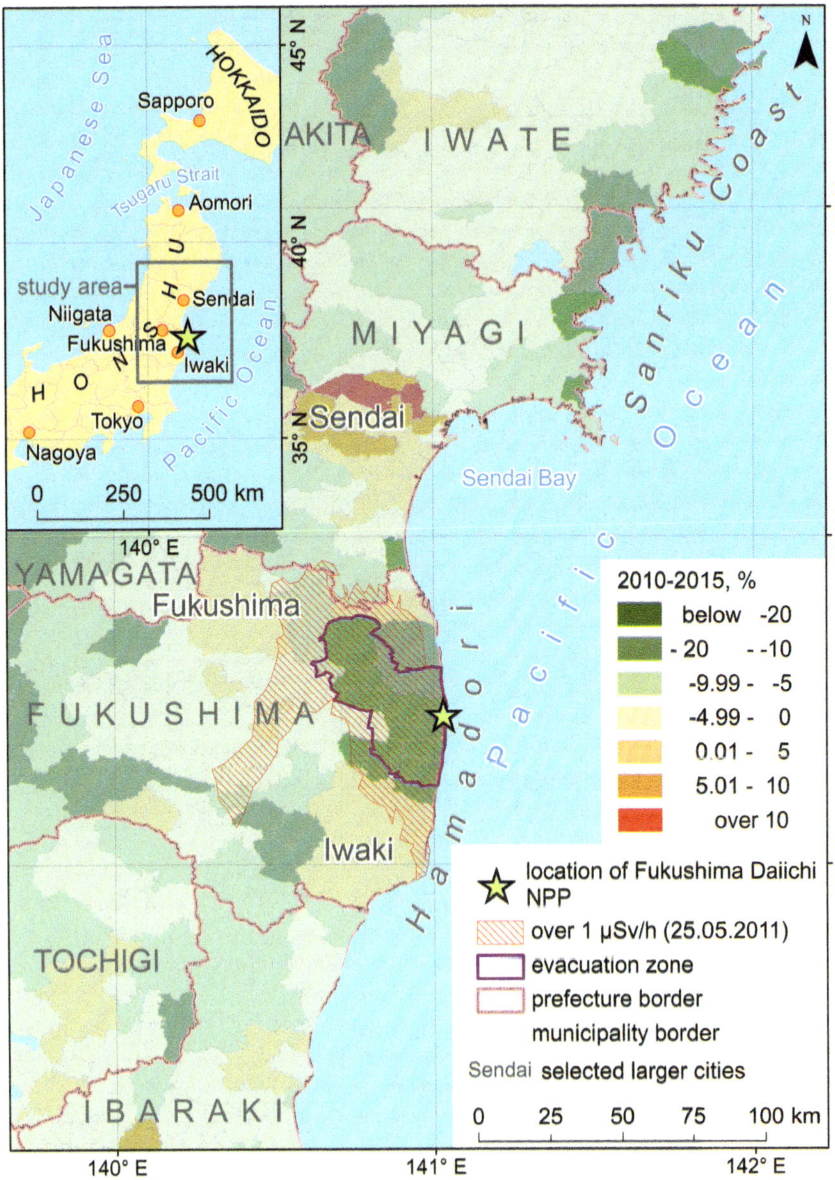

Fig. 2.5 Total population change by municipalities (Author and cartography by Karácsonyi, calculation based on 2010, 2015 Japanese census populations)

Table 2.7 Proportions of internal migrants by destination (based on 2010, 2015 Japanese census populations)

Destination	Unit	Current age group (years old)								
		5–14	15–29	30–39	40–49	50–59	60–69	70–79	80+	5+
From Tomioka Town										
Outside Fukushima Pref	%	**31.9**	**35.7**	**30.4**	**28.7**	23.6	23.1	19.2	20.2	26.9
Inside Fukushima Pref	%	68.1	64.3	69.6	71.3	**76.4**	**76.9**	**80.8**	**79.8**	73.1
Iwaki City (43 km)	%	38.6	36.9	40.3	39.7	**46.4**	**41.9**	**42.7**	**43.7**	41.3
Koriyama City (77 km)	%	13.3	14.9	13.5	16.0	16.8	**20.3**	**25.1**	**23.1**	17.6
Fukushima City (95 km)	%	**3.3**	2.4	**4.3**	2.0	1.8	2.4	1.6	**3.5**	2.6
Aizuwakamatsu City (124 km)	%	**2.0**	0.9	**1.6**	**1.4**	0.8	0.4	0.5	0.6	1.0
Total	N	1,137	1,842	1,564	1,795	1,888	2,029	1,255	1,121	12,631
From Okuma Town										
Outside Fukushima Pref	%	**31.5**	**33.2**	**32.0**	**28.7**	20.8	23.3	16.8	20.5	26.4
Inside Fukushima Pref	%	68.5	66.8	68.0	71.3	**79.2**	**76.7**	**83.2**	**79.5**	73.6
Iwaki City (50 km)	%	36.7	37.6	36.4	37.3	**45.4**	**42.1**	**43.4**	**41.7**	40.0
Koriyama City (68 km)	%	7.2	**8.4**	**8.4**	7.3	**8.7**	**8.9**	**8.6**	**8.8**	8.3
Fukushima City (95 km)	%	2.0	2.0	2.1	**3.1**	**3.3**	2.0	1.1	**2.8**	2.3
Aizuwakamatsu City (116 km)	%	**13.1**	10.7	9.1	**13.5**	11.2	**14.3**	**20.4**	**16.1**	13.1
Total	N	1,105	1,363	1,273	1,227	1,378	1,480	852	782	9,460
From Futaba Town										
Outside Fukushima Pref	%	**52.6**	**49.6**	**48.8**	**41.6**	32.9	39.2	30.4	34.3	40.7
Inside Fukushima Pref	%	47.4	50.4	51.2	58.4	**67.1**	**60.8**	**69.6**	**65.7**	59.3
Iwaki City (58 km)	%	27.0	25.5	26.0	**29.8**	**36.0**	27.6	**31.6**	**32.3**	29.5

(continued)

Table 2.7 (continued)

Destination	Unit	Current age group (years old)								
		5–14	15–29	30–39	40–49	50–59	60–69	70–79	80+	5+
Koriyama City (73 km)	%	6.6	8.8	8.7	9.0	**10.8**	**11.1**	**13.4**	8.9	9.8
Fukushima City (85 km)	%	1.7	3.1	3.3	2.9	**6.2**	**6.4**	**7.8**	**7.7**	5.1
Aizuwakamatsu City (120 km)	%	**1.9**	1.0	1.1	**2.0**	0.6	0.5	**1.4**	**1.6**	1.2
Total	N	534	718	642	652	776	996	629	685	5,632
From Namie Town										
Outside Fukushima Pref	%	**40.0**	**38.8**	**36.5**	**30.9**	23.6	23.9	20.5	22.1	28.8
Inside Fukushima Pref	%	60.0	61.2	63.5	69.1	**76.4**	**76.1**	**79.5**	**77.9**	71.2
Iwaki City (63 km)	%	**14.1**	13.0	**15.5**	**14.4**	**17.3**	13.8	10.7	12.3	14.0
Koriyama City (75 km)	%	**7.7**	7.2	**7.9**	6.9	**8.1**	7.2	6.0	6.9	7.3
Fukushima City (79 km)	%	12.7	14.5	11.6	**17.0**	**18.3**	**17.2**	**22.9**	**20.2**	17.0
Aizuwakamatsu City (126 km)	%	**2.9**	1.0	**2.7**	**1.4**	0.5	0.8	1.2	1.3	1.3
Total	N	1,499	2,207	1,868	1,936	2,567	2,973	1,994	1,864	16,908
From Iitate Village										
Outside Fukushima Pref	%	**12.4**	**12.7**	**9.4**	**7.3**	3.4	3.0	3.7	4.8	6.6
Inside Fukushima Pref	%	87.6	87.3	90.6	92.7	**96.6**	**97.0**	**96.3**	**95.2**	93.4
Iwaki City (101 km)	%	0.0	**0.6**	0.2	0.2	0.5	0.2	0.2	0.0	0.3
Koriyama City (64 km)	%	0.7	2.0	**0.9**	0.7	**1.1**	0.5	0.9	0.6	0.9
Fukushima City (35 km)	%	**69.3**	**68.1**	60.5	**65.6**	58.5	62.0	**62.8**	57.0	62.4
Aizuwakamatsu City (104 km)	%	**1.4**	**0.6**	**1.2**	0.2	0.0	0.0	0.0	0.0	0.3
Total	N	476	624	469	477	754	838	575	691	4,904

Values above the average (= age group: 5+) are presented in bold
Distance in brackets is based on the shortest path between two municipal offices calculated by Google Maps

concerning the nuclear accident at school. In this way, resettlement after the nuclear accident, unfortunately, has been accompanied by various hardships in the new environment for evacuees. Figure 2.6 shows those municipatities where evacuees were

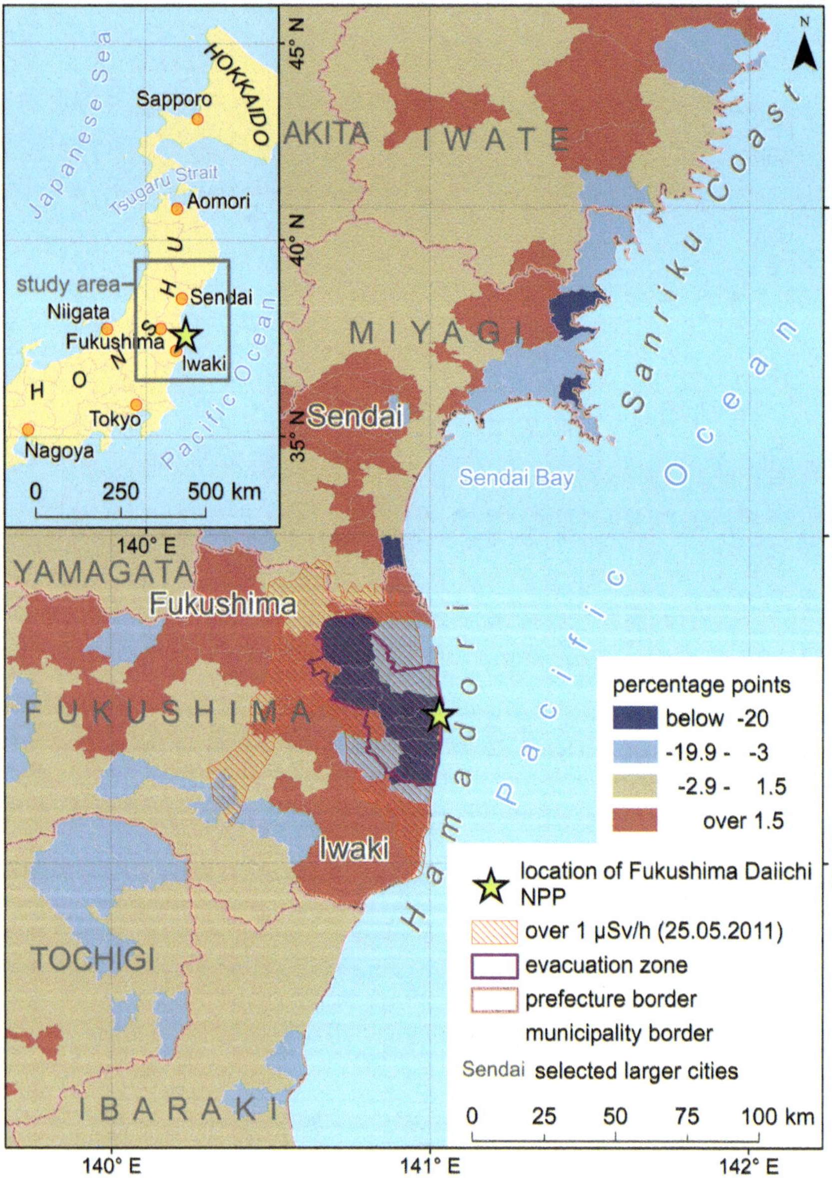

Fig. 2.6 Changing population dynamics after 2011 (Difference between population change of 2005–2010 and 2010–2015) (Author and cartography by Karácsonyi, calculation based on 2005, 2010, 2015 Japanese census populations)

hosted (red colour), changing the demographic trends between 2010 and 2015 in these settlements.

In Fukushima, affected municipalities are planning several resettlement sites outside evacuation areas, but they seem to be temporal, not permanent resettlements. This is because firstly, some evacuees' wish to return to their hometowns, and secondly, permanent resettlement sites managed by the affected municipality in the receiving municipality are practically difficult under the current Japanese local government system. There are several issues to solve, such as whether dual resident registration can be admitted or not and how to share taxes and public services among municipalities (Tsunoda 2015). Therefore, in principle, affected municipalities are making every effort to enable residents to return to their homes through decontamination. The evacuation order has already been lifted for some evacuation areas, and radiation levels dropped as a result of the natural degradation and decontamination. In the spring of 2017, the evacuation order excepting the highest contaminated area was lifted for the majority of evacuation areas in Iitate Village, Kawamata Town, Namie Town and Tomioka Town. Approximately, 32,000 people lived there before the accident (Asahi 2017b). The total evacuation areas being lifted until April 2017 was 70% of the initial evacuation areas issued immediately after the accident. Yet, many people decided not to return. They not only worried about the radiation level, but basic amenities (e.g. shops, hospitals) and employment opportunities are limited. For example, in Naraha Town, the evacuation order was lifted for all evacuation areas, which covered 80% of the municipality. According to official town records (Naraha Town 2016), only 781 people, which is equivalent to 10.6% of the total population before the accident, have returned. Among them, the proportion of people aged 65 years and over reached 53% (the elderly ratio in 2010 was 24%).

Young people resettled outside Fukushima Prefecture, while many of the migrants who returned are elderly people. This trend accompanies geographical separations of generations in rural areas where young families and their parents traditionally lived together. In addition, since the 1970s, nuclear industries had attracted many migrants from the outside. Many of the first generations reached or are reaching the age of retirement (65 years old). Thus, it is a difficult question whether such people who migrated from the outside previously and settled for a long time will choose to return again. Pre-disaster population structure characterised by a mixture of rural and industrialised areas makes the estimation of future demographic trends difficult in Fukushima-affected areas.

Using spatial mobile statistics, we mapped the average of hourly population per 500 m grid cell in June 2016 and compared it to 2010 population census data to analyse the changes (Fig. 2.7). Comparing the two maps, we found that the geographic distribution of grid cells with high population density almost remained the same outside the evacuation areas between two time periods, while inside, they disappeared almost completely. In particular, there is no peak in population density distribution in the town centre proximity to railway stations, implying that densely populated residential areas no longer exist after the accident. However, there are several grid cells with somewhat higher population density in areas close to the nuclear power plant in Okuma Town and Tomioka Town. This population distribution does not well

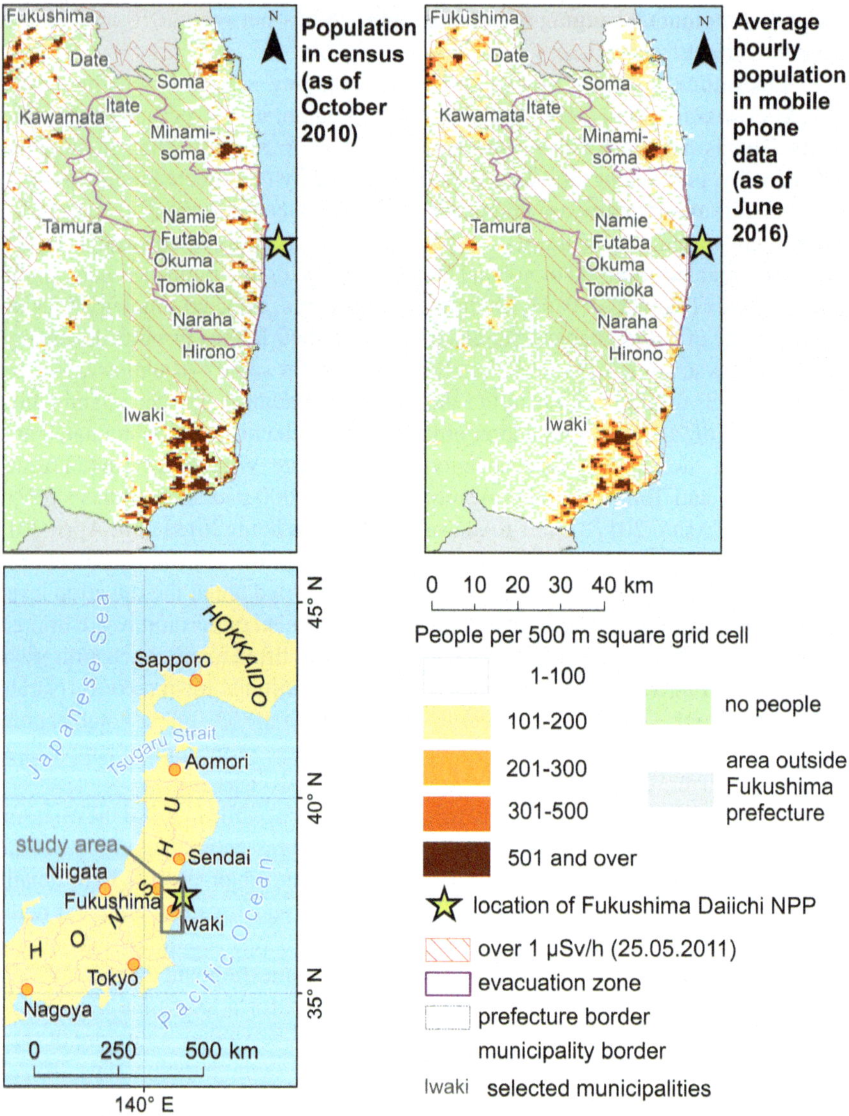

Fig. 2.7 Population distributions in census and mobile phone data (Author Hanaoka cartography by Hanaoka and Karácsonyi)

overlap with the census one, suggesting that most of the people are temporal visitors such as engineering and technical workers at nuclear power plants.

Eighty percent of the area of Naraha Town, located within the 20 km radius from the nuclear power plant, was previously included in the evacuation areas. As we discussed above, the evacuation order was lifted for these areas in September 2015, and the total number of grid cells with population density above 100 people also increased from 32 grid cells in June 2015 to 41 grid cells in June 2016, among 472 grid cells across Naraha Town. This result may suggest that people already returned in these areas or people can enter them more frequently than before, perhaps in preparation for returning. The Japanese population census is conducted every five years, and it is not able to capture such temporal dynamics of population changes. Using mobile spatial statistics will help to continuously monitor and explore the reconstruction progress after the evacuation areas are lifted.

2.6 Discussion and Recommendations

The mass displacement after a nuclear disaster, rather than the radiation itself has a much more significant impact on deteriorating health, natural reproduction and economic performance of the affected population. Based on the literature and the results of this study, these consequences can be summarised as follows:

(1) In both Fukushima and Chernobyl cases, the regional impact of the accident resulted in a dramatic loss of population in the contaminated areas and accelerated the concentration of populations in adjacent major cities through evacuation. Evacuated communities were traumatised and destroyed, facing challenges that result from not having any spatial or social attachment in the recipient areas (see also Chap. 11). The receiving communities faced population growth and a radically changed composition by the arriving evacuees, whose integration often ended up in social segregation and marginalisation.
(2) A strong spatial shift towards urbanisation can also be observed because urban centres provide a better chance for socio-economic recovery and re-integration (job opportunities, more extensive social network) for the evacuees, even for former rural residents (Voloshin et al. 1996; IOM 1997). Carson et al. (Chap. 5) emphasise that other types of disasters can also increase urbanisation because urban areas providing people more opportunities to cope with the consequences.
(3) Large migration shifts are significant during the first 5–10 years after the disaster, when a large number of evacuated population are on the move, often staying temporarily at one location and going through multiple migration steps until settling down. The population trends can take a totally different direction for this decade in certain areas. This challenges the local housing market, service sector and government policy. Later, these shifts are less and less significant. Some of the people remain, but the majority of the people (mostly young families) start a new life mostly in urban areas out of the affected region.
(4) After a decade, however, positive migration balance has been observed in certain areas of the affected region, mainly resulting from clean-up workers, scientists or even tourists and settlers along with the returning elderly. High variation in short-term population numbers can be registered in the regions with extremely low population density caused by the disaster.

Population immigration trends are influenced by employment opportunities. Sectors vulnerable to radiation such as agriculture, fishing, food industry are in decline (IOM 1997; UNDP 2002b), while others requiring a lesser sized or a rotating labour force, such as forestry or nuclear waste storage facilities are maintained. New jobs in health care, engineering, science and construction are also represented.

(5) The returning migration or remaining in place is more common in the case of elderly people. Firstly because they have a higher attachment to the place and because of their age, they are less flexible to start a new life somewhere else. Their expected lifelong exposure is also much lower than for the adolescent population. Therefore, they are more likely to accept the risks associated with living in these polluted areas. In-migration to contaminated areas can also be observed in Belarus recently. However, this results from government policy to attract new settlers, rather than the return of the elderly population.

Unlike the Chernobyl case, in Fukushima, permanent resettlement sites outside the affected area were not organised, and instead, extensive decontamination work and higher accepted radiation thresholds make evacuees legally able to return their original homes within six years or so after the nuclear accident. This was also possible because of lower level of Cs^{137} contamination and lack of high risk trans-uranium elements. Not to mention that according to some experts (see Hjelmgaard 2016), after the ^{131}I and ^{137}Cs phases (IOM 1997), a third phase related to ^{241}Pu–^{241}Am decay has begun in Chernobyl with health consequences yet to be realised and understood. Given the differences in radio-ecological conditions, reconstruction policy and the time framework, Fukushima may demonstrate demographic consequences that are different from the Chernobyl case. In contrast, the return rate is very low which causes a large drop in population density similarly to Chernobyl.

According to World Nuclear Association (2018), nuclear accidents have a low and decreasing probability. Even though previously unforeseen circumstances always could cause accidents in the nuclear industry (Labaudiniere 2012), bring forth the need for evacuation or even permanent displacement of large populations. There are certain points that should be considered when developing resettlement and redevelopment policy for possible future disaster-stricken regions based on the consequences of past displacements. Most of the following points were not considered in post-Chernobyl but were followed in post-Fukushima mass displacement showing a clear policy improvement and better situation-adapted decision-making.

(1) A well-planned short-term evacuation is necessary during the rapid-onset phase ("^{131}I phase") of the disaster when a high number of temporary or permanent housing, as well as financial and social aid, is needed to cope with the consequences. This is the most significant challenge within 1–2 years following the disaster. Chernobyl, and to some extent Fukushima, evacuation measures failed during this phase.

(2) A clear distinction between temporary and permanent displacement as well as planning and straightforward communication of community futures accordingly is strongly required during the slow-onset phase ("^{137}Cs-phase" from 2

to 50 years onwards following the disaster). The permanent mass displacement should be seen only as the ultimate solution to cope with the consequences of a nuclear disaster. It can cause much larger and longer demography shifts than the distortion caused by a low-level radiation exposure resulting in health problems. People should be provided with reliable information about the threat and assistance if they decide to move. On the other hand, people should be allowed to take the risk if they decide to do so, but compensation is necessary.

(3) The intent to move, stay or return is strongly age-specific, and the return migration should be supported accordingly. Infrastructure redevelopment should be planned in accordance with the demographic shift towards aging populations such as more senior homes, hospitals rather than nursery. Low return rates, falling population density and shifting settlement system towards population concentration and urbanisation are also common consequences. Thus, infrastructure regeneration, reconstruction (roads, railway links) should target these areas.

In summary, the most significant lesson from this study is that a poorly planned mass displacement can cause a larger economic loss than the disaster itself. There were clear differences between the management of Chernobyl and Fukushima disasters in this regard. The post-Chernobyl policies that drove population displacement lacked previous experiences on effect of radiation on human health in large populations. The administrative rigidity of the zoning and lack of financial sources for infrastructure-reconstruction and redevelopment (see also Chap. 11) accelerated further the total population and economic loss in Chernobyl. The knowledge and experience derived from the Chernobyl case helped the decision-making in Fukushima. Townships were opened up for returners in recent years in the Fukushima disaster area is possible because of the lower level of radiation and lack of emitted trans-uranium isotopes. The long-lasting existence of Chernobyl zone contributed significantly to the false view to explain all demographic consequences by the invisible radiation threat presented in and around the "death zone". This delivered a negative image of the entire area, making regional redevelopment even more difficult.

References

Abbott, P., Wallace, C., Beck, M. 2006. Chernobyl: Living with risk and uncertainty. *Health, Risk & Society*, 8(2):105–121.

Asahi 2016. Genpatsu hinan fueru teiju: baisho susumi fudosan shotoku nanasenken cho (in Japanese) (Resettlements of evacuees from the nuclear accident are increasing: Compensation progressed and house acquisitions exceeded 7,000 cases). *Asahi Newspaper*, Morning 21st February 2016, p.1.

Asahi 2017a. Hinan kenmin 8 man nin kiru (in Japanese) (The number of evacuees from Fukushima Prefecture dropped below 80,000 people). *Asahi Newspaper*, Morning 21st February 2017, p.27.

Asahi 2017b. Genpatu hina, konshun 4 choson 3.2 man nin kaijo: konnan kuiki nao 2.4 man nin (in Japanese) (Evacuation from the nuclear accident, 4 municipalities will be lifted this spring and the total number of evacuees are 32,000 people. Still 24,000 evacuees of "difficult-to-return" zone). *Asahi Newspaper*, Morning 28th February 2017, p.1.

Balonov, M., Crick, M., Louvat, D. 2010. *Update of Impacts of the Chernobyl Accident: Assessments of the Chernobyl Forum (2003–2005) and UNSCEAR (2005–2008)*. Proceedings of Third European IRPA Congress 2010 June 14–16, Helsinki, Finland.

Baranovsky, M. O. 2010. *Silski depresivni teritorii Polissia: osoblivosti rozvitku ta sanacii* (in Ukrainian) (Rural depressive areas in Polissia: development characteristics and improvement). Nizhin Gogol State University, Nizhin.

Baverstock, K, Wiliams, D 2006. The Chernobyl Accident 20 Years On: An Assessment of the Health Consequences and the International Response. *Environmental Health Perspectives*, 114(9):1312–1316.

Brenot J., Charron S., Verger P. 2000. Mental health effects from radiological accidents and their social management. *10th international congress of the International Radiation Protection Association* Hiroshima, Japan. pp. 10–165. Accessed on-line 27.11.2019: https://www.irpa.net/irpa10/cdrom/00633.pdf.

Cernea, M. 2004. Impoverishment risks, risk management, and reconstruction: A model of population displacement and resettlement. *UN Symposium on Hydropower and Sustainable Development*, 27–29 October: Beijing, China. Accessed on-line on 27.11.2019: https://commdev.org/pdf/publications/Impoverishment-Risks-Risk-Management-and-Reconstruction.pdf.

European Commission 2011. Recent scientific findings and publications on the health effects of Chernobyl. Summary report. *Radiation protection No. 170*. Luxembourg. Accessed on-line on 27.11.2019: https://ec.europa.eu/energy/sites/ener/files/documents/170.pdf.

Fukushima Prefecture. 2017. Accessed on-line on 27.11.2019: https://www.pref.fukushima.lg.jp/uploaded/life/256620_601558_misc.pdf.

Greenpeace 2006. *The Chernobyl Catastrophe. Consequences on Human Health*. Greenpeace, Amsterdam. Accessed on-line on 27.11.2019: https://www.sortirdunucleaire.org/IMG/pdf/greenpeace-2006-the_chernobyl_catastrophe-consequences_on_human_health.pdf.

Hatch, M., Ron, E., Bouville, A., Zablotska, L., Howe, G. 2005. The Chernobyl Disaster: Cancer following the Accident at the Chernobyl Nuclear Power Plant. *Epidemiologic Reviews* 27(1):56–66.

Hjelmgaard, K. 2016. Exiled scientist: 'Chernobyl is not finished, it has only just begun'. USA Today. Accessed online on 28.11.2019: https://www.usatoday.com/story/news/world/2016/04/17/nuclear-exile-chernobyl-30th-anniversary/82896510/.

Hugo, G. 2008. Migration, Development and Environment. *IOM Migration Research Series*. Geneva.

IAEA 2006. *Chernobyl's Legacy: Health, Environmental and Socio-Economic Impacts and Recommendations to the Governments of Belarus, the Russian Federation and Ukraine*. IAEA, Vienna. Accessed on-line on 27.11.2019: https://www.iaea.org/Publications/Booklets/Chernobyl/chernobyl.pdf.

IAEA-WHO-UNDP. 2005. Chernobyl: *The True Scale of the Accident*. Accessed on-line on 27.11.2019: https://www.who.int/mediacentre/news/releases/2005/pr38/en/.

IOM 1997. *Ecological Migrants in Belarus: Returning Home after Chernobyl?* IOM, Geneva. Accessed on-line on 27.11.2019: https://publications.iom.int/system/files/pdf/return_chernobyl_0.pdf.

ICRP 2009. Application of the Commission's Recommendations to the Protection of People Living in Long-term Contaminated Areas After a Nuclear Accident or a Radiation Emergency. *ICRP (International Commission on Radiological Protection) Publication* 111. Ann. ICRP 39 (3).

Imanaka, T. 2016. Chernobyl and Fukushima: Comparison of accident process and radioactive contamination. *Kagaku* 86(3):252–257.

Ioffe, G. 2007. Belarus and Chernobyl: Separating seeds from chaff. *Post-Soviet Affairs*, 23(2):1–14.

Isoda, Y. 2015. Fukushima daiichi gennshiryoku hatudensho jiko ni yoru hinansha no seikatu to sentaku teki ido: jinteki shihon ron ni moto duku "Okuma machi fuko keikaku chomin ankeito no bunseki (in Japanese) (Refuge life of evacuees from the Fukushima nuclear accident and their selective migration: Analysis of Okuma Town Survey based on human capital theory.) In: ed. Yoshihara, N., Nihei, Y., Matsumoto, M.: *Records of the Victims' Refuge Lives in the Great East Japan Earthquake*, Rikka Press, pp. 544–561.

Jaworowski Z. 2010. Belarus to Repopulate Chernobyl Exclusion Zone. *Nuclear Update*, 2010 (Summer):46–47.

Kajita, S. 2014. Establishment of nuclear power stations and reconfiguration of local society and economy: A case study of Tomioka, Fukushima Prefecture. *Geographical Review of Japan Series A*, 87(2):108–127.

Khomra A. U. 1989. Rural depopulation trends in the Ukrainian SSR: The delimitation and spatial differentiation. In: ed. Stasiak, A., Mirówski, W. *The process of depopulation of rural areas in Central an Eastern Europe*. Polish Academy of Sciences, Institute of Geography and Spatial Organisation, Warsaw pp. 173–182.

Labaudiniere, M. S. 2012. *Three Mile Island, Chernobyl, and the Fukushima Daiichi Nuclear Crises*. BA, Boston College.

Lakiza-Sachuk, N. N., Omelianets, N. I., & Pyrozhkov, S. I. 1994. The significance of the socio-demographic consequences of the Chernobyl disaster for Ukraine: Current situation and prospects. Working Paper No. 20. National Institute for Strategic Studies, Kiev.

Lavell, C., Ginnetti, J. 2013. Technical Paper: The risk of disaster-inducted displacement. Central America and the Caribbean. Internal Displacement Monitoring Centre (IDMC). Geneva. Norwegian Refugee Council.

Lehman, H., Wadsworth, J. 2009. The Impact of Chernobyl on Health and Labour Market Performance in the Ukraine. Discussion Paper No. 4467. Institute for the Study of Labor, Bonn.

Lehman, H., Wadsworth, J. 2011. The Impact of Chernobyl on Health and Labour Market Performance. CEP Discussion Paper No 1052. Centre for Economic Performance.

Libanova, E. M. 2007. Chornobylskaya Katastrofa: 25 Rokiv Potomu (in Ukrainian) (Chernobyl Disaster: 25 Years After) *Demografiya ta Socialna Ekonomica*, 2(16):3–18.

Linge, I., Melikhova, I., Pavlovski, O. 1997. Medico-demographic criteria in estimating the consequences of the Chernonyl accident. In: One decade after Chernobyl: Summing up the consequences of the accident. Poster presentations — Volume 1 International Conference 8–12 April 1996, Vienna. Accessed on-line on 27.11.2019: https://www.iaea.org/inis/collection/NCLCollectionStore/_Public/28/073/28073807.pdf.

Lochard, J. 1996. Psychological and social impacts of post-accident situations: Lessons from the Chernobyl accident – International Congress on Radiation Protection, Vienna. Accessed on-line on 27.11.2019: https://www.irpa.net/irpa9/cdrom/VOL.1/V1_10.PDF, pp. 105–111.

Malko, M. V. 1998. Social Aspects of the Chernobyl Activity in Belarus – Research Activities about the Radiological Consequences of the Chernobyl NPS Accident and Social Activities to Assist the Sufferers by the Accident, Kyoto. Accessed on-line on 27.11.2019: https://www.rri.kyoto-u.ac.jp/NSRG/reports/kr21/kr21pdf/Malko3.pdf, pp. 24–37.

Marples, D. R. 1993. A Correlation Between Radiation and Health Problems in Belarus? *Post-Soviet Geography*, 34(5):281–292.

Marples, D. R. 1996. *Belarus: From Soviet Rule to Nuclear Catastrophe*. New York, St. Martin's Press.

Mesle, F., Poniakina, S. 2012. Prypiat died – long live Slavutych: Mortality profile of population evacuated from Chornobyl exclusion zone. Accessed on-line on 27.11.2019: https://cdn.uclouvain.be/public/Exports%20reddot/demo/documents/MeslePoniakina.pdf.

Ministry of Environment. 2017. Decontamination. Accessed on-line on 27.11.2019: https://josen.env.go.jp/en/decontamination/.

Naraha Town 2016. Accessed on-line on 27.11.2019: https://www.town.naraha.lg.jp/information/genpatu/001261.html.

NAIIC 2012. The Official Report of the Fukushima Nuclear Accident Independent Investigation Commission. National Diet of Japan Fukushima Nuclear Accident Independent Investigation Commission (NAIIC). Accessed on-line on 27.11.2019: https://warp.da.ndl.go.jp/info:ndljp/pid/3856371/naiic.go.jp/en/report/.

Nuclear Emergency Response Headquarters 2015. For Accelerating the Reconstruction of Fukushima: From the Nuclear Disaster (revised). Accessed on-line on 27.11.2019: https://www.meti.go.jp/earthquake/nuclear/kinkyu/pdf/2015/0612_02.pdf.

OECD 2002. *Chernobyl: Assessment of Radiological and Health Impacts. 2002 Update of Chernobyl: Ten Years On.* Nuclear Energy Agency. Organisation for Economic Co-operation and Development. Accessed on-line on 27.11.2019: https://www.oecd-nea.org/rp/pubs/2003/3508-chernobyl.pdf.

Omel'anets, N.I., Miretski, G.I., Saurov, M.M., Torbin, V.F. 1988. Medical demographic consequences of the Chernobyl accident. Medical aspects of the Chernobyl accident. A technical document issued by the International atomic energy agency. IAEA-Tecdoc-516. Vienna. Accessed on-line on 27.11.2019: https://www-pub.iaea.org/books/IAEABooks/771/Medical-Aspects-of-the-ChernobylAccident-Proceedings-of-the-All-Union-Conference-Kiev-USSR-11-13-May-1988, pp. 315–325.

Omelianets, N., Bazyka, D., Igumnov, S., Loganovsky, K., Prysyazhnyuk, A., Stepanova, E., Afanasev, D. 2016. *Health Effects of Chernobyl and Fukushima: 30 and 5 years down the line.* Greenpeace, Brussels.

Oliver-Smith, A. 1996. Anthropological research on hazards and disasters. *Annual Review of Anthropology*, 1996(25):303–328.

Oliver-Smith, A. 2009. Climate change and population displacement: Disasters and diasporas in the twenty-first century. In: Crate, S. A., Nuttal, M. eds. *Anthropology and climate change: from encounters to actions.* Left Coast Press, Walnut Creek, CA.

Oliver-Smith, A. 2013. Catastrophes, Mass Displacement and Population Resettlement. In: Bissel, R. *Preparadness and Response for Catastrophic Disasters.* CRC Press, Taylor&Francis Group, Boca Raton, London, New York. pp. 185–224.

Oyabu, Y., Terada, M., Yamaguchi, T., Iwasawa, S., Hagiwara, J. and Koizumi, D. 2013. Evaluating reliability of mobile spatial statistics. *NTT DOCOMO Technical Journal*, 14(3):16–23.

Peplow, M. 2006. Counting the dead. *Nature,* 440(20):982–983.

Reconstruction Agency 2016. Gensiryoku hisai jichitai ni okeru jumin iko chosa kekka (in Japanese) (Results of questionnaire survey about the residents in municipalities affected by the nuclear accident). Accessed on-line on 27.11.2019: https://www.reconstruction.go.jp/topics/main-cat1/sub-cat1-4/ikoucyousa/.

Robinson, C. 2003. Overview of Disaster. In Robinson, C., Hill, K. eds. 2003. *Demographic Methods In Emergency Assessment. A Guide For Practitioners.* Center for International Emergency, Disaster and Refugee Studies (CIEDRS) and the Hopkins Population Center; Johns Hopkins University Bloomberg School of Public Health, Baltimore, Maryland.

Rolevich, I. V., Kenik, I. A., Babosov, E. M., Lych, G. M. 1996. Report for Belarus. In: One decade after Chernobyl. Summing up the Consequences of the Accident. Proceedings of an International Conference Vienna, 8–12 April 1996. IAEA.

Rumyantseva G.M., Drottz-Sjoberg B.-M., Allen P. T., Arkhangelskaya H. V., Nyagus A. I., Ageeva L. A., Prilipko V. 1996. The influence of social and psychological factors in the management of contaminated territories. In: Karaoglou A., et al., eds. *The radiological consequences of the Chernobyl accident.* Minsk, European Commission, pp. 443–452.

Shestopalov, V.M., Naboka, M.V., Bobylova, O.A., Isakova, T.I. 1996. Influence of Ecological factors on the morbidity among the Chernobyl accident sufferers. In: "Ten Years after the Chernobyl Catastrophe" Proceedings of International Conference, UNESCO, Minsk. pp. 101–109.

Steinhauser, G., Brandl, A., Johnsom, T.E. 2014. Comparison of the Chernobyl and Fukushima nuclear accidents: a review of the environmental impacts. *Science of the Total Environment,* 470–471(2014):800–817.

Team in Charge of Assisting the Lives of Disaster Victims, Cabinet Office 2013. Kikan konnan chiiki ni tsuite (in Japanese) (About areas where it is expected that the residents have difficulties in returning for a long time). Accessed on-line on 27.11.2019: https://www.mext.go.jp/b_menu/shingi/chousa/kaihatu/016/shiryo/__icsFiles/afieldfile/2013/10/02/1340046_4_2.pdf.

TEPCO 2011. Investigation Committee on the Accident at the Fukushima Nuclear Power Stations of Tokyo Electric Power Company (TEPCO). Interim Report. Accessed on-line on 27.11.2019: https://www.cas.go.jp/jp/seisaku/icanps/eng/interim-report.html.

Terada, M., Nagata, T. and Kobayashi, M. 2013. Population estimation technology for mobile spatial statistics. *NTT DOCOMO Technical Journal,* 14(3):10–15.

TORCH 2006. *The other report on Chernobyl.* The Greens in the European Parliament. Berlin, Brussels, Kiev.

TORCH 2016. *An independent scientifc evaluation of the health-related effects of the Chernobyl nuclear disaster.* GLOBAL 2000/Friends of the Earth, Vienna

Tsunoda, H. 2015. Genpatsu jichitai no chogai komyunithi koso to jichitai saiken no kadai (in Japanese) (Planning a new community outside a municipality affected by the nuclear accident and issues for reconstructing the community) Kenkyu to Hokoku, Jichiroren Institute of Local Government 107. Accessed on-line on 27.11.2019: https://www.jilg.jp/_cms/wpcontent/uploads/2015/02/6d5b33ffbe0caa1faf28c3d7e50147e2.pdf.

Tykhyi, V. 1998. Chernobyl Sufferers in Ukraine and Their Social Problems: Short Outline. Research Activities about the Radiological Consequences of the Chernobyl NPS Accident and Social Activities to Assist the Sufferers by the Accident, Kyoto. Accessed on-line on 27.11.2019: https://www.rri.kyoto-u.ac.jp/NSRG/reports/kr21/kr21pdf/Tykhyi.pdf, pp. 235–245.

UN 2002. *The Human Consequences of the Chernobyl Nuclear Accident, A Strategy for Recovery.* United Nations. Accessed on-line on 10.06.2018: https://www.un.org/ha/chernobyl/docs/report.pdf.

UNDP 2002a. *The Human Consequences of the Chernobyl Nuclear Accident. A Strategy for Recovery.* Report by UNDP and UNICEF. Accessed on-line on 27.11.2019: https://www.iaea.org/sites/default/files/strategy_for_recovery.pdf.

UNDP 2002b. Belarus National Report. UNDP, Minsk. Accessed on-line on 27.11.2019: https://chernobyl.undp.org/english/docs/blr_report_2002.pdf.

UNDP 2002c. Chernobyl Recovery and Development Program. Program Outcome Evaluation. UNDP, Kyiv. Accessed on-line on 27.11.2019: https://erc.undp.org/evaluation/documents/download/646.

UNSCEAR 2000. *Sources and Effects of Ionizing Radiation. Vol. II. Annex J. Exposures and effects of the Chernobyl accident.* UN, New York. Accessed on-line on 27.11.2019: https://www.unscear.org/docs/publications/2000/UNSCEAR_2000_Annex-J.pdf.

Veselkova, I.N., Zemlyanova, E.V., Silina, Z.D. 1994. Some demographic tendencies in the Russian Federation. *Health Care of the Russian Federation,* 3(1994):30–33.

Voloshin, V., Gukalova, I., Reshetnyk, V. 1996. Geographic aspects of the socio-economic consequences of the Chernobyl catastrophe in Ukraine, in UNESCO 1996 Ten Years after the Chernobyl Catastrophe Proceedings of International Conference, Minsk, Belarus. pp. 189–197.

WHO 2006. *Health Effects of the Chernobyl Accident and Special Health Programmes.* Report of the UN Chernobyl Forum. Geneva. Accessed on-line on 27.11.2019: https://www.who.int/ionizing_radiation/chernobyl/who_chernobyl_report_2006.pdf.

WHO 2016. 1986–2016: *Chernobyl at 30. An update.* Accessed on-line on 27.11.2019: https://www.who.int/ionizing_radiation/chernobyl/Chernobyl-update.pdf?ua=1.

World Nuclear Association (2018). *Safety of Nuclear Power Reactors.* Accessed on-line on 27.11.2019: https://www.world-nuclear.org/information-library/safety-and-security/safety-of-plants/safety-of-nuclear-power-reactors.aspx.

Chapter 3
Estimating Migration Impacts of Wildfire: California's 2017 North Bay Fires

Ethan Sharygin

Abstract This chapter examines a cluster of wildfire conflagrations that hit northern California during October 2017, which resulted in significant loss of housing units (6874 residential structures destroyed or damaged). To assess the magnitude of the migration response and network of destinations, a method to estimate migration drawing from a proxy universe of households with students enrolled in public schools was proposed, using data on school exits and re-enrollments from a longitudinal student database. The analysis finds that a small minority of households affected by the fires moved out of the area. Out of nearly 7800 persons displaced by the central fire complex in one city, this study estimated fewer than 1000 changed neighborhoods; of those, fewer than 500 moved out of Sonoma County. These findings are applicable to other wildfires and localized disasters where a substantial portion of housing is lost but public infrastructure in the region remains intact.

Keywords Wildland-urban interface · Wildfire · Natural hazards · Migration · Administrative data

3.1 Introduction

This chapter examines a cluster of wildfire conflagrations that hit northern California during October 2017, which resulted in significant loss of housing units (6874 residential structures destroyed or damaged). The magnitude of migration response was unknown, as was the timing and the network of possible destinations. This chapter provides a standard theoretical framework for understanding fire disasters: their frequency and locations most at risk. It also presents an overview of the 'October 2017 Fire Siege' and the conditions which precipitated the fire followed by a discussion into possible approaches to estimating population impacts. A method is proposed to estimate migration drawing from a proxy universe of households with students enrolled in public schools using data on school exits and re-enrollments from a

E. Sharygin (✉)
Population Research Center, Portland State University, Portland, OR, USA
e-mail: sharygin@pdx.edu

D. Karácsonyi et al. (eds.), *The Demography of Disasters*,
https://doi.org/10.1007/978-3-030-49920-4_3

longitudinal student database. The analysis finds that a small minority of households affected by the fires moved out of the area. Out of nearly 7800 persons displaced by the central fire complex in one city, this study estimated that fewer than 1000 people changed neighborhoods; of those, fewer than 500 moved out of Sonoma County. These methods and findings are applicable to policies and policy research regarding future wildfires and localized disasters where a substantial portion of housing is lost but public infrastructure in the region remains intact.

Section 3.1 of this chapter discusses the growth of wildfire risk in California over the last and next century. Section 3.2 provides background on the Central Sonoma-Lake-Napa Unit (LNU) Complex fires, in particular the largest, the Tubbs fire in Sonoma County. Sections 3.3 and 3.4, the size of the displaced population is estimated, and previous research on estimating migration discussed. The 'school enrollment proxy method,' a new method of estimating migration, is described along with the results. The closing section outlines limitations of this approach and identifies directions for future research.

3.2 Background

3.2.1 Wildfire Hazard in California

Disasters are a major part of the Californian experience. Throughout American expansion into California, earthquakes, floods, fire, and other catastrophes have played an important role in defining life in the state. A tragic standard was set by the 1906 San Francisco earthquake and ensuing fire that followed, which destroyed 28,000 buildings, displaced at least 200,000 residents, and killed between 500 and 3000 people. Flooding in Los Angeles in 1938 destroyed 5600 buildings and caused over 100 deaths. Earthquakes at Loma Prieta in 1989 and Northridge in 1994 injured thousands and caused billions of dollars of damage. Wildfires in Berkeley during 1923 and then Oakland in 1991 destroyed thousands of residences and emphasized wildfire as a constant threat in urban and suburban parts of the state.

Disasters can be broadly classified into two types of looming threats. Gradual-onset disasters include droughts, sea level rising, global warming, while sudden-onset disasters would include fires, floods, and earthquakes. Sudden-onset threats are especially important for applied demographers to anticipate and prioritize due to their disruptive and chaotic aftermath and the essential role that demographic data can play in the government response. Sudden-onset disasters have immediate consequences, including mass mobilizations of people and resources.

This paper presents a case study from a series of wildfires in late 2017 that displaced more than 10,000 people in Sonoma County in the State of California. This event presented an exceptional challenge to account for the population of each county and city as of January 1, 2018 and to estimate the size of the affected population and the relocation patterns of people displaced by the fires.

While migration responses to wildfire are the focus of the present research, a broader agenda on disaster demography should consider sequelae of disasters and differential vulnerability (see Chaps. 4, 6 and 10). In the example of wildfires, smoke plumes may have a significant public health impact. Equity concerns are also important. Climate change is expanding the geography of risk and changing the demographic profile of the affected populations (see also Chap. 8). No longer limited to remote areas or vacation homes, vulnerability to wildfire in particular, may be reversing from a progressive to a regressive hazard (Davies et al. 2018; Baylis and Boomhower 2018).

Even though California is one-third forested, many of these forests are in poor health due to a combination of factors that including climate change and inadequate, misguided, or uncoordinated forest management policy. Abatzogulou and Williams (2016) found that climate change was responsible for half of the increase in aridity in Western US forests between 1966 and 2016 and that without climate change the cumulative area burned by fires would be just half of what has occurred between 1984 and 2016. Climate change is expected to continue to increase the frequency and intensity of fires (Keyser and Westerling 2017; Moghaddas et al. 2018).

More than 40% of California housing stock is in areas at heightened risk of fire damage, where housing abuts forest, referred to as the wildland–urban interface (WUI; Hammer et al. 2007). The hills, forests, and peripheries of cities and towns are desirable areas, both for builders and potential residents attracted to their natural amenities. Housing growth in WUI areas in California grew by approximately 20% each decade during 1990–2010, compared to 10% each decade for non-WUI areas even though the WUI accounts for just 10% of land area (Radeloff et al. 2018). This has contributed to more problematic forests due to fire suppression efforts and also exposed an increasing number of residents to increased fire risks.

3.2.2 Conditions of the 2017 Fire Season and the LNU Complex Fires

The year 2017 marked a record fire season in California, with 436 significant fires that burned more than 630,000 ha (CDFFP 2019). Yet, the records set during the fire season of 2017 have already been surpassed. Fires during 2018, notably the Carr fire near the city of Redding and the Camp fire which razed the city of Paradise, were the deadliest (93 confirmed fatalities) and most destructive (650,000 ha) in the state's history. This increase has necessitated the development of a robust toolset for estimating demographic impacts of disasters.

The LNU Complex fires comprised of the Tubbs, Nuns, Atlas, and several other fires deserves special mention. These three conflagrations occurred during October 2017 and in close to each other in the WUI. Nauslar et al. (2018) state that these fires 'pushed the bounds of conventional fire wisdom with the extreme rates of spread, size, and timing.' They affected four counties, but Sonoma County bore the brunt of the population and housing impacts. They affected four counties, but Sonoma County bore the brunt of the population and housing impacts.

Sonoma County is located north of San Francisco, along the California coast of the Pacific Ocean. The county has a great deal of climatic variation, according to elevation and proximity to the ocean. Parts of the county close to the ocean remain cool and moist year-round and can be foggy most days from late afternoon until morning. Inland areas have a more typically Mediterranean climate, with hot, dry summers and cool, wet winters. The Northern Coastal Range of mountains runs on either side of the Sonoma Valley. Its abundance of variable microclimates lends itself well to viticulture, with California's first wineries established in the area in the 1850s. Santa Rosa is the largest city in Sonoma County and was home to 178,000 of the county's 504,000 residents in 2017 (CDOF 2018).

The region had suffered the effects of severe drought since 2011—possibly the worst drought in more than 1000 years (Griffin and Anchukaitis 2014). These conditions contributed to a record high total of 129 million dead trees in the state as of late 2017 (USFS 2018; Guerin 2017; Buluç 2017). In total, the state's tree mortality survey counted 147 million tree deaths during 2010–2018. A normal average would be 1 million tree deaths each year. However, 62 million deaths were recorded in 2016 at the peak of the drought.

As of October 2016, 83% of the state was in drought, and 21% of the state was in 'exceptional' drought status, the most severe category. The first months of 2017 saw high rainfall, which mitigated drought conditions. Most of the precipitation of the 2016–17 season fell during January and February, marking the state's wettest winter and the second-wettest year since records began in 1895 (Di Liberto 2017). The rain fell so fast that it caused flooding throughout the state and damaged the nation's tallest dam, Oroville Dam. The state's drought declaration had been in place since 2014 and was lifted in April 2017. The immense volume of rain was a welcome reprieve to depleted reservoirs but also promoted the growth of shrubs and grasses that typically wither during the hot, dry summer and become fuel for more intense fires (Dudney et al. 2017). Recent research found that the connection between wet weather and fire activity in California has broken down completely during the second half of the twentieth century, as the effect of growth and fuel buildup during rainier seasons created greater fire hazards within the context of increased annual temperatures (Wahl et al. 2019).

The state's seasonal wind patterns are another factor in making California especially susceptible to fires late in the year. Late summer winds, known regionally as the Santa Ana or Diablo winds, can occasionally bring fast hot winds that exacerbate fires. These winds may grow more intense as a result of climate change, although the effect is debated (Jin et al. 2015; Mass and Ovens 2019). These late summer or fall winds were a factor in all of the state's most destructive WUI fires and have long been recognized as 'a wind of bad augury […] at this season of the year' (Russell and Boyd 1923).

It was during one such windy evening on October 8, 2017 that a small fire ignited near the town of Calistoga, a city of approximately 5000 in Napa County. The state had effected a 'red flag warning' which alerts the public to heightened fire hazard conditions. Warm air at 32 °C combined with strong winds with peak gusts of 175 km/h. Weather stations in the hills around Santa Rosa recorded windspeed exceeding the ninety-ninth percentile as well as relative humidity below the second percentile (Nauslar et al. 2018). At the 9:44 pm local time, a resident of Calistoga reported a fire in the vicinity of Tubbs Road. Within six hours of the report, numerous fires in the region had grown into major conflagrations, collectively termed the LNU Complex (Fig. 3.1). The Tubbs fire, while not the largest of these, was the most destructive. It moved out of the WUI into urban Santa Rosa and surrounding communities.

Embers carried by strong winds ignited cascading spot fires that complicated rescue efforts by making it impossible to track the extent of the fire during the period of its most rapid expansion overnight. The fire spread uncontrolled during the night and reached the city limits of Santa Rosa by 1 am on October 9. Although

Fig. 3.1 Central LNU Complex fires in Sonoma County (Author: Sharygin, cartography by Karácsonyi)

Table 3.1 Characteristics of Sonoma County, Santa Rosa City and Tubbs fire perimeter

Geography	Total population	*N* Households			Owner-occupied share of housing	Median household income
		Total	With 1 or more child (<18)	With 1 or more senior (65+)		
Sonoma	500,943	190,058	54,398 (27%)	62,770 (33%)	0.603	71,769
Santa Rosa	174,244	64,709	19,641 (30%)	19,958 (31%)	0.531	67,144
Tubbs Fire Zone	33,431	12,590	3,514 (28%)	4,774 (38%)	0.727	98,479

it was not fully contained until October 31, the urban destruction caused by the fire peaked within 24 h before winds died down. The Tubbs fire eventually burned almost 15,000 ha and destroyed 5636 structures, of which 4651 were residential. Collectively, the LNU Complex fires consumed more than 5000 housing units in Sonoma, Napa, and Solano counties (Hawks et al. 2017).

The following analysis focuses on impacts in Sonoma County, where the Tubbs fire displaced the greatest number of residents. The city of Santa Rosa lost 3081 housing units, representing approximately 5% of the city's housing stock (CDOF 2018). The city was already facing a housing shortage before the fire: with an overall vacancy rate of just 4.2% in 2017 (1.6% vacancy for rental units), and with 6.5% of housing units defined as 'overcrowded' (more than one resident per room).[1] Table 3.1 provides additional characteristics of Sonoma County, the city of Santa Rosa, and the fire-devastated areas.

The census block groups affected by the Tubbs fire had fewer renters as a share of households and higher average income. Households in the fire zone were also more likely to have at least one person in the household over the age of sixty-four, but the difference in the share of households with children was not found to be statistically significant.

3.3 Estimating Population Impacts From Housing Data

3.3.1 Role of Housing in Population Estimates

This section will investigate ways in which the statistics of destruction presented in the previous section could be translated to population and migration data necessary to plan a recovery effort. The State of California Demographic Research Unit produces

[1]U.S. Census Bureau; ACS 2017 5-year SF, table DP04. Retrieved 1 April 2019 from https://factfinder.census.gov.

population estimates for July and January of each year. The methods of accounting for population change include housing data that enters into the calculation in several ways. Annual estimates for January 1 are produced via complimentary top-down and bottom-up estimation steps. The headline state population is estimated using a cohort component method (Swanson and Tayman 2012), in which the total population is determined by the last known population[2] accounting for births, deaths, and migration. Net migration is composited from different datasets for different age groups that include school enrollment; tax returns, driver's licenses, and immigration data; and pension and health insurance data.

County population estimates are produced by ensemble averaging (Clemen 1989). The first piece of the ensemble is a composite method in which different methods are used to estimate the size of the population of different age groups (Bogue and Duncan 1959). For example, births and school enrollment inform estimates of the size of the child population; driver's licenses, deaths, and tax information for the adult population, and administrative records including pension and health insurance data are the primary source of information about change in the population age 65 and older. The second piece is a 'ratio correlation' regression-based method, which predicts the county's total population as a function of covariates such as the county birthrate, housing stock, and labor force (Schmitt and Crosetti 1954). The third piece of the ensemble is a cohort component model, in which births, deaths, and net migrants from federal administrative data are included as measured by the U.S. Census Bureau. These three models are combined into the model (as equally weighted averages, although the weights could be specified differently) to produce a single county estimate, which is translated into a county share of the total state population and applied to the state total number estimated above.[3]

The populations of each of California's 539 city and county jurisdictions that include 482 cities, unincorporated parts of 57 counties, as well as the city and county of San Francisco, were estimated using a housing unit method (Swanson and Tayman 2012), which relates the total population to the number of housing units, the persons per household, and the vacancy rate. An advantage of the housing unit method is that, as an accounting identity, the only source of error is in the estimation of the

[2]Generally, this would be the last decennial Census, post adjustments.

[3]A small set of 14 counties with population below 65,000 is incremented (births) or decremented (deaths) from the last decennial census only, without an attempt to estimate net migration.

parameters. However, the parameters can be extremely challenging to monitor and update due to data constraints, especially for small areas. Change in the population that live in group quarters[4] can be accounted for exclusively from administrative data.

The housing unit method and the ratio correlation method both rely on housing stock to adjust population counts. In the case of the housing unit method, the effect of loss in housing stock could be offset by updated vacancy and density constants. However, in the wake of a sudden-onset disaster, there may be no data with which to update the persons per household or vacancy rates, even if the data on housing unit change are rapidly updated and very high quality. The ratio correlation method can be even more vulnerable, depending on whether housing is part of the equation. A large shock in the form of housing stock loss would mechanically produce a significant drop in the population of Santa Rosa and Sonoma County. Qualitative accounts from reportage on the fire suggested that many displaced people stayed in proximity to the fire area, anticipating a return to their land after cleanup and reconstruction. The higher income and homeownership rates in the area are consistent with this notion.

3.3.2 *Estimating the Number of Displaced Persons*

To estimate the size of the displaced population in Sonoma County, we tested two approaches. In the immediate aftermath, without data on the precise location or addresses of destroyed housing, we interacted the total number of destroyed housing units inside the county boundaries with the persons per household most recently estimated for the city, weighted by the occupied share of housing.

In late October 2017, the state fire agency published a report on damage from the Tubbs Fire (Hawks et al. 2017), in which fire affected addresses and land parcel numbers were published along with assessments of the extent of the damage to the building and the building type (residential, commercial, or outbuilding). From this list, we generated a distribution of damaged and destroyed residential structures by census block group, which was used to weight block group level household size estimates from the American Community Survey (ACS), an annual household survey.[5] Results for estimated population displacement using these two methods are presented below in Table 3.2. Despite the more careful use of block group specific housing tenure and vacancy rate data, the two alternatives are within 1% of each other; for this study, we adopted the simpler calculation that resulted in the count of 11,521.

---

[4]Group quarters include residential care facilities, hospitals, school dormitories, military barracks, prisons and jails, and other shared living arrangements.

[5]Geocoding of fire addresses was performed using the Census Bureau geocoding service, with manual geocoding of 296 records that returned no matches.

Table 3.2 Estimates of the population displaced by the Tubbs fire: Sonoma County, October 2017, **a** 2017 DOF housing and population estimates (Table E-5). **b** 2013–17 American community survey (Tables B25002, B2510)

a

Geography	Residential structure loss	Vacancy rate	Persons per household	Displaced persons
	A	*B*	*C*	$A * (1 - B) * C$
Sonoma County	1,569	0.083	2.59	3,726
Santa Rosa City	3,081	0.038	2.63	7,795
Total	4,650			11,521

b

Block group	Residential structure loss		Rental share of occupied housing	Occupied units	Persons per household		Displaced persons
	Freq.	Percent			Owned	Rented	
	A		*B*	*C*	*D*	*E*	$D * C * (1 - B) + E * C * B$
152100.1	131	2.8	0.39	395	1.74	1.70	226
152400.1	169	3.6	0.11	947	2.50	1.36	402
152400.3	397	8.5	0.47	942	1.80	1.30	538
152400.4	642	13.8	0.15	980	2.90	2.54	1,828
152400.5	503	10.8	0.00	630	2.89	0.00	1,454
152502.1	2	0.0	0.35	614	2.65	3.62	6
152600.1	500	10.8	0.13	570	2.24	4.20	1,180
152600.5	169	3.6	0.16	526	2.12	1.41	241
152701.1	276	5.9	0.18	461	2.53	3.49	649
152701.3	35	0.8	0.19	387	2.50	1.83	83
152702.2	98	2.1	0.67	481	2.16	2.78	252
152702.3	315	6.8	0.06	368	2.16	2.13	665
152702.5	8	0.2	0.31	216	2.84	2.36	22
152801.1	219	4.7	0.58	531	2.55	2.49	551
152801.2	401	8.6	0.19	412	2.78	3.97	1,206
152801.3	275	5.9	0.35	403	2.13	4.15	740
152801.4	166	3.6	0.18	250	2.60	4.72	496
152801.5	207	4.5	0.22	296	2.67	2.38	540
152905.1	92	2.0	0.44	482	2.21	2.43	212
152906.1	1	0.0	0.40	1,326	2.80	2.95	3
153807.2	21	0.5	0.09	1,016	2.93	2.50	57
154100.4	23	0.5	0.49	357	2.57	4.19	63
Total	4,650	100.0					11,414

3.3.3 Estimating Migration

Having prepared an estimate of the population displaced by the loss of their home, we then needed to devise a system that could accurately model where people moved. In doing so, we analyzed how many remained in the same city, moved elsewhere in the county or state, or left the state altogether. We considered the literature from other efforts to assess disasters, as well as the methods used by other U.S. states.

An inspiring account came from the state of Florida in the aftermath of Hurricane Andrew in 1992, where a publicly funded telephone and field survey provided updated information on vacancy rates, persons per household, and population in transitory locations such as hotels and motels. This survey data provided rapid and accurate new inputs to the classical housing unit method (Smith 1996). However, costs and logistical challenges mean that this approach has not gained traction. Also, there are reasons why it might not succeed in all contexts. For instance, the accuracy of the approach depends on how many likely destinations of displaced people are captured. In the case of Hurricane Andrew, or indeed the Tubbs fire, this approach showed great promise because the housing losses were substantial but small relative to the regional housing capacity. In other cases where these conditions do not hold, a survey may not be practical or effective. A new data collection effort may not always be feasible, but there are many other private and public data sources were considered as shown in Table 3.3.

The US Census Bureau has worked with the US Federal Emergency Management Agency (FEMA) to collect data on persons registered with the agency to receive disaster assistance. In the case of wildfire, the extent of federal operations is determined by whether the federal government declares a 'major disaster,' 'emergency,' or a 'Fire Management Assistance Declaration.' The 2018 Camp Fire was declared a major disaster; however, the 2017 LNU Complex fires were only given the latter designation, limiting the extent of federal assistance and the accuracy of federal data. Residents affected by the LNU Complex fires had limited time to register with FEMA for grants or loans that support uninsured or underinsured residents. In addition, many homeowners may have opted to rely on private insurance. In Sonoma County, FEMA approved just 3200 registrations, and ultimately, only 119 households received temporary housing relocation assistance, leaving major gaps in the coverage of FEMA data for this disaster (Morris 2017; Schmitt 2019).

The US Postal Service provides change of address (COA) data with names and addresses of individuals, families, and businesses who have filed a change of address for mail delivery through a service called NCOALink. Postal service address data are the backbone of the US Census Bureau's Master Address File, a putative master list of all living quarters in the USA. The NCOALink product is the most comprehensive source of household moves, but it is not a panacea, due to several limitations. The data are licensed for the purpose of updating mailing lists, and for privacy reasons cannot be queried without an extant mailing list which includes a name and address. From cadastral datasets, we generated a database of names and addresses associated with properties in the burn area, but the names of those with legal title to affected land

Table 3.3 Sources of data on migration

	Data source	Coverage	Notes
1	FEMA Assistance	Registered persons or households	Requires national disaster declaration; confidential data
2	USPS NCOALink	Registered changes of address	Requires names and addresses; no data for temporary moves
3	ID, drivers license, or vehicle registrations	Drivers or ID holders; vehicle owners	Delayed registration of address changes. Potentially longitudinal, but confidential data difficult to access
4	Public health insurance/pension data	Beneficiaries	Varying coverage and rapidity of updates according to each program. Potentially longitudinal, but confidential data difficult to access
5	Traffic measurements	Drivers	Difficulty of parsing traffic data and relating to total population
6	Social media (Twitter, Facebook)	Users	Without adjustment, limited representativeness of selected population; difficulty of parsing data
7	Wastewater	Households using municipal sewer system	Missing population living in transitional locations or septic tanks; difficulty collecting and parsing/interpreting data
8	Solid waste	Households using municipal solid waste collection	Missing population living in commercial buildings (e.g. apartments and hotels); difficulty collecting and parsing/interpreting data; indirect relationship to disaster migration
9	Historical migration patterns (census/survey)	All persons	Geographic scale may be limited; historical data may not be representative of disaster migration patterns
10	Mobile phone movement	Mobile phone users	Rapid updates, but expensive, proprietary data source
11	Remote sensing	All structures	Potentially high cost of imagery and other sensor data; indirect relationship to disaster migration

(continued)

Table 3.3 (continued)

	Data source	Coverage	Notes
12	Utilities (e.g. electric/gas)	Utility customers	Missing population living in transitional locations; difficulty of accessing and parsing/interpreting data. Limited longitudinal data
13	HUD Point-in-Time Count	Homeless	Conducted yearly or biennially
14	Voter registration rolls	Registered voters	Delayed registration of moves. Potentially longitudinal, but confidential data difficult to access. Representativeness issues
15	Credit Bureau Headers	Persons with credit files	Expensive, proprietary data; licensing issues
16	Tax data	State or federal taxpayers	Delayed registration of moves and publication of data. Generally not available at small geographies
17	Public education enrollment	Public school students	Longitudinal, but potentially limited representativeness; confidential data

parcels will not give complete coverage. For example, cadastral datasets will include apartment and rental housing addresses and landlords' names, but not the names of individual tenants. In addition, historical metadata are not attached to moves. In other words, a search of COA within the past 18 months would return the most recent address only, not a sequence of moves if more than one move occurred. Another limitation is that temporary COA, for example, a hold mail or temporary change of address order may be filed with the postal service for up to one year, but these temporary orders are not searchable via NCOALink. People change residence for many reasons, and a blanket query of addresses in a disaster perimeter will overstate out-migration (as well as missing possible moves into undamaged housing in the area). For these reasons, NCOALink would be valuable only when a specific list of destroyed addresses and associated names are available.

Government programs such as public pensions, medical insurance, cash or in-kind transfer programs, and other entitlements may be a source of migration data. Their coverage will vary according to the socioeconomic characteristics of the area affected by the disaster, and the ability or willingness of the agencies involved to share data with the appropriate geographic specificity. These data were not available for this study, but in the future, they should be explored further. Agencies that collect such data ought to give thought to ways in which they could be made available for demographic studies without violating laws regarding privacy and disclosure of sensitive information. There are a variety of administrative data sources that may be useful resources, but they share the drawback that the populations that participants in these programs may be few and highly selected relative to the total population. Still, some of these programs may enable estimation of population migration more rapidly than other sources, and it may be possible to statistically control for some aspects of population selectivity.

We pursued leads from the available data sources, and research into the use of new data is an area of continuing activity. For the LNU Complex fires, we eventually gravitated toward the use of public education enrollment data. We considered that public education is more representative than other government programs of the general population, while still capturing data in a timely manner as relocated families re-enroll their children in school. California, like many other US states, received federal grant funding to produce a student longitudinal data system (SLDS). In California, the SLDS program is called CalPADS and assigns individual identifiers to each student. The program records all enrollment activity such as transfers to other public schools or transfers out of the public school system (to private schools or out of state). Bias still exists: the behavior of families with children enrolled in Santa Rosa schools may not be representative of the decisions made by people in other living arrangements (living alone or with others in households without children, or in group quarters). On balance, these data offered a superior balance of timeliness, completeness, and representativeness compared to the alternatives.

3.4 Student Enrollment Proxy Method For Estimating Migration

Previously, in Sect. 3.2, the displaced population (N) was estimated at 11,521. To model new locations for this population, we focused this study on developing a methodology to use longitudinal student enrollment data in order to calculate a rate of migration of students in Sonoma County and extrapolate from their behavior to the wider population. To this end, the following data were collected.

1. Fire perimeter (GIS; or, geocoded housing loss registry).
2. Census block group boundaries (GIS)
3. Census block group population and housing characteristics
4. School attendance zones (GIS)
5. Student enrollment data by campus, including:
    a. Total enrollment in October 2017 by grade (or age).
    b. Enrollment changes by school campus of origin and destination.

Fire perimeter polygons were downloaded from the state's fire agency. Census block group boundaries and data were collected from IPUMS-NHGIS (Manson et al. 2018). School attendance boundaries were obtained from the School Attendance Boundary Survey (SABS), a program of the national department of education and the US Census Bureau (Geverdt 2018). We used the SABS polygons to determine which

Table 3.4 Student transfers ($N = 211$) by destination: Sonoma County school districts; 1 Oct-31 Dec 2017

Many (>20)	*N*	(%)	Some (5–20)	*N*	(%)	Few (<5)	N	(%)
San Jose—San Francisco—Oakland CSA			*Sacramento-Arden Arcade-Yuba City CSA*			*Bakersfield MSA*		
Total	138	**65.4**	Total	18	**8.5**	Kern	3	**1.4**
Alameda	3	1.4	Sacramento	10	4.7	*Fresno-Madera CSA*		
Contra Costa	4	1.9	Placer	2	1.0	Fresno	2	**1.0**
San Francisco	1	0.5	Nevada	4	1.9	*Stockton MSA*		
Santa Clara	1	0.5	El Dorado	2	1.0	San Joaquin	2	**1.0**
Napa	3	1.4	*Los Angeles-Long Beach-Riverside CSA*			*Modesto MSA*		
Marin	10	4.7	Total	13	**6.2**	Stanislaus	2	**1.0**
Solano	8	3.8	Los Angeles	7	3.3	Visalia-Porterville MSA		
Sonoma; of which	108		Riverside	1	0.5	Tulare	1	**0.5**
Other city	51	24.2	San Bernardino	3	1.4	*Salinas MSA*		
Santa Rosa	56	26.5	Ventura	2	1.0	Monterey	1	**0.5**
Balance of County	1	0.5	*Clearlake μSA*			*San Luis Obispo-Paso Robles MSA*		
			Lake	16	**7.6**	San Luis Obispo	1	**0.5**
			Redding MSA			*Red Bluff μSA*		
			Shasta	8	**3.8**	Tehama	1	**0.5**
			Ukiah μSA					
			Mendocino	5	**2.4**			

Source CalPADS
Totals may not be equal due to rounding

schools' attendance areas included census block groups that experienced housing losses from the fires, by calculating the percentage overlap between the fire perimeter and school attendance zones. In Sonoma County, we ultimately selected 41 schools, spread over six school districts. We received from the state education department a tabulation from CalPADS data of school level enrollment change data for these districts, whose coverage areas included all block groups where a significant amount of housing was lost. The data we received consisted of total student transfers that occurred between October 1, 2017 and December 31, 2017 that were reported to the department as of March 2018. The report includes the number of transfers for every unique origin–destination pair of schools, along with the destination school city and county summarized in Table 3.4.[6]

[6]Due to a processing error, we did not receive out of state transfers, only transfers to other schools in California.

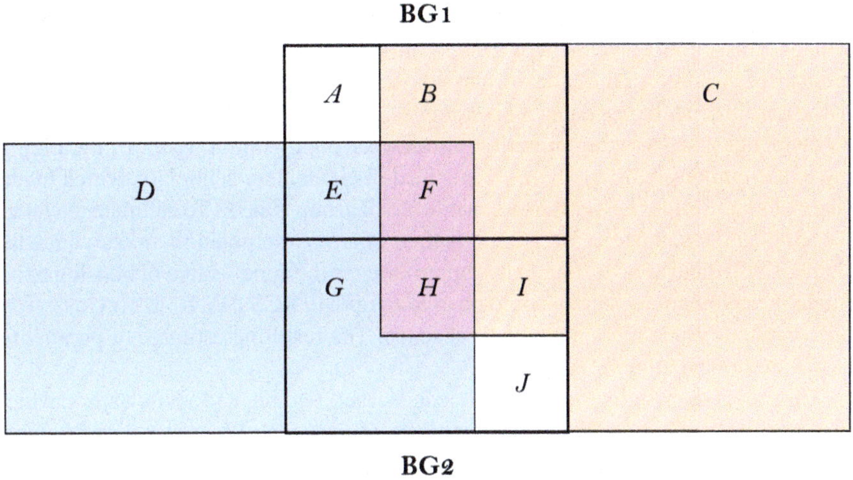

Fig. 3.2 Calculating migration universe for school enrollment proxy model. *Note* labels refer to the entire area inside the intersection of fire/school/block group boundaries

In total, the data indicated that there were 211 transfer students out of schools serving homes in the burned area. The majority of the 211 transfer students did not move far: over half of withdrawn students re-registered at other schools in Santa Rosa or other jurisdictions in Sonoma County. Other significant destinations (circa 5% of transfers) were Lake, Marin, and Sacramento counties. The next step was to follow up by identifying the population at risk of these moves, to estimate a migration rate for the general population based on the total number of displaced students.

I proposed a *school enrollment proxy method*, in which the percentage overlap is calculated between each school's attendance zone, census block group, and fire area shown in Fig. 3.2.

In Fig. 3.2, the area $B + C + F + H + I$ (written hereafter as *BCFHI*) is bounded by the fire area, and *DEFGH* is defined by the school attendance boundary. Two census block groups are represented above where BG1 $=$ *ABEF*, and BG2 $=$ *GHIJ*. Area *A* is in neither the school attendance zone nor the fire area. *B* is the fire area, but not the school attendance zone. *E* and *F* are in the school attendance zone, but only *F* is within the fire area. The total number of households inside areas *F* and *H* was estimated using a GIS analysis. This revealed households whose migration trends can be followed in the school enrollment data. The above is an extreme example. In most cases, there was an insignificant amount of area in a given block group that was not captured by the school attendance zone. For this reason, the simplifying assumption was made such that $H \approx HI$ and $F \approx FB$.

To estimate the population at risk of migration, we further needed to assume which portion of the total enrollment of each school lived in the households in the intersecting areas. To do this, we used student enrollment data at each school by grade, which are regularly published online by the education department (students' ages are withheld from these data for privacy reasons).[7] At the block group level, we compared with data on school enrollment by grade from the latest five-year ACS to estimate the number of children in the block group whose household moves should be captured in the enrollment change data.[8] For example, if school boundary *DEFGH* represents an elementary school, the migration universe or population at risk of migration is the total number of children in BG1 enrolled in elementary school, weighted by the share of households in BG1 that are inside block group zone *F*. To estimate the latter, we adopted the equal area weighting method of areal interpolation, under which the share of BG1 land area in zone *F* also approximates the zone's share of housing units. Student enrollment by school could be used to update ACS data from previous years or to estimate student counts in their absence. The resulting estimate of population at risk is shown below in Table 3.5.

This novel approach relies on a geocoded structure damage/loss inventory, which may not be readily available for other types of disasters. Perimeter data are more readily available for a variety of disaster events. However, a perimeter-only method without structure losses by block group greatly overestimates housing units lost since it implicitly assumes that burned area in any given block is equally likely to have consumed housing as not (which is unrealistic and usually counter to the goal of fire suppression). We tested the effect of generating a perimeter-based estimate by comparing the housing loss estimates from perimeter-only method to the known total housing loss. This process identified all block groups that intersected the fire perimeter (46 compared to only 22 block groups identified in the geocoded loss data). Interacting the share of the area of each of these block groups inside the fire perimeter with the housing stock produces an overestimate of 10,490 housing units lost. Weighting the estimated impacted students by the ratio of an independent standard results in what is likely an undercount of the displaced student population ($N = 1293$), but a useful point of comparison (Table 3.6).

The migration proportion of Sonoma students can now be calculated by:

$$\text{Migration rate} = N \text{ movers}(211)/N \text{ at risk } (1740) = 12.13\%$$

We assume that all transfers were the result of fire related displacement, and that there were no moves in due to the fire, which may result in an overestimate of net migration attributable to the fire. In estimating the student population, the equal area weighted interpolation assumption implies homogeneity within the block group. Results will be biased if block group average characteristics do not accurately describe the characteristics of lost housing units. For example, we implicitly assume that destroyed homes were equally likely to contain students as standing homes in the same block group.

There are caveats in the approach as it is currently formulated. A significant source of uncertainty is whether the rate of migration we estimated for students

[7]CDE Data Report Office. https://www.cde.ca.gov/ds/sd/sd/.

[8]ACS 5-year 2013–17. "School enrolment by detailed level of school for the population 3 years and over" (B14007).

Table 3.5 Estimating population at risk of migration from ACS survey data

Block Group	Enrolled students, by grade			Housing lost (share of total)	Student population at risk of move
	K-6	7–8	9–12		
	A	*B*	*C*	*D*	$(A + B + C) * D$
1521001	60	5	18	0.332	28
1524001	84	82	202	0.178	66
1524003	0	0	0	0.366	0
1524004	231	63	142	0.655	286
1524005	52	62	241	0.798	283
1525021	228	72	114	0.003	1
1526001	112	0	57	0.828	140
1526005	60	14	36	0.229	25
1527011	109	21	53	0.521	95
1527013	94	21	59	0.090	16
1527022	123	10	39	0.204	35
1527023	72	8	69	0.838	125
1527025	101	8	8	0.037	4
1528011	88	0	6	0.412	39
1528012	152	18	45	0.973	209
1528013	198	28	21	0.647	160
1528014	48	0	37	0.664	56
1528015	96	66	21	0.699	128
1529051	43	20	24	0.191	17
1529061	392	167	209	0.001	1
1538072	312	112	323	0.019	15
1541004	91	38	77	0.052	11
Total					1,710

applies equally to the non-student population. Households without children may be more or less likely to relocate depending on their age, housing tenure, length of time in the area, and employment. Of those that migrate, they may be less likely than households with children to remain in the region. Migration of families with children could also be underestimated, for example, children may have commuted from new residences back to their original schools to avoid the stress of changing schools. If households without children in school are more likely to relocate, then these results are biased toward underestimating the impact of the fire on out-migration.

To generate new estimates for January 1, 2018, we follow the usual steps for the housing unit method and county estimates ensemble. However, we deviated from usual practice in that we keep the housing lost from the disaster in the housing stock when the models are run. After generating counterfactual estimates this way,

Table 3.6 GIS-only method for displaced student population from fire zone

Block Group	Share of BG Area inside Fire perimeter	Housing units	Student population (K-12)
	A	*B*	*C*
1501001	0.450	610	64
1501003	0.034	272	38
1502023	0.745	450	115
1502024	0.545	803	200
1503032	0.002	504	135
1503063	0.489	902	257
[...]			
	Independent estimate	Modelled housing loss	Weighted exposure
		$B * A$	$C * A * (D/E)$
1501001		275	13
1501003		9	1
1502023		335	38
1502024		438	48
1503032		1	0
1503063		442	56
[...]			
Total	$D = 4650$	$E = 10{,}490$	1,293

we account for the disaster impacts separately. We then define the population at risk of migration for each jurisdiction where population estimates are needed. In this case, example results are shown below in Table 3.7 for the city of Santa Rosa. We apply the migration probability to the population at risk of migration and then assign migrants to new destinations according to the distribution of destinations in the student enrollment change data.

3.5 Conclusion

The results suggest that most Santa Rosa residents did not move out of the city: according to our analysis, only 695 or 9% of displaced persons crossed jurisdictional boundaries in this case and most often to nearby cities or unincorporated areas in the same county. The estimation framework is intended to be generalizable to any origin–destination dyad data, as long as the geographic data can be used to estimate a population at risk. The methodology we presented may be suited to timely assessment of the population impacts of disasters that occur in countries with a well-developed public infrastructure. The effort we made would not have been possible without

Table 3.7 Net migration estimates for fire displaced population of Santa Rosa, Sonoma County

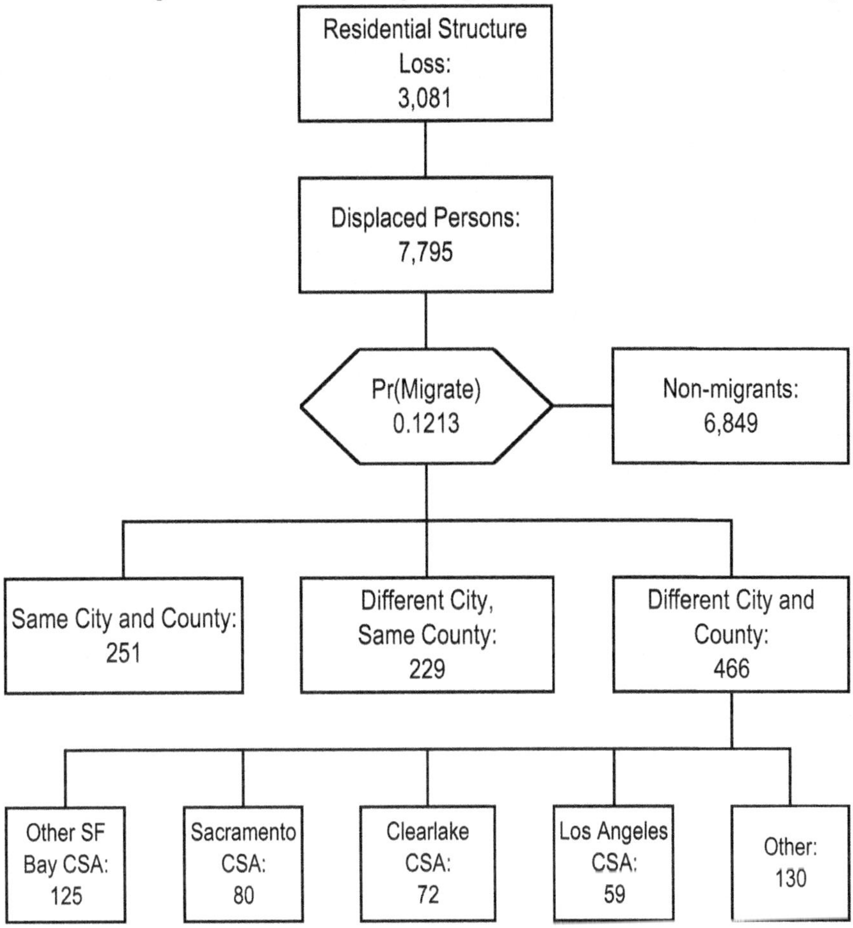

the ability of government institutions to produce and share reliable data about the emergency, as well as high quality small area data from the ACS and longitudinal records from the public education system.

We propose that the housing stock destroyed by disasters should be handled carefully by population estimation methods that rely on housing stock. One solution is to keep the destroyed housing in the model as if it was still existing and then to separately estimate and account for migration as a post-estimation adjustment (similar to the treatment of GQ population). This would improve the accuracy of the ratio correlation or other regression-based models. The housing unit method could use this approach also, but a simpler solution would be to adjust vacancy rates and persons per household to accommodate the lost housing units while keeping the population within the jurisdiction. While we anticipate that this approach can be useful for studying many disaster types, more effort should be put into developing a typology of methods that work for a broader variety of disaster types and in other state and country contexts.

Future work on this methodology should focus on identifying systematic differences between the general and proxy population, such as students, that can be represented with data available to the model. Because our primary concern was adjusting for the very high bias toward overstatement of the number of migrants in the housing unit and ratio correlation methods, we considered these acceptable tradeoffs.

The area of Sonoma County that burned in the Tubbs fire in 2017 bore great resemblance to a fire of 53 years previous—the Hanly Fire of 1964. Like the Tubbs Fire, Hanly started in the hills near Calistoga; propelled by strong winds, it reached Santa Rosa city in less than 12 hours (Kovner 2013). It caused much less destruction because the WUI at that time was relatively sparsely inhabited. The fire did not halt continued growth of the population or expansion of housing continually deeper into the WUI. As the wildfire hazard increases throughout the state in coming years, continued research into methods of estimating population impacts will become increasingly in demand.

References

Abatzogulou J, Williams AP. 2016. "Impact of anthropogenic climate change on wildfire across western US forests." PNAS 113(42) 11770-11775. https://doi.org/10.1073/pnas.1607171113.

Baylis P, Boomhower J. 2018. "Moral Hazard, Wildfires, and the Economic Incidence of Natural Disasters". SIEPR Working Paper 18–044. Retrieved 1 April 2019 from https://siepr.stanford.edu/sites/default/files/publications/18-044.pdf.

Bogue D, Duncan B. 1959. "A composite method for estimating postcensal population of small areas by age, sex and color." Vital Statistics Special Report, Vol 47 No 6. Washington, DC: National Center for Health Statistics.

Buluç L, Fischer C, Ko J, Balachowski J, Ostoja S. 2017. "Drought and Tree Mortality in the Pacific Southwest Region." A synopsis of presentations and work group sessions from the Science and Management Symposium: "Lessons Learned From Extreme Drought and Tree Mortality in the Sierra Nevada: How Can Past Events Inform Our Approach Forward?". Symposium held July 2017, Sacramento, CA. Retrieved 1 April 2019 from https://www.fs.fed.us/psw/topics/tree_mortality/california/documents/DroughtFactSheet_R5_2017.pdf.

CDOF. California Department of Finance (CDOF). 2018. E-5 Population and Housing Estimates for Cities, Counties and the State — January 1, 2011-2018. Sacramento, California, May 2018.

CDFFP. California Department of Forestry and Fire Protection (CDFFP/CalFire). Preliminary statistics from Annual Wildfire Activity Statistics Report. Retrieved 1 December 2019 from https://www.fire.ca.gov/incidents/2017/.

Clemen RT. (1989) "Combining forecasts: A review and annotated bibliography." *International Journal of Forecasting* 5(4): 559–583.

Davies I, Haugo R, Robertson J, Levin P. 2018. "The unequal vulnerability of communities of color to wildfire". PLOS ONE. https://doi.org/10.1371/journal.pone.0205825.

Di Liberto T. 2017. "Very wet 2017 water year ends in California". *NOAA Climate News*: October 10, 2017. Retrieved 1 April 2019 from https://www.climate.gov/news-features/featured-images/very-wet-2017-water-year-ends-california.

Dudney J, Hallett L, Larios L, Farrer E, Spotswood E, Stein C, Suding K. 2017. "Lagging behind: Have we overlooked previous-year rainfall effects in annual grasslands?" *J. Ecol.* 105:484–495.

Geverdt D. 2018. *School Attendance Boundary Survey (SABS) File Documentation: 2015–16 (NCES 2018–099).* US Department of Education. Washington, DC: National Center for Education Statistics. Retrieved 1 April 2019 from https://nces.ed.gov/programs/edge/SABS.

Griffin D, Anchukaitis K. 2014. "How unusual is the 2012–2014 California drought?", *Geophys. Res. Lett.* 41:9017–9023. https://doi.org/10.1002/2014GL062433.

Guerin E. 2017. "Drought kills 27 million more trees in California". *KPCC*. Retrieved 1 April 2019 from https://www.scpr.org/news/2017/12/12/78804/drought-kills-another-27-million-trees-in-californ/.

Hammer RB, Radeloff VC, Fried JS, Stewart SI. 2007. "Wildland-urban interface housing growth during the 1990s in California, Oregon, and Washington." *Int J Wildland Fire* 16:255–265.

Hawks S, Goodrich B, Fowler C. 2017. *Tubbs Incident Central LNU Complex (CALNU 010045) Damage Inspection Report.* Sacramento, CA: California Department of Forestry and Fire Protection (CDFFP/CalFire).

Jin YF, Goulden M, Faivre N, Veraverbeke S, Sun F, Hall A, Hand M, Hook S, Randerson J. 2015. "Identification of two distinct fire regimes in Southern California: implications for economic impact and future change". *Env Res Lett* 10(9). https://doi.org/10.1088/1748-9326/10/9/094005.

Keyser A, Westerling A. 2017. "Climate drives inter-annual variability in probability of high severity fire occurrence in the western United States". Environmental Research Letters 12(6). https://doi.org/10.1088/1748-9326/aa6b10.

Kovner G. 2013. "Redwood Empire fire history remains visible in wild spots". *Santa Rosa Press-Democrat, September 14, 2013.* Retrieved 1 April 2019 from https://www.pressdemocrat.com/news/2217177-181/redwood-empire-fire-history-remains.

Manson S, Schroeder J, Van Riper D, Ruggles S. 2018. *IPUMS National Historical Geographic Information System: Version 13.0* [Database]. Minneapolis: University of Minnesota. 2018. https://doi.org/10.18128/D050.V13.0.

Mass C, Ovens D. 2019. "The Northern California Wildfires of 8–9 October 2017: The Role of a Major Downslope Wind Event". *Bull Am Met Soc.* Feb 2019: 235–256. https://doi.org/10.1175/BAMS-D-18-0037.1.

Morris JD. 2017. "Northern California fire victims must register for FEMA, SBA assistance by Monday". *Santa Rosa Press-Democrat*, December 9, 2017. Retrieved 1 April 2019 from https://www.pressdemocrat.com/news/7732075-181/northern-california-fire-victims-must.

Moghaddas J, Roller G, Long J, Saah D, Moritz M, Star D, Schmidt D, Buchholz T, Freed T, Alvey E, Gunn J. 2018. "Fuel treatment for forest resilience and climate mitigation: a critical review for coniferous forests of California." California Natural Resources Agency. Publication number: CCCA4-CNRA-2018–017.

Nauslar N, Abatzogolou J, Marsh P. 2018. "The 2017 North Bay and Southern California Fires: A Case Study". *Fire* 1(1), 18; https://doi.org/10.3390/fire1010018.

Radeloff V, Helmers D, Kramer H, Mockrin M, Alexandre P, Bar-Massada A, Butsic V, Hawbaker T, Martinuzzi S, Syphard A, Stewart S. 2018. "Rapid growth of the US wildland-urban interface raises wildfire risk". *PNAS* 115 (13):3314-3319. https://doi.org/10.1073/pnas.1718850115.

Russell G, Boyd A, eds. 1923. *The Berkeley fire: dedicated to the people of Berkeley who have proved themselves great hearted in giving, courageous in losing, and clear eyed in building toward a safer future.* Berkeley, CA: Lederer, Street & Zeus Co.

Smith SK. (1996): Demography of Disasters: Population Estimates after Hurricane Andrew. Population Research and Policy Review, vol. 15, pp. 459–477.

Schmitt R, Crosetti A. 1954. "Accuracy of the ratio-correlation method for estimating postcensal population." *Land Economics* 30:279–281.

Schmitt W. 2019. "FEMA extends housing aid for fire survivors as Santa Rosa offers home loans to eligible Coffey Park fire victims". *Santa Rosa Press Democrat*, March 21, 2019. Retrieved 1 April 2019 from https://www.pressdemocrat.com/news/9412262-181/fema-extends-temporary-housing-for.

Swanson D, Tayman J. 2012. *Subnational Population Estimates*. The Springer Series on Demographic Methods and Population Analysis. Springer Netherlands. https://doi.org/10.1007/978-90-481-8954-0.

U.S. Forest Service (USFS). 2018. "2018 Tree Mortality Aerial Detection Survey Results". Retrieved 1 April 2019 from https://www.fs.usda.gov/Internet/FSE_DOCUMENTS/fseprd609295.pdf.

Wahl E, Zorita E, Trouet V, Taylor A. 2019. "Jet stream dynamics, hydroclimate, and fire in California from 1600 CE to present". *PNAS* 116(12):5393-5398. https://doi.org/10.1073/pnas.1815292116.

Chapter 4
The 2010 Catastrophic Forest Fires in Russia: Consequence of Rural Depopulation?

Tatiana Nefedova

Abstract Catastrophic forest fires hit the European part of Russian Federation during summer 2010 as result of a two-month-long period with temperatures above the average by 10 °C coupled with an unusually long drought period. Even though forest fires are usual for Russia, these are typically located in the sparsely populated Asian part of the country. As a consequence of the 2010 summer fires in the densely populated European part of Russia, the mass media poured forth reports of burning forests, villages, victims, about lost crops by fires. The situation was further dramatized by the fact that the extreme smoke from these fires reached Moscow. Seventeen million people lived in the regions where the state of emergency has been declared, and another 10 million in Moscow suffered from smoke. Over one-third of the population of the Russian Federation lived in those regions where the fires were very intense in summer 2010. Still, the main question remained after the disaster which I attempt to answer in this chapter; was the heat the only cause?

Keywords Russia · Forest fire · Forestry · Rural depopulation · Land use

4.1 Introduction

Catastrophic forest fires hit the European part of Russian Federation during summer 2010 as result of a two-month-long period with temperatures above the average by 10 °C coupled with an unusually long drought period. Even though forest fires are usual for Russia, these are typically located in the sparsely populated Asian part of the country. As a consequence of the 2010 summer fires in the densely populated European part of Russia, the mass media poured forth reports of burning forests, villages, victims, about lost crops by fires. The situation was further dramatized by the fact that the extreme smoke from these fires reached Moscow (see similar case in

Nefedova's work was done on the topic of the state task of the Institute of Geography of Russian Academy of Sciences Nr. 0148-2019-0008.

T. Nefedova (✉)
Institute of Geography, Russian Academy of Sciences, Moscow, Russia
e-mail: trene12@yandex.ru

D. Karácsonyi et al. (eds.), *The Demography of Disasters*,
https://doi.org/10.1007/978-3-030-49920-4_4

Chap. 10). Seventeen million people lived in the regions where the state of emergency has been declared, and another 10 million in Moscow suffered from smoke. Over one-third of the population of the Russian Federation lived in those regions where the fires were very intense in summer 2010. Still, the main question remained after the disaster which I attempt to answer in this chapter; was the heat the only cause?

Chapter 3 by Sharygin highlighted how a California bush fire influenced outmigration from affected areas; while in contrast, this case study attempts to highlight a vice versa causal relationship: Could demography be a root cause for a forest fire disaster? In this study, I used media analysis and my existing knowledge and experience on spatial organisation of Russian countryside (Nefedova 2003; Nefedova et al. 2001), the rural demographic problems (Ioffe et al. 2004, 2006; Nefedova 2013) and the land use and agricultural change (Nefedova 2017; Nefedova and Pallot 2007) in the European part of Russia to explain the root causes of the 2010 catastrophic fires.

4.2 Data About the 2010 Russian Forest Fires and the Course of the Disaster

According to the estimation by ADSR (2010), there were 55 thousand excess deaths as the consequence of extreme heat, smoke and fire in Russia in 2010. Although the death toll published by the annual report of ADSR (2010) seems overestimated, there is no reliable direct data available in Russia on the fires and their consequences. The area of fires in Russia are normally underestimated by the official statistics. There is an objective reason for it—in a huge and sparsely populated country, not all forests are available for monitoring and protection. Besides, some officials intentionally understate the areas and consequences. So, according to the Ministry of Emergencies and the Federal Forestry Agency (*Rossleshoz*), fires affected 1.5 million ha of forests in 2010. Independent institutions suggest 6 million ha were affected, while, according to the Global Fire Monitoring Centre, which used satellite image data, it was 10 million ha. So, the difference in the sizes of affected areas reported by different agencies reached 5–10 times. For Europe, or the USA, it is never more than 20%.

Another problem is how to estimate the scale of the disaster. Either by the number of hot spots? Or according to the total area of affected forests? Or by the number of the devastated villages, or of the human victims? All these could be considered, while there is no exact data available on each of them. I analysed publications about fires in mass media. The intensity of fires may be estimated according to the frequency of reported areas, villages and other facts concerning fires. Such a digest is not quite full and reliable; however, it reflects the chronicle of the 2010 forest fire disaster.

The initial reports about local fires in April and May were not terrifying. In spring, residents usually burn old grass which results in local fires in the suburbs of Moscow and neighbouring regions. Also, as usual, forest fires began to occur in the Siberian part of Russia. The first signs of the catastrophe in European Russia appeared in June 2010, when extensive fires were reported in the so-called Black Earth Zone (Fig. 4.1),

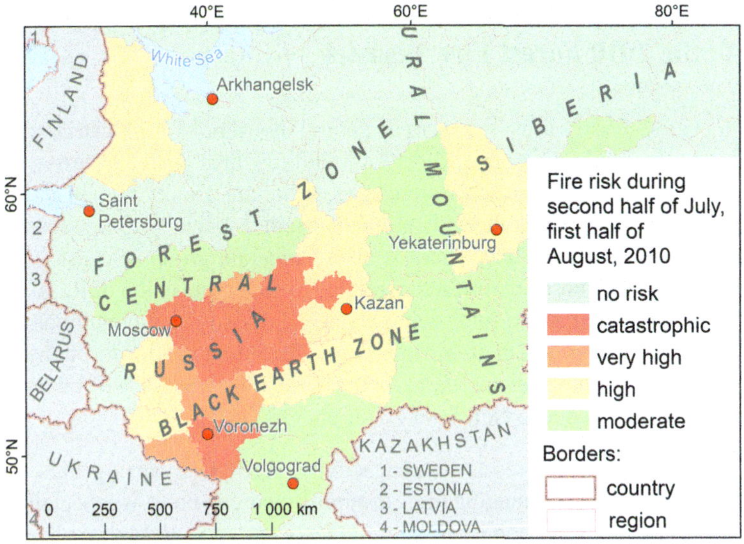

Fig. 4.1 Fire risk during summer 2010 in Russia (Author Nefedova, cartography by Karácsonyi)

the southern part of the core regions European Russia, where the hot weather had been observed first. In early July, large fires have passed through the Ural and Volga Region in the southeast part of European Russia, moving towards the core Russian areas.

Since the mid-July, fires had been extending to the east from Moscow, burning not only the forest but also dried peat bogs fires (Nefedova 2010). At the time, initial media reported burnt-down villages and associated deaths. Over 19 thousand hot spots were reported in early August 2010. The area affected by fire, according to the official data, covered 0.5 million ha. However, the Federal Forestry Agency and the Ministry of Emergencies pointed out that this was less than in the year 2009 by a factor of 2 or 3. Their arguments were right, except the fact that this summer was cooler in Siberia, where fires usually extended in previous years. President Dmitry Medvedev, who at that time was acting president of Russian Federation, decreed a state of emergency in seven regions of the Russian Federation on 7 August 2010 (Fig. 4.1).

Entire villages were devastated by the fires during summer in 2010. In total, three thousand houses from over 150 settlements burnt down and some people were lost and died. Proceeding these events, the government allocated construction funds for new buildings to support the victims of the fires who became homeless. Large cities, including Moscow, suffered heavy smoke from these fires. By mid-August, the situation began to stabilise. The number of eliminated fires began to exceed the number of new ones. By the end of August, after the rains started, the situation in most regions of Central Russia improved significantly. However, it remained difficult even in September that year in the Urals and in the south of Siberia.

4.3 Institutional Causes of Fires in Russia, in Particular to the 2010 Forest Fire Disaster

The authorities classify fires by causes as natural, household or industrial fires. This classification is artificial because 90–95% of all forest fires, which occurred in 2010, are caused by human factors. The root causes of fire are usually human activities as well as the problems within the institutional system, especially those areas responsible for regulation and policy. There is also a separate group of causes, which is seldom mentioned. These are the factors related to the organisation of Russian populated space, in particular the rural depopulation and associated population concentration into large cities and their surroundings. The human and institutional factors are summarised in the following paragraphs, while a new factor will be introduced in the next section (see also Nefedova 2013, pp.338–346).

(1) The main cause of fires is human activity. Activities causing fires, such as burning the old grass, making fires, dropping cigarette butts, is especially significant in the areas of people concentration, for instance, in Moscow's and other cities' suburbs. Sometimes individuals or enterprisers even provoke forest fires purposefully to conceal illegal or excessive woodcuts. This is possible only because the penalties for breaching fire-prevention rules are too lenient in Russia. In addition, officials lack initiative and funding. They have not taken basic fire prevention measures; for example, ploughing around villages and maintaining ponds. Some officials misrepresent the information about fires, concealing real danger and so on.

(2) Many people thought that the 2010 catastrophic forest fires are caused by the reform of legislation and of forest management in 2007–2009 (Gricyuk 2010). In Russia, with its vast forest areas, these changes have had grave consequences. According to this reform, a new Forestry Code was issued by the federal government, partly based on the European and Canadian best practices. In frame of the new Code, the Federal Service of Forest Wardens was eliminated, and all forest management was handed over to regional authorities who employed a small number of forest inspectors. Before the legislation change, fire security in forests was provided by 70 thousand foresters, but according to the new legislation, supervision of forests is charged to 12 thousand people who are engaged in a considerable amount of paperwork leaving almost nobody in the forests to detect fires. According to this new legislation, logging concession enterprises not only harvest timber, but they are also responsible for protecting and restoring forests. However, they control only 13% of forest areas. Moreover, small enterprises have neither means nor desire to do so. As for large companies, they prefer to pay small penalties for not implementing forest protection, than to spend a significant amount of money on implementing them. Air protection of forests was also handed over to regions, according to the new Code, after which its performance has decreased dramatically. It became impossible to transfer forces quickly from one region to another.

When the disaster occurred in 2010, President Medvedev dismissed the head of Federal Forestry Agency and transferred the Agency into direct subordination to the government and has charged to correct the Forestry Code. The legislation about the supervision and control in the Forest Code has been amended hastily during December 2010, and the Federal Service for Forest Wardens has been restored with a limited budget (Kuzminov 2011). Because of the limited budget, the amendment of the Forest Code happened only on paper according to Kuzminov.

4.4 Demographic Root Causes of the 2010 Forest Fire Disaster

As mentioned previously, there is another root cause of the 2010 forest fire disaster in Russia, which is the organisation of Russia's demographic space. The deep stratification of the Russian space was seldom mentioned during the discussion of the 2010 fire disaster (see, for example, articles of Rossiyskaya Gazeta journal in the period of May–August 2010), especially the rapid rural depopulation of periphery areas of the Non-Black Earth or Forest Zone and the suburbanisation in urban cores.

Rural population in European Russia has sharply decreased and concentrated in cities as a result of urbanisation during the twentieth century (see Nefedova et al. 2001). The rural population decline coupled with a negative social selection because young and most active population left villages for cities. The zone of the rapid rural depopulation between 1959 and 2017 (Fig. 4.2) has affected almost the entire Russian

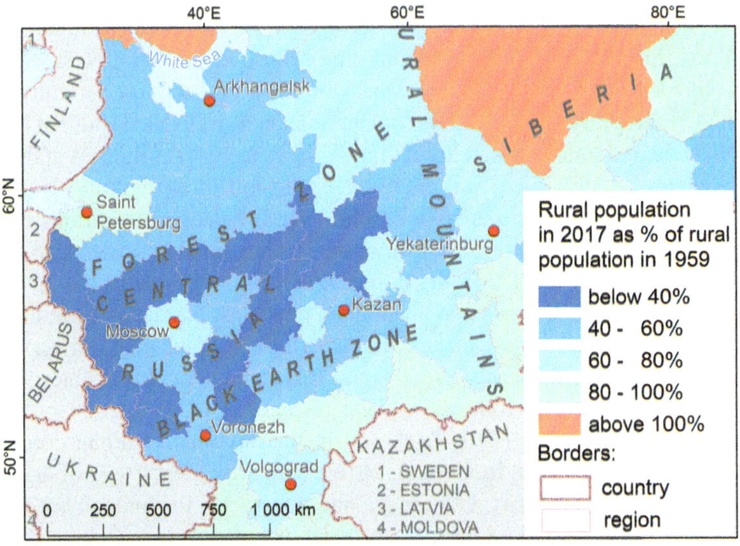

Fig. 4.2 Rural depopulation in Russia (Author Nefedova, cartography by Karácsonyi)

core areas. In the 1990s, the trends of migration have changed, so not only large cities but also suburbs started to attract rural population from peripheral districts. At the same time, the urban population from large cities leaked out to the suburbs as a result of suburbanisation as well (Nefedova and Treivish 2019). Hence, the contrast between suburbs and periphery areas has been increasing. The largest gap exists in the northern Forest Zone between peripheries and urban–suburban cores. In the periphery areas of the Forest Zone, not only villages but also towns are characterised by strong outmigration and depopulation. As an illustration, even in the Core of European Russia, only 100 km out of Moscow, you can find periphery areas without passable roads. At the same time, public bus services in many areas were eliminated due to economic insufficiency. On the other hand, the lack of roads is partially compensated by recent technical developments of other communication facilities (mobile phones or the internet), but the territory equipped with these facilities is also limited. As a result, these peripheries are forming a social and economic "desert" within urban core areas of large cities.

Further, due to globalisation and post-Soviet socio-economic development, the contrasts between cities and remote rural areas became more striking. Population density is higher, people are more active, and large enterprises are more successful in suburban zones of urban cores. Those districts lying beyond these suburban zones have been coming under social and economic depression, which resulted in the shrinking of agricultural land. It is also hard to attract agricultural businesses to areas without sufficient human resources. Neither large nor small farms are strong and numerous enough to deal with huge areas of abandoned land. The majority of peripheral villages are populated by old women and strongly drinking men. They are not capable of implementing even basic fire protection measures.

Consequently, destiny of rural territories is closely connected with agriculture and forestry. Agriculture in European Russia strongly varies in space and determined by natural conditions and by distance from cities (Ioffe et al. 2004; Nefedova 2012, 2017). Only 14% of Russia's landmass has climate favourable for agriculture, such as the highly fertile Black Earth Zone (or Chernozem Belt) located south of Moscow. The collective and state farms in Soviet times were strongly subsidised in the Forest Zone. Because of this, there was a strong expansion of agriculture to the north during the Soviet times, into areas with unfavourable natural conditions and rural depopulation. Furthermore, this expansion was accompanied by large-scale drainage of peat bogs typically located in the Forest Zone. In the course of the 2010 forest fires disaster, these drained out and abandoned peat bogs burned and smouldered.

During the transition period, in the 1990s with the end of subsidies for unprofitable farms, the arable land decreased greatly, and Russian's agricultural enterprises fell into depression. This was followed by the restoration of production only in the favourable southern Black Earth Zone and in the environs of urban cores in the 2000s (Nefedova 2013, Meyfroidt et al. 2016). However, great losses of arable land were typical for the majority of Forest Zone due to land abandonment that saw recovering wildlife expanding through formerly populated areas (Fig. 4.3). Hence, the traditional rural landscape of the Forest Zone of Central Russia with villages surrounded by fields was transformed. Young forests come nearer to villages as

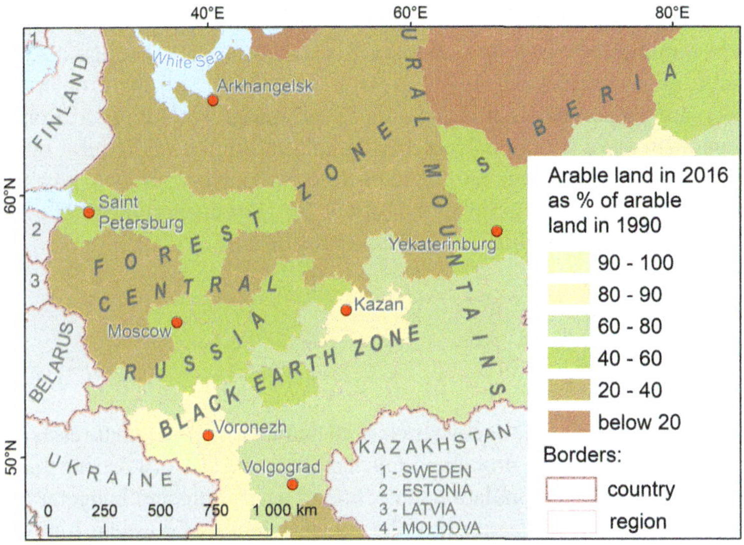

Fig. 4.3 Land use change in Russia (Author Nefedova, cartography by Karácsonyi)

possible "bridges" for fires. Few trees managed to grow on these abandoned lands, so these strips are covered by very high dry grass leading to the ever-greater danger of fire.

On the contrary, using timber resources is a more promising way for the Forest Zone than agriculture, but the land used by the timber industry has also retreated. As a result of the large state-owned logging enterprises transformation in the 1990s, the Soviet forest infrastructure was wrecked. Timber harvesting moved to forests close to urban core areas with passable roads. As a result, the intensity of logging in these areas and the danger of fire have increased.

During the same time, a large part of agricultural land was transferred into construction land in suburban zones of urban core areas. However, Russian suburbanization is different from the Western equivalent because it is characterised by the so-called *dacha*s, which are secondary homes belonging to urban dwellers and only inhabited during the summer. Only 20% of new-built houses in Moscow suburban zone are used as permanent residences. Also, more than 70% of urban dwellers have such secondary house. The concentration of Moscow residents in the suburban zone is very high, and these dacha settlements are very crowded during summer, yet, they face a lack of infrastructure, such as garbage removal. The risk of wildfire increases the when people burn their garbage.

Beside these suburban zones, remote *dacha*s are existing as well. As an illustration, the *dacha* zones of Moscow and Saint Petersburg have already merged despite the 700 km distance between the two cities. Remote *dachas* in rural depopulation areas are completely different where summer residents double or multiply the local population. They prevent houses and villages from complete destruction, but cannot restore agriculture (Nefedova and Pallot 2013; Nefedova and Pokrovskiy 2018). However, coming for one or two months a year, these urban dwellers take care of their rural houses and try to cut dry grass around them. Still the problems of fire protection are not solved. There is a lack of local manpower willing to work (aging and alcoholism are a common problems) in spite of high level of rural unemployment. Additionally, one can mention the insufficient financing of local governments, especially those located in peripheral districts. So, they cannot maintain even basic infrastructure for permanent local or summer residents.

4.5 Conclusions

In conclusion, in the struggle against fires, both human activities and forest legislation are important. However, the problem is much more complicated, covering many other issues, including rural depopulation, a decline of agriculture and budgetary support of local communities. Understanding the transformation of Russian rural space is also important. According to Greenpeace Russia's data, there were around 3 million hectare forests on fire in 2012, but some experts (see Delivoria 2019) estimate these fire areas up the 10 million ha. However, the total area on fire was larger in 2012 than in 2010. The fires had not caused smoke in large cities, so the situation was less echoed by the media. In spring 2019, a special firefighting system was introduced in 25 regions of Russia, but only in those areas where it can potentially threaten large cities. In other regions, such as the sparsely populated Siberia, nothing has changed. Hence, the root causes of the 2010 fire disaster still exists. Coupling this with more extreme weather conditions due to climate change could cause a similar disaster at any time again.

References

ADSR 2010, *Annual Disaster Statistical Review 2010 – The numbers and trends*. Centre for Research on the Epidemiology of Disasters, Brussels.

Delivoria, K. 2019, Gorit ono ognem, Ploshchad lesnih pozharov kazhdiy god rastet. (In English: It burns with fire. The firest fire area growing every year.) *Versiya*. Accessed ont he 2/10/2019: https://versia.ru/ploshhad-lesnyx-pozharov-v-rossii-kazhdyj-god-rastyot.

Gricyuk, M, 2010, Pozharni srok (In English: Fire time), *Rossiyskaya Gazeta* Nr. 5186 (106), 19th May, 2010, Accessed on the 2/10/2019: https://rg.ru/2010/05/19/les.html.

Ioffe, G, Nefedova, T, Zaslavsky, I, 2004, From Spatial Continuity to Fragmentation: The Case of Russian Farming. *Annals of the Association of American Geographers*, vol. 94, no. 4, pp. 913–943.

Ioffe, G, Nefedova, T, Zaslavsky, I, 2006, *The End of Peasantry? The Disintegration of Rural Russia*. Pitt Series in Russian and East European Studies. Pittsburgh: University of Pittsburgh Press.

Kuzminov, I, 2011. *Organizacionnye innovacii v lesnom sektore Rossii i socialno-ekonmicheskie effekti ih vnedreniya* (In English: Organisational innovation in the forestry sector of Russia and the socio-economic effect of its introduction). Innovacionnie in integracionnie processi v regionah i stranah SNG, Moscow, Media-Press.

Meyfroidt, P., Schierhorn, F., Prishchepov, A., Müller, D., and Kuemmerle, T., 2016, Drivers, constraints and trade-offs associated with recultivating abandoned cropland in Russia, Ukraine, and Kazakhstan, *Global Environ. Change*, vol. 37, pp. 1–15.
Nefedova, T, 2003, *Selskaya Rossiya na perepute.* (In English: Rural Russia in crossroads) Novoje. IG RAN. Moskva. 403 p.
Nefedova, T, 2010 Goryachee leto 2010 g.: Hronika i prociny secnih pozharov. (In English: Hot summer in 2010: chronicle and causes of forest fires). *Geografiya*, Vol. 917, No. 21, pp. 4–11.
Nefedova T, 2012, Major Trends for Changes in the Socioeconomic Space of Rural Russia. *Regional Research of Russia*, Vol. 2, No. 1, pp. 41–54.
Nefedova, T, 2013, *Desyat aktualnyh voprosov o selskoy Rossii. Otvety geographa* (In English: Ten Topical Issues about Rural Russia. A geographer's viewpoint). URSS. Moscow. 452 p.
Nefedova, T, 2017, Twenty-Five Years of Russia's Post-Soviet Agriculture: Geographical Trends and Contradictions, *Regional Research of Russia*, Vol. 7, No. 4, pp. 311–321.
Nefedova, T, Polian, P, Treivish, A, 2001, *Gorod in Derevnya v Evropeiskoy Rossii: sto let peremen.* (In English: Town and village in European Russia – Changes of one hundred year). OGI. Moscow. 560 p.
Nefedova, T, Pallot, J, 2007, *Russia's Unknown Agriculture: Household Production in Post-Soviet Russia.* Oxford University Press.
Nefedova T., Pallot J. 2013. The Multiplicity of Second Home Development in the Russian Federation: A Case of "Seasonal Suburbanization". In: *Second Home Tourism in Europe: Lifestyle Issues and Policy Responses* / Roca Z. (ed.). Farnham, UK – Berlington, USA: Ashgate Publishing, 2013. pp. 91–121.
Nefedova T, Pokrovskiy N, 2018, Terra Incognita of the Russian Near North: Counter-urbanization in Today's Russia And the Formation of Dacha Communities, *European Countryside.* Vol. 10 No. 4, pp. 673–692.
Nefedova, T, Treivish, A, 2019, Urbanization and Seasonal Deurbanization in Modern Russia. *Regional Research of Russia*, 2019, Vol. 9, No. 1, pp. 1–11.

Chapter 5
Disruptions and Diversions: The Demographic Consequences of Natural Disasters in Sparsely Populated Areas

Dean B. Carson, Doris A. Carson, Per Axelsson, Peter Sköld, and Gabriella Sköld

Abstract The *Eight Ds* model (Carson and Carson 2014) explains the unique characteristics of human and economic geography for sparsely populated areas (SPAs) as *disconnected*, *discontinuous*, *diverse*, *detailed*, *dynamic*, *distant*, *dependent* and *delicate*. According to the model, SPAs are subject to dramatic changes in demographic characteristics that result from both identifiable *black swan* events and less apparent *tipping points* in longer-term processes of demographic change (Carson et al. 2011). The conceptual foundations for this assertion are clear. Populations in SPAs can experience large and long-term impacts on the overall demographic structure as a result of decisions by a relatively small number of people. High levels of migration and mobility cause constant shifts in the demographic profile and prime SPAs to adapt to many different demographic states (Carson and Carson 2014). The Northern Territory of Australia, for example, experienced previously unseen waves of pre-retirement aged migrants in the past decade or so (Martel et al. 2013) as evidence of *detailed* but important changes to past trends. However, while dramatic demographic changes are conceptually possible and occasionally observable, there have been few attempts to examine the conditions under which such changes are likely to occur or not to occur. This is an important question particularly in relation to black swan events such as natural disasters because effective disaster management policy and planning is at least partially dependent on understanding who is affected and in what ways (Bird et al. 2013).

D. B. Carson (✉)
CQUniversity Australia, Rockhampton, Australia
e-mail: d.carson@cqu.edu.au

D. A. Carson · P. Axelsson · P. Sköld · G. Sköld
Umeå University, Umeå, Sweden
e-mail: doris.carson@umu.se

P. Axelsson
e-mail: per.axelsson@umu.se

P. Sköld
e-mail: peter.skold@umu.se

D. Karácsonyi et al. (eds.), *The Demography of Disasters*,
https://doi.org/10.1007/978-3-030-49920-4_5

Keywords Flood · Cyclone · Famine · Sparsely populated · Eight Ds

5.1 Introduction

The *Eight Ds* model (Carson and Carson 2014) explains the unique characteristics of human and economic geography for sparsely populated areas (SPAs) as *disconnected, discontinuous, diverse, detailed, dynamic, distant, dependent* and *delicate*. According to the model, SPAs are subject to dramatic changes in demographic characteristics that result from both identifiable *black swan* events and less apparent *tipping points* in longer-term processes of demographic change (Carson et al. 2011). The conceptual foundations for this assertion are clear. Populations in SPAs can experience large and long-term impacts on the overall demographic structure as a result of decisions by a relatively small number of people. High levels of migration and mobility cause constant shifts in the demographic profile and prime SPAs to adapt to many different demographic states (Carson and Carson 2014). The Northern Territory of Australia, for example, experienced previously unseen waves of pre-retirement aged migrants in the past decade or so (Martel et al. 2013) as evidence of *detailed* but important changes to past trends. However, while dramatic demographic changes are conceptually possible and occasionally observable, there have been few attempts to examine the conditions under which such changes are likely to occur or not to occur (see also Chap. 6). This is an important question particularly in relation to black swan events such as natural disasters because effective disaster management policy and planning is at least partially dependent on understanding who is affected and in what ways (Bird et al. 2013).

The purpose of this chapter, therefore, is to begin the process of identifying the conditions under which dramatic demographic responses to natural disasters in SPAs might occur. In the process, we introduce two new 'Ds' with which to describe the nature of demographic change. We propose that natural disasters such as cyclones, floods, earthquakes, bushfires, landslides, avalanches and crop failures present the potential to *disrupt* or to *divert* demographic development. Disruption involves a temporary break in demographic development after which the pattern apparent before the disaster resumes. Diversion involves the transition to a new pattern clearly and substantially distinct from the pre-event state. Disruption and diversion can be considered as end points on a scale with attributes such as the length of time taken to revert to pre-event development patterns and the number and type of demographic characteristics that change post-event along with the severity of the change determining where an event might be positioned on that scale.

The chapter focuses on two apparently dissimilar natural disasters which are the *Great Deprivation* crop failure(s) in northern Sweden in 1867–8 and the flooding of the Katherine-Daly river basins in the Northern Territory of Australia in 1998 caused by Cyclone Les. While the Great Deprivation affected large areas of Norway, Sweden and Finland, this chapter will focus on the parish of Gällivare in Sweden's

Norrbotten County due to the region's prominent role in the economic and demographic development of the north at the time through forestry and mining. Figures 5.1 and 5.2 shows the locations of Gällivare parish and the Katherine-Daly river basins.

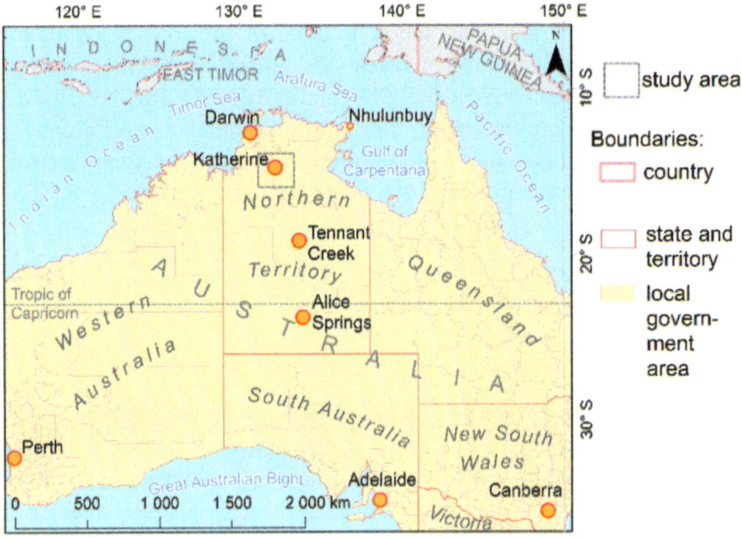

Fig. 5.1 Location of case site Katherine-Daly region, Australia (Cartography by Karácsonyi)

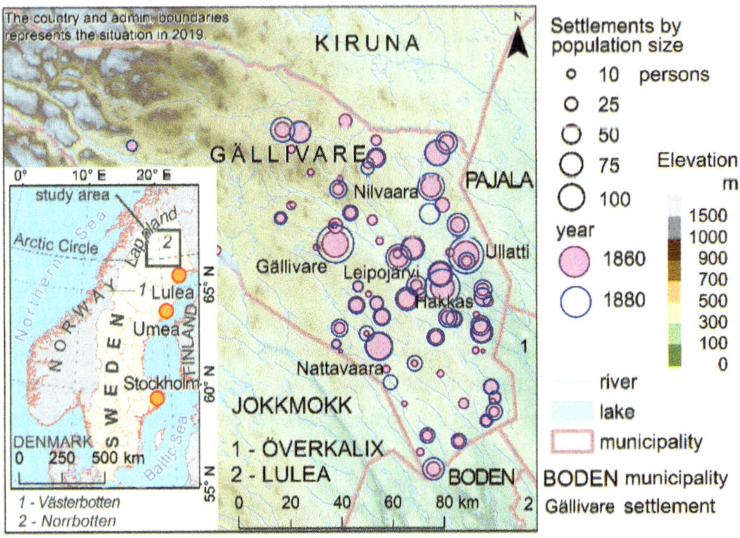

Fig. 5.2 Settlements in Gällivare, 1860 and 1880 (Authors: B. Carson, A. Carson, Axelsson, P. Sköld, G. Sköld, cartography by Karácsonyi)

Research has not been published about the impacts of the Great Deprivation on Gällivare, but Nordin's (2009) analysis of long-term demographic change in this and other northern parishes shows a brief (two or three years) slowing of population growth around the time of the crop failures amid an extended period of population growth before and after the event. This is primae facie evidence (at least at the gross population level) of a demographic disruption. In contrast, grey and academic literature relating to the Katherine-Daly region reveals a dramatic change in gross population trends (from substantial increase in the 1980s and early 1990s to a long period of stagnation persisting to the present time), along with substantial (and persistent) changes in the proportion of Indigenous people and women in the population (Harwood et al. 2011). This is primae facie evidence of a demographic diversion.

5.2 The Challenges of Identifying Demographic 'Impacts'

Attributing demographic changes in SPAs to black swan events is difficult precisely because of the dynamic nature of populations in those areas. Their susceptibility to dramatic change exists, even in the absence of such events. In the case of migration impacts, for example, high levels of migration already occurring in SPAs might be difficult to separate from migration specifically related to a natural disaster (Adamo and Sherbinin 2011). Likewise, demographic transitions in fertility and mortality can occur relatively quickly but sporadically in SPAs (Taylor 2011), such that *bubbles and craters* (Martel et al. 2011) are expected characteristics of distributions of demographic characteristics of SPAs in most circumstances.

Analyses of demographic change are also impacted by the modifiable areal, modifiable temporal and modifiable social unit problems (MAUP, MTUP, MSUP) (Koch and Carson 2012). The MAUP is concerned with the problem of spatial scale when analysing statistical (e.g. population) data, as results can change according to the levels of scale applied to the analysis. It is difficult to ascertain at what spatial scale the demographic impacts of natural disasters can be best understood. On the one hand, natural disasters may have physical impacts that are highly localised as is the case with landslides and avalanches, and sometimes bushfires and flooding and even violent storm events. On the other hand, public perceptions often attach these events to broad geographical ideas such as *the north, the Arctic* or *the tropics*. While researchers must be careful in recognising the physical boundaries of event impacts, they should also not underestimate the power of perceived risk boundaries in influencing demographic behaviour (Ford et al. 2006). In this chapter, we are interested in somewhat local impacts, which bring the additional challenge of deciding what demographic characteristics to observe, and how to represent them. Most demographic rates are designed for application to relatively large populations of 100,000 persons or more while here, we are working with populations much smaller. In the absence of well-established demographic analysis techniques for small populations (Taylor 2014), we have restricted our analysis to broad indicators that include gross populations and percentage of persons with particular characteristics.

Similar to the MAUP, the MTUP refers to the impact that different temporal dimensions can have on the outcomes of data analysis. This is important in the context of demographic change following natural disasters, whereby Ford et al. (2006) noted that demographic change may be a long-term consequence of exposure to the risk of natural disasters as much as a short-term response to a single event. In fact, the expectation of experiencing natural disasters in SPAs may make demographic structures quite resilient to a single major event, and changes in demographic behaviour may be a cumulative effect of relatively minor events.

This last statement also reveals the challenges of the MSUP, which refers to the problem that the outcomes of data analysis may differ according to the social units selected for analysis. McLeman's (2010) research in Canada showed that specific individuals may migrate out of northern communities as a result of extreme weather events, but migration flows may persist with out-migrants being replaced by in-migrants with similar characteristics. In this way, the demographic behaviours of individuals might be impacted by the event, but the demographic behaviours of the population as a whole remain relatively stable. Bailey (2011) noted, for example, that communities with high pre-existing demographic volatility are likely to experience high demographic volatility as a result of a natural disaster. This is not the only problematic dimension of the MSUP, with several researchers identifying that the differential impacts of events on various social groups (defined by ethnicity, social status, gender and even religion) can be hidden in higher level analyses (Ellis 2009).

These *unit problems* are apparent in trying to summarise the impacts of the Great Deprivation as reported in the academic literature. That literature is replete with apparent contradictions and counter-intuitive observations. The Great Deprivation was a severe shortage of food resulting from consecutive crop failures in the Nordic countries in 1867 and 1868. The 1867 crop failures were caused by a long, cold and wet winter, while the 1868 failures were caused by drought. The crop failures affected at least Finland, Norway and Sweden (Nelson 1988). Finland, during the Great Deprivation, has been the focus of much attention regarding demographic impacts. This is because of estimates that state nearly 10% of the population there died as a direct result of the food shortage (Jantunen and Ruosteenoja 2000). Debate continues as to the impacts of the Great Deprivation across the entire region, but it is apparent that impacts differed both between and within countries. It has been common, for example, to attribute the substantial increase in Swedish (and Norwegian) emigration to North America at the end of the nineteenth century to the Great Deprivation (Akenson 2011), although emigration from Finland to North America was comparatively rare. More recent research, however, suggests that either the Great Deprivation did not have such an immediate impact on Swedish emigration (with the substantial increase coming at least a decade after the event), or emigration was largely from southern parts of Sweden (Alestalo and Kuhnle 1987). Likewise, while direct Finland–America migration was limited, it has been estimated that up to 20% of Finns who migrated to other parts of Scandinavia (mostly northern Sweden) at the time of the Great Deprivation ultimately on-migrated to America (Newby 2014). This simple example reveals aspects of the MAUP (different migration patterns observable at different spatial scales), MTUP (different patterns observable depending on the ‘lag’ times allowed between the Great Deprivation and emigration) and MSUP (the behaviours of people in northern Sweden have been hidden within the analyses of the total Swedish experience (Doblhammer et al. 2013), just as the behaviour of Finnish on-migrants may have been hidden by both time and the relatively small size of the population within northern Sweden).

Migration responses to the Great Deprivation are not the only subject of contention in the literature. While there appears to be consensus that fertility rates decreased during the immediate Great Deprivation period (typically considered to be 1867–70 allowing for recovery time) but recovered shortly thereafter, the impacts on life expectancy and mortality continue to be debated. Hayward et al. (2013) claim lower life expectancies for people born around the time of the Great Deprivation, while Saxton et al. (2013) claim that life expectancy impacts were very small. The MSUP might be important here, with specific population groups such as landless labourers, Sami (Nordin and Sköld 2012), rural dwellers, and Finnish migrants believed to have been more dramatically affected than others.

Overall, the published evidence from the Great Deprivation suggests a demographic disruption lasting between three and five years and affecting fertility, mortality and migration. These conclusions are tempered by observations of substantial spatial and social variation, particularly for mortality, migration and possibly also life expectancy. These presumed impacts of the Great Deprivation are generally consistent with the observations made about other ‘famine’ events in western and northern Europe (O Grada 2007). In general, local impacts were less where there were higher industrialisation and urbanisation. Industrialisation and urbanisation generally reduce reliance on local food sources, increase access to public and private sources of famine relief, and provide the resources for individuals and families to use adaptation strategies such as long-term and short-term migration and changes of occupation (Isacson et al. 2013).

The Great Deprivation is seen as the last of the European famines to result from natural events (although food scarcity has since occurred as a result of war and economic depression). The nineteenth century saw other years of crop failure (notably 1809, 1832–33 and 1857–58 (Bengsston and Dribe 2002)), with industrialisation and urbanisation changing food supply chains and reducing the reliance of communities on local food production. Nevertheless, crop (and stock) failures as a result of droughts, floods and other natural events continue in many SPAs in industrialised (and post-industrial) nations, and their demographic impacts continue to be debated. While the risk of food scarcity is low, there are still likely to be substantial demographic impacts linked to economic and social consequences of bad seasons. Drought, cyclone, flood and fire events have been linked to out-migration (particularly youth out-migration), increased suicide rates, and urbanisation in SPAs in Australia, Canada and the USA (Hogan et al. 2014; McLeman and Smit 2006).

The specific literature on the Katherine-Daly cyclone-induced flood event of 1998 will be reviewed later in this chapter, but similar events have been researched from a demographic perspective. Cyclones and floods have the potential to physically destroy towns and settlements, although the typical response (in the past century at least) has been to rebuild those towns and settlements once the event has passed.

Perhaps the most famous rebuilding exercise in SPAs was associated with Cyclone Tracy. This cyclone hit Darwin (population of 40,000) in Australia's Northern Territory on Christmas Eve 1974, destroying over 90% of the housing and general building stock of the city. Over half of the city's residents were evacuated, with the population only returning in substantial numbers more than a year later once rebuilding had commenced. The population growth that was being experienced in Darwin before the cyclone quickly resumed by the late 1970s and even accelerated based on substantial in-migration associated with rebuilding and with the transition from dependent Territory to self-governing Territory status in 1978. However, it has been estimated that less than half of the population residing in the city in 1978 had also been at the time of the cyclone (Britton 1981). While the collective demographic characteristics (particularly age and sex distribution) looked quite similar, they were the characteristics of a new population, and the individual demographic impacts in terms of both out and in-migration were dramatic.

Cyclone Tracy may also have had different impacts on different population groups, with recent research (Haynes et al. 2011) suggesting much higher return migration rates of Indigenous people, and in-migration of Indigenous people from nearby communities which were also affected by the cyclone but which had less access to recovery and rebuilding resources. In a review of the literature on Indigenous migration responses to cyclones and other severe weather events, however, Carson et al. (2013) argued that the Cyclone Tracy experience may not be typical (nor may it be atypical) given that so many factors influence individual and collective behaviours. The review suggested that migration responses could differ depending on the extent to which: migration was forced [by evacuation policies, for example (Taylor and Freeman 2010)]; alternative residential locations were available nearby; resources were available to support both the out-migration and return migration; economic and cultural livelihoods were tightly linked to specific locations; and individual decisions were influenced by social and cultural group membership.

The literature on famines, cyclones and floods in SPAs suggests that demographic impacts are difficult to predict and dependent on a range of factors which need to be understood in their spatial, temporal and social contexts. Immediate impacts may be somewhat predictable based on exposure to risk (such as fatalities) and organisation of disaster response (such as forced evacuation), but longer-term impacts are subject to many intervening factors, including the difficulties associated with isolating the impacts of a single event from the ongoing dynamic and detailed nature of demographic change (see discussion on long-term population shifts in Chap. 2 as well). This argument can be further supported through the examination of the experiences of Gällivare and the Katherine-Daly area linked to the Great Deprivation and Cyclone Les respectively. While these cases are separated by 130 years, 14,000 km and a plethora of social, economic and political attributes, they demonstrate some of the ways in which demographic disruptions and diversions may emerge from natural disasters in SPAs.

5.3 Gällivare and the Great Deprivation

While northern Sweden is often considered to have been slow to industrialise when compared with other parts of Europe (Engberg 2004), there is evidence that Gällivare was well advanced by the 1860s (Godet 2009). Industrialisation in Sweden tended to come first to rural primary sectors such as forestry and mining. This trend coincided with the dismantling of guilds and opportunities for private enterprise in manufacturing, thus, limiting urban industrialisation at least until after the liberal reforms of 1846 (Ryner 2003). During this time, Gällivare was in the Timber Frontier region, and the forestry sector experienced relatively rapid industrialisation after the introduction of the first steam-powered timber mills in the 1840s (Maas and van Leeuwen 2002). Gällivare was also at the forefront of mining development in northern Sweden, with commercial ore deposits near Malmberget (five kilometres north of Gällivare town itself) being extracted in the early nineteenth century. By the 1860s, British investors had secured funding to construct the first railway in Norbotten from Malmberget to the Lule River. Construction commenced in 1865, but was abandoned in 1867 due to financial strain (Godet 2009). There is no specific evidence that the Great Deprivation played a role in halting this initial construction, and, in fact, large projects such as railway construction would likely have been embraced by government as a means of providing employment for displaced rural workers during this time. The Swedish government did take over construction the railway, but poor planning and engineering meant that a working line did not eventuate in the region until 1888. Nevertheless, mining developments were already occurring by the 1860s, bringing with them new extraction and transportation technologies (Dewsnap 1981).

Likewise, while it was the impact of the railway on mining and forestry which led to the major economic and demographic development of Gällivare in the 1890s and early twentieth century, the signs of forestry industrialisation were already in place by the 1860s. Northern Sweden as a whole was already exporting timber by 1860, and the period of government regulation and management of forests as an export resource began in 1866 when England removed import barriers for Swedish timber (Axelsson 2014). During the second half of the nineteenth century, and particularly during the 1860s, there was a substantial consolidation of forestry ownership, with both government and private ownership focusing on larger landholdings. By 1872, new settlements were being actively discouraged in Sami pasture areas, which impacted the larger western parts of Gällivare, and led to a focus of development in the eastern quarter. Gällivare village itself had been settled in the early eighteenth century and became the centre of the Gällivare Parish formed in 1742. The Parish was dominated by small settlements up until the turn of the twentieth century, with just three villages having more than 100 population in 1860 (Gällivare, Ullatti and Nattavaara). Even by 1880, there were just six villages larger than 100 population (now including Hakkas, Leipojärvi and Nilivaara). Figure 5.2 shows the gross population development for Gällivare Parish between 1860 and 1880. The size of each dot represents the *settled* population of the village, with Gällivare Village in 1860 equating to 105 persons.

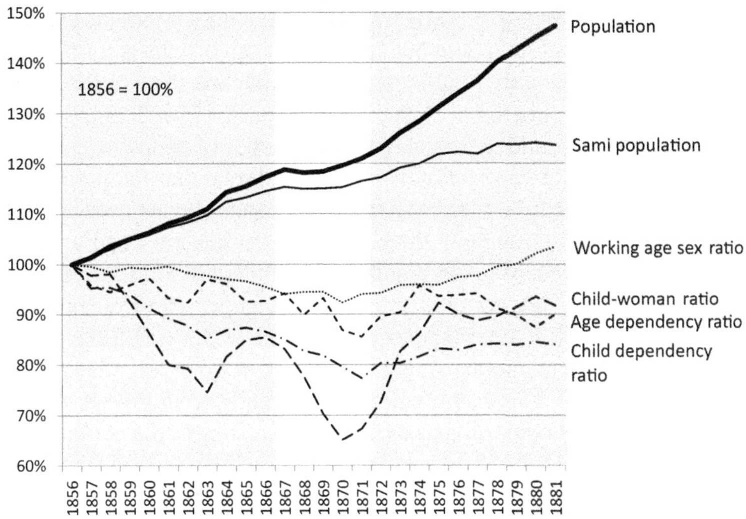

Fig. 5.3 Demographic Indicators for Gällivare, 1856–1881

Note that many Sami people (about 400 in both 1860 and 1880) were excluded from village population lists as they were still considered to be nomadic.

By 1865, Gällivare Parish had an estimated population of 2650 people, including nearly 1500 Sami people. There were 97 men for every 100 women, although this rate was much higher among non-Sami (104 men for every 100 women) population. Over one-third of the population (Sami and non-Sami) were aged under 15 years, and just five per cent (seven per cent of Sami) were aged 65 years and over. Annual in-migration rates for the past 10 years had averaged five per 1000 population, with out-migration averaging seven per 1000. Nearly two-thirds of the male population were employed in forestry or agriculture (and often in both) (Bäcklund 1988).

The short-term impacts of the Great Deprivation are apparent in a range of demographic indicators (see Fig. 5.3). These indicators are derived from data provided by the Demographic Data Base (DDB) at Umeå University. The DDB contains digitised parish records and links data on individuals identified in these records for most of the nineteenth century. Additional data were drawn from the 1890 and 1900 Swedish Census provided by the North Atlantic Population Project (NAPP). Figure 5.3 standardises indicators according to the 1856 value. The figure highlights the period 1867–1871, which has been considered the critical impact period (Isacson et al. 2013). Population growth stalled very early in this period but had resumed by the end. The Parish population was 2700 in 1866 and increased only to 2720 by 1869, but then increased to 2782 by 1872 and continued to grow by between 1.5 and 2.0% per annum thereafter. The Sami population had been increasing rapidly up until the early 1860s but remained at 1500 people for the entire critical period. There was a brief period of growth again between 1871 and 1878 (to 1600 people), but the Sami population declined again to about 1500 people by 1900 (Karlsson 2013).

The other indicators in Fig. 5.3 experienced decline throughout the entire critical period. The child–woman ratio (number of children aged 0–4 years for every woman aged 15–44 years) experienced the most dramatic decline, from about 85% of its 1856 value in 1866, to just 65% of the 1856 value in 1870. The ratio rose very quickly thereafter. The child dependency ratio (number of people aged 0–14 years for every person aged 15–64 years) also declined throughout the period. The age dependency ratio (number of people aged 65 years and over for every person aged 15–64 years) fluctuated throughout the period, but the general trend was a decline (indicating fewer older people). The small spike in this ratio in 1869 may be related to the 'dip' in working-age sex ratio (number of men per 100 women in the ages 15–64 years) in 1869–70. There is some evidence here of working-age males leaving Gällivare, probably via temporary migration to seek work in areas less affected by the famine (Dribe 2003). This also suggests only a small excess impact of the famine on mortality for older people (Edvinsson 2014). Interestingly, the period considered in Fig. 5.3 includes the 1857–58 famine, and there is some small evidence of decline in variables of similar type (but lower intensity) for those years.

The impact of the Great Deprivation on migration to and from Gällivare is difficult to ascertain. There was negative net migration (in the order of 10–15 persons per year) for four of the five years between 1867 and 1871, but negative net migration was not unusual and had occurred in 8 of the preceding 11 years and would occur in four of the following 10 years despite substantial population growth in that period. Population turnover was lower in the Great Deprivation critical period (10 per 1000 population) than the preceding (13 per 1000) or proceeding (12 per 1000) five-year periods. Of the 90 people who did leave Gällivare between 1867 and 1871, none were recorded as moving to North America, and just three moved to Norway. The remainder moved to other northern parishes within Sweden such as Råneå (20 migrants) and Överkalix (20 migrants). The destinations of migrants for 1867–1871 were similar to destinations in the preceding and proceeding five-year periods, with the exception that Norway was less popular in 1867–1871 (there were 14 migrants to Norway 1862–1866 and 13 in 1872–1876). There were 49 in-migrants to Gällivare between 1867 and 1871, with the majority again coming from other northern parishes where Överkalix and Jokkmokk accounting for more than half of all migrants. None were from Finland.

However, migration events may not have been well captured in the data, particularly short-term moves to find work or famine relief. The DDB does record 14 cases of people who moved out of Gällivare between 1867 and 1871 and moved back to Gällivare later. Only three of these were working-aged males (and one other male aged 14 years), while seven were working-aged females. Consequently, the migration events recorded during the period do not account for the relative decrease in the adult male population. Instead, there was an excess of adult male deaths (105 males died during the period compared to 92 females). In contrast, the period 1877–1881 saw 104 adult female deaths and 69 adult male deaths.

Mortality did not increase in Gällivare during the 1867–68 famine, crude mortality rates were 33 and 28 per 1000, but in perspective of the entire nineteenth century, these years mark a start of a high-mortality period that lasted until the end of the century (Sköld and Axelsson 2008). The dominant cause of death during the famine resulted from infectious diseases. Smallpox broke out in 1867 and measles followed in 1868. It is difficult to relate this to the starvation and suffering related to the famine, especially since these epidemics were nationwide. The Sami experienced a limited mortality increase around 1867–68 that was followed by similar peaks in the following decades. These changes were not dramatic when seen through a long-term perspective as they were not remarkable. Infant mortality actually decreased in Gällivare during the 1860s. The limited immediate impact on Sami populations may be explained in part by their capacity to use migration (a large proportion of Sami were at this time still nomadic) as a response to the famine threat (Sköld 1997).

After the Great Deprivation, the Gällivare population became more male (the total sex ratio rose steadily until the end of the nineteenth century, along with the working-age sex ratio), and less Sami. Gällivare also became more urbanised, with nearly three-quarters of the *settled* population living in villages with populations over 50 people in 1880, compared with just 47% in 1860. Furthermore, the five largest villages contained less than 25% of the population in 1860, which rose to approximately 30% in 1880. However, this urbanisation was not apparent by 1870 with 47% of the population in larger villages and 23% in the five largest villages. So the Great Deprivation may likely have had a lesser impact than the processes of industrialisation in both the forestry and mining sectors, despite the more readily available famine relief in the larger towns (Engberg 2004).

5.4 Katherine-Daly and Cyclone Les

The Katherine-Daly river catchment area is home to about 21,000 people and consists largely of two broad demographic zones. The town of Katherine is a predominantly non-Indigenous (70%) urban centre of 10,000 inhabitants whose location coincides with the main river crossing as well as the intersection of the only north–south and east–west sealed road transport routes in northern Australia. The surrounding region has over 85% of its population identifying as Indigenous, and features at least ten large (more than 200 people) 'Indigenous communities'—population centres located on Indigenous land and subject to specific legislation and regulations about who can live there.

Both Katherine town and the Katherine-Daly region have been well aware of the risk of flood events, with the site of the town shifted several times between initial European settlement in the late nineteenth century and the final siting that occurred once the railway bridge intended as part of the north–south railway line (not completed until 2004) had been constructed. The Indigenous inhabitants had long since adopted practices of temporary and longer-term migration to adapt to changing environmental conditions (Bird et al. 2013). While Katherine, like other large Northern Territory towns, was founded largely as a result of the Overland Telegraph line from Adelaide to Darwin, it quickly developed an entrepreneurial culture which distinguished it from the heavy government dependence typical of

Northern Territory regional and local economies (Carson 2011). Katherine was one of the main service centres for the cattle industry in the north of Australia, was the site of experimental pastoral (cattle and sheep) and agricultural (peanuts and citrus) activities and had relatively large mining and manufacturing sectors in the 1980s and 1990s. The Katherine Research Station was opened in 1956 to inform agricultural and pastoral development.

The Katherine-Daly region was also at the forefront of tourism development in the north of Australia in the 1970s and 1980s. One of the first northern national parks was established at Katherine Gorge (Nitmiluk) in 1962. Tourism ventures originating in Katherine and focusing on Nitmiluk and Kakadu National Parks were very important contributors to the local and regional economy by the late 1980s when Nitmiluk National Park was handed back to its traditional Aboriginal owners (Berzins 2007). Faulkner and Vikulov (2001) report that Nitmiluk received nearly a quarter of a million visitors in 1997, and over 175,000 visitors stayed overnight in Katherine town, where the majority of accommodation infrastructure existed. While many visitors were self-drive, there were large organised tour markets, including 45,000 international visitors. Large tour companies such as Travel North were formed and based in Katherine, and large motels (many constructed in the late 1970s and early 1980s) dominated the accommodation market.

By the time of the January 1998 flood, however, there were signs that Katherine's development was slowing down. Attempts to sustain scheduled commercial air transport (subsidised heavily by the government) had largely failed. Travel North had been broken up into two companies, with some operations relocated to Darwin. Darwin had also supplanted Katherine as a major accommodation centre for people visiting Kakadu National Park, and even Nitmiluk National Park (Schmallegger and Carson 2010). Overall, visitor numbers to Outback Australia had begun a decline that continues to this day (Taylor et al. 2015). For the Katherine tourism region, visitor numbers have continued to decline since Faulkner and Vikulov's (2001) paper claimed that a 'bounce back' was likely. Visitor numbers did increase substantially in 2000 and 2001 (by approximately 15% compared with 1999), but declined even more dramatically between 2002 and 2003 (by over 30%) and continued to decline at least until 2013 (when visitor numbers were less than two-thirds of the 1999 benchmark).

While detailed data prior to 1996 are difficult to obtain, there is some anecdotal evidence at least, that other private sector activities such as mining and manufacturing in Katherine were also in decline before 1998 (Desert Knowledge Australia 2005). The Mount Todd mine, for example, was already in the process of winding down operations by 1997. Population growth had already plateaued by the 1996 Census at around 9500 people in the town, and 19,000 in the Katherine-Daly region as a whole. The balance between Indigenous and non-Indigenous residents of the town had also been shifting, with steady migration of Indigenous people from the surrounding region to the town observed at least since the early 1970s (Taylor 1989) and may have increased in the mid-1990s as a result of less support from government for Indigenous people living in very small communities in very remote areas (Altman 2006). The early impact of this migration had somewhat been hidden by the major expansion of the Tindal Airforce Base in the late 1980s, which brought nearly 2000 mostly non-Indigenous (and male) residents to the town.

The broad demographic experience of the town of Katherine in response to Cyclone Les and associated flooding in 1998 has studied by Harwood et al. (2011). Their work focused on the period from 1971 to 2006 reached some broad conclusions:

- The town experienced rapid population growth from the late 1970s (inspired by the establishment of a new military installation and the growth of the tourism industry) until the flooding, when population growth stagnated;
- There was a substantial change in economic geography, with the private sector and government administrative employment declining, and public services employment (education and health) increasing;
- There was a substantial change in the age distribution between 1996 and 2001, with an increase in the proportion of children aged under 15 years and people aged over 45 years at the expense of younger working-age adults;
- There was a substantial increase in the proportion of Indigenous people (from 18 to 25% of the population) and in the proportion of women in the population (from about 45% to nearly 50%).

Harwood et al. (2011) focused on the town itself, although the floods affected the larger Katherine-Daly river catchment area. They did suggest that the increase in Indigenous population could be partially explained by the continuing migration of people from more remote parts of the area to the main town. While this migration may have increased after the flood (evidence is hard to find), the changing proportions of Indigenous and non-Indigenous people in town was likely more due to out-migration and a 'failure to return' (p. 317) by non-Indigenous people.

Harwood et al. assumed that the flood event was responsible for the immediate demographic change, and suggested, as did Faulkner and Vikulov (2001), that such impacts would likely be temporary. Katherine was expected to be well positioned to *bounce back* from this event. Partly, because the town's familiarity with flood events (big floods also occurred in 1957 and 1974, and since Cyclone Les there have been flood events in 2006 and 2014). Additionally, rapid recovery should arise because of Katherine's advantageous geographical position. Finally, because of the 'entrepreneurial spirit' which had underpinned rapid development during the 1980s and early 1990s, and the economic diversity and human capital which had accrued as a result. The expectation of rapid and full recovery (and even increased growth) existed despite 98% of businesses being affected by flood damage (James 2009).

Both Harwood et al. (2011) and Faulkner and Vikulov (2001) may have overestimated the capacity of the town in particular to recover economically and demographically as quickly as observed in Gällivare. Katherine town did not return to 1996 population levels until 2007, and its growth since has been predicated on the continued expansion of the public sector workforce, including substantial local government employment as local government offices for the surrounding region became located in Katherine town after local government reforms in 2007 (Michel and Taylor 2012). Public sector employment in health, welfare, education and protective services also increased dramatically as a result of the 2007 Northern Territory Emergency Response, an Australian Government programme designed to combat reported high levels of domestic violence and child abuse in Aboriginal communities (Taylor and Carson 2009).

In contrast to the town, population growth in the surrounding region averaged 2% per annum from 1999 until 2013, with 1997–1998 gross population growth (72 persons) being low, but not unusually so. For the surrounding area, growth between 2003 and 2004 was estimated at 82 persons, between 2006 and 2007 (coinciding with another flood event) at 78 persons, and 2010 and 2011 at just 23 persons. In both Katherine and the region as a whole, Indigenous population growth outstripped non-Indigenous population growth, and the region became more female as well as more concentrated in the very young and the older age groups (Brokensha and Taylor 2014). No doubt the flood event was significant in the embedding of the new demographic trends reported by Harwood et al. (2011). However, there is some evidence that Katherine (particularly the town itself) was *primed* for such a demographic shift even before the 1998 flood event. The disaster may have served to exacerbate the new trends, rather than to create them.

5.5 Explaining the Differences

On the surface at least, Katherine appears to have experienced dramatic long-term consequences of Cyclone Les and the 1998 flood, while Gällivare suffered some substantial immediate impacts and relatively quickly returned to a pattern of demographic development which had been emerging by the mid-1860s. However, beyond the direct impact on individuals (which even so is difficult to assess), it is problematic to ascribe specific outcomes to the natural hazards. The dynamic nature of populations and economies in sparsely populated areas is well captured in these two case examples. For Gällivare, the impending mining and timber booms linked to northern settlement strategies, the introduction of industrial technologies, the opening up of global markets (particularly the UK) and the arrival of rail transport were the major drivers of demographic change in this period. There may be hints of correlations between Great Deprivation-induced activities such as the failed British attempt to construct a railway line or the increased Swedish government attention to forest regulation, but the literature has not teased these out to this point. For Katherine, the impending tourism 'bust' and the diversion of political attention and infrastructure investment to other locations in the north (principally Darwin) were likely very significant drivers of demographic change by the mid-1990s. It is likely that the processes of change would have occurred more slowly without the flood, but it is not by any means clear that the ultimate outcomes would have been much different.

The Gällivare story stands as an exception to some of the perceptions about the Great Deprivation—that it affected the north more than the south due to lack of industrialisation and that it resulted in substantial international emigration—but reinforces some others—that impacts were short lived, and that component effects (births, deaths, migration), while noticeable, were generally small. This reinforces the idea that spatial scale is important when examining the demographic impacts of natural disasters. Gällivare's experience must be somewhat unique given its economic, social and political resources, and, at least when it comes to economic circumstances, is likely to be importantly unique.

The spatial scale at which the 1998 flood disaster in the Northern Territory of Australia is examined has also proven important. A demographic diversion is clear when the town of Katherine alone is examined, but the broader region appears to have been less impacted. The town experience was more dramatic because of the starting economic and demographic conditions—not only that these were so different to the surrounding region, but because new trends which the flood would exacerbate in the short term had already been flagged. Consequently, those trends have persisted at least until the present day.

This research also includes an interesting comparison of affect—in Gällivare, the principal threat was to human life. Property was not much impacted by the bad seasons (although some land was not able to be used productively for a couple of years). In contrast, human life was not substantially threatened by the 1998 flood, but there was substantial property damage. This could explain the failure of Katherine to recover as expected, although property damage and the need to rebuild have elsewhere been considered drivers of Darwin's rapid recovery from Cyclone Tracy (Britton 1981). Loss of human life in a context of high mobility may actually be less important than the loss of physical infrastructure.

The research has highlighted the ways in which the dynamic processes of demographic change that typify sparsely populated areas make it difficult to identify the causes of specific changes, even where black swan events occur. Dynamic change makes it more possible for sparsely populated areas to *absorb* immediate demographic shocks within what are anyway often short-term demographic cycles. It also makes it easier to exaggerate trends which may have only been barely apparent prior to the event. 'Timing is everything', but interpreting the specific effects of the timing of a disaster presents challenges. In both cases, had the event been just a few years earlier, the impacts may have been vastly different. Resulting in perhaps more dramatic impacts for Gällivare than for Katherine. Timing influences the adaptations that are available for populations—are there economic alternatives? Are there migration opportunities? Are there systems in place to coordinate disaster response? The answers to these questions can change within even very short periods of time.

Similarly, the perception of impact of a disaster event can be altered by the temporal scale at which those impacts are viewed. It is only at a relatively fine scale (single years in this analysis) that the *shock* that Gällivare suffered can be seen in the surviving data. Perhaps examination at an even finer scale (seasons or months) could reveal even sharper impacts (Fellman and Eriksson 2001). In the case of Katherine et al. (2001) and Harwood et al. (2011) inability to *stretch* the timeframe of analysis sufficiently both prior to and after the event led to a misinterpretation of the economic and demographic context, and consequently a misunderstanding of at least the medium term economic and demographic potential.

The third attribute of a temporal scale is the relationship between the disaster event being examined and preceding and proceeding events of similar type. The annual modelling of demographic characteristics presented for Gällivare, for example, poses some interesting questions about the relationship between the 1857–58 famine and the Great Deprivation a decade later. While it is commonly assumed that people living in sparsely populated areas have heightened *risk capacity* as a result of frequent exposure (or expected exposure) to disaster events, the importance of the frequency of those events and the potentially different reactions of different social and economic groups are not well known. The slowdown in Indigenous population growth in the Katherine-Daly region around the time of the 2006 flood is also worthy of further investigation.

A lot more work additionally needs to be done on examining the social scale issues that have been flagged in this research. The experiences of women of the Great Deprivation in Gällivare may actually have been important in the long-term influencing both urbanisation (if even a few women moved to larger towns when their partners died) and masculinisation (if future generations of women were discouraged from in-migrating or encouraged to out-migrate) (see also Chap. 9). The evidence linking these processes and the event is currently unclear, and other methods (analysis of oral and written personal histories, for example) are likely to prove much more useful in creating such links than quantitative demographic analysis. The same can be said for understanding the role of the Great Deprivation and the Katherine floods in changing the behaviour of Indigenous people, an issue to which we have already been alerted in relation to Cyclone Tracy in Darwin in 1974 (Haynes et al. 2011).

In conclusion, the study of disaster events in sparsely populated areas is a task made particularly complex by the *Ds* of human geography in those areas. We can postulate that disaster events can lead to temporary disruptions in demographic development when existing trends are so dynamic and powerful as to readily absorb temporary shocks. We can also postulate that disaster events can lead to longer lasting demographic diversions when economic, political, social and demographic conditions are somewhat *primed* for change in any case (see also Chap. 6). We may not be able, however, to make any definitive statements of cause and effect except at the very fine spatial and social scale (essentially the attribution of individual behaviour). Rather, disaster events should be considered as one part of an ever present and dynamic set of influences on the human geography of sparsely populated areas.

References

Adamo, S. B. and A. de Sherbinin (2011). The impact of climate change on the spatial distribution of the population and on migration Population Distribution, Urbanisation, Internal Migration and Development: An International Perspective. P. D. United Nations Department of Economic and Social Affairs, United Nations: 161–195.

Akenson, D. H. (2011). Ireland, Sweden and the great European migration, 1815–1914. Montreal, McGill-Queen's University Press.
Alestelao, M. and S. Kuhnle (1987). "The Scandinavian Route: Economic, Social, and Political Developments in Denmark, Finland, Norway, and Sweden." International Journal of Sociology **16**(3/4): 3–38.
Altman, J. C. (2006). In search of an outstations policy for Indigenous Australians. Working Paper. Canberra, Centre for Aboriginal Economic Policy Research, Australian National University. **34**.
Axelsson, E. (2014). The Forest as a Resource: conflicts in the northern Sweden wooded land in the 19th century. 54th European Regional Science Congress. St. Petersburg.
Bailey, A. J. (2011). "Population geographies and climate change." Progress in Human Geography **35**(5): 686–695.
Bengtsson, T. and M. Dribe (2002). New evidence on the standard of living in Sweden during the 18th and 19th centuries. Lund Papers in Economic History. Lund, Sweden, Department of Economic History, Lund University. **82**.
Berzins, B. (2007). Australia's northern secret: Tourism in the Northern Territory, 1920s to 1980s. Sydney, B. Berzins.
Bird, D., J. Govan, H. Murphy, S. Harwood, K. Haynes, D. Carson, S. Russell, D. King, E. Wensing, N. Tsakissiris and S. Larkin (2013). Future change in ancient worlds: Indigenous adaptation in northern Australia. Gold Coast, National Climate Change Adaptation Research Facility.
Britton, N. (1981). Darwin's Cyclone Max. An exploratory investigation of a natural hazard sequence on the development of a disaster subculture. Disaster Investigation Report. Townsville, James Cook University. **4**.
Brokensha, H. and A. Taylor (2014). The demography of the Territory's 'Midtowns': Katherine. Darwin, Australia, Northern Institute, Charles Darwin University.
Bäcklund, D. (1988). I Industrisamhällets Utkant: småbrukets omvandling i Lappmarken 1870–1970. Umeå Studies in Economic History. Umeå, Sweden, Umeå University. **8**.
Carson, D. (2011). "Political economy, demography and development in Australia's Northern Territory." Canadian Geographer **55**(2): 226–242.
Carson, D., P. Ensign, R. Rasmussen and A. Taylor (2011). Perspectives on 'demography at the edge'. Demography at the Edge: Remote human populations in developed nations. D. Carson, R. Rasmussen, P. Ensign, L. Huskey and A. Taylor. Farnham, United Kingdom, Ashgate Publishing Ltd.: 3–20.
Carson, D. B., D. Bird, L. Bell and P. Yuhun (2013). "Migration as an Adaptation to Climate Change for Remote Indigenous Communities: What might we expect?" Inner Asia Studies in the Humanities **2**: 92–111.
Carson, D. B. and D. A. Carson (2014). "Local Economies of Mobility in Sparsely Populated Areas: cases from Australia's Spine." Journal of Rural Studies **36**: 340–349.
Desert Knowledge Australia (2005). Our Outback: Partnerships and Pathways to Success in Tourism. Alice Springs, Desert Knowledge Australia.
Dewsnap, R. (1981). "The history of the Lappland iron ore fields." Minerals and Energy - Raw Materials Report **1**(1): 64–69.
Doblhammer, G., G. J. ven den Berg and L. H. Lumey (2013). "A re-analysis of the long-term effects on life expectancy of the Great Finnish Famine of 1866–68." Population Studies **67**(3): 309–322.
Dribe, M. (2003). "Dealing with economic stress through migration: Lessons from nineteenth century rural Sweden." European Review of Economic History **7**: 271–299.
Edvinsson, R. (2014). Pre-industrial population and economic growth: was there a Malthusian mechanism in Sweden? Stockholm, Stockholm University.
Ellis, M. (2009). "Vital statistics." The Professional Geographer **61**(3): 301–309.
Engberg, E. (2004). "Boarded out by auction: poor children and their families in nineteenth-century northern Sweden." Continuity and Change **19**(3): 431–457.
Faulkner, B. and S. Vikulov (2001). "Katherine, washed out one day, back on track the next: A post-mortem of a tourism disaster Tourism Management." Tourism Management **22**(4): 331–344.

Fellman, J. D. and A. W. Eriksson (2001). "Regional, temporal and seasonal variations in births and deaths: the effects of famines." Social Biology **48**(1–2): 86–104.

Ford, J. D., B. Smit and J. Wandel (2006). "Vulnerability to climate change in the Arctic: A case study from Arctic Bay, Canada." Global Environmental Change **16**: 145–160.

Godet, L. (2009). Transforming Gällivare: the emancipation of destruction. Masters of Architecture, Ecol Polytechnique Federale de Lausanne.

Harwood, S., D. Carson, E. Marino and N. McTurk (2011). Weather hazards, place and resilience in the remote Norths. Demography at the Edge: Remote human populations in developed nations. D. Carson, R. Rasmussen, P. Ensign, L. Huskey and A. Taylor. Farnham, United Kingdom, Ashgate Publishing Ltd.: 307–320.

Haynes, K., D. K. Bird, D. Carson, S. Larkin and M. Mason (2011). Institutional response and Indigenous experiences of Cyclone Tracy. Gold Coast, Australia, Report for the National Climate Change Adaptation Research Facility.

Hayward, A. D., I. J. Rickard and V. Lummaa (2013). "Influence of early-life nutrition on mortality and reproductive success during a subsequent famine in a preindustrial population." Proceedings of the National Academy of Sciences **110**(34): 13886–13891.

Hogan, A., D. Carson, J. Cleary, D. Donnelly, K. Houghton, R. Phillips and R. Tanton (2014). Community Adaptability Tool: Securing the wealth and wellbeing of rural communities. Canberra, Australia, Rural Industries Research and Development Corporation. **14/041**.

Isacson, M., M. Morell, I. A. Flygare, T. Lennartsson, A. Dahlström and M. Eriksson (2013). Resilience of peasant households during harvest fluctuations and crop failures in Nineteenth century Sweden – a micro study based on peasant diaries. First EURHO Rural History Conference. Bern, Switzerland.

James, G. (2009). Katherine, Northern Territory. Risks from Climate Change to Indigenous Communities in the Tropical North of Australia. D. Green, S. Jackson and J. Morrison. Canberra, Department of Climate Change.

Jantunen, J. and K. Ruosteenoja (2000). "Weather Conditions in Northern Europe in the Exceptionally Cold Spring Season of the Famine Year 1867." Geophysica **36**(1–2): 69–84.

Karlsson, L. (2013). "Indigenous life expectancy in Sweden 1850–1899: Towards a long and healthy life?" Demographic Research **28**(16): 433–456.

Koch, A. and D. B. Carson (2012). Spatial, Temporal and Social Scaling in Sparsely Populated Areas—geospatial mapping and simulation techniques to investigate social diversity. GI_Forum 2012. Salzburg, Austria.

Maas, I. and M. H. D. van Leeuwen (2002). Occupational careers of the total male labor force during industrialization: the example of nineteenth-century Sweden. Origins of the Modern Career. D. Mitch, J. Brown and M. H. D. van Leeuwen. Aldershot, Ashgate: 258–277.

Martel, C., D. Carson, E. Lundholm and D. Muller (2011). Bubbles and Craters: Analysing Ageing Patterns of Remote Area Populations. Demography at the Edge: Remote human populations in developed nations. D. Carson, R. Rasmussen, P. Ensign, L. Huskey and A. Taylor. Farnham, United Kingdom, Ashgate Publishing Ltd.: 107–124.

Martel, C., D. Carson and A. Taylor (2013). "Changing patterns of migration to Australia's Northern Territory: Evidence of new forms of escalator migration to frontier regions?" Migration Letters **10**(1): 101–113.

McLeman, R. (2010). "Impacts of population change on vulnerability and the capacity to adapt to climate change and variability: a typology based on lessons from "a hard country"." Population and Environment **31**: 286–316.

McLeman, R. and B. Smit (2006). "Migration as an adaptation to climate change." Climate Change **76**: 31–53.

Michel, T. and A. Taylor (2012). "Death by a thousand grants? The challenge of grant funding reliance for local government councils in the Northern Territory of Australia." Local Government Studies **iFirst**: 1–16.

Nelson, M. C. (1988). Bitter Bread: the famine in Norrbotten 1867–1868. Uppsala, Uppsala University.

Newby, A. G. (2014). ""Neither do these tenants or their children emigrate": famine and transatlantic emigration from Finland in the nineteenth century." Atlantic Studies: global currents **11**(3): 383–402.
Nordin, G. (2009). Äktenskap i Sapmi: giftermålsmönster och etnisk komplexitet i kolonisationens tidevarv, 1722–1895. PhD, Umeå Universitet.
Nordin, G. and P. Sköld (2012). "True or false? Nineteenth-century Sápmi fertility in qualitative vs. demographic sources." The History of the Family **17**(2): 157–177.
O Grada, C. (2007). "Making famine history." Journal of economic Literature **45**(1): 5–38.
Ryner, J. M. (2003). Capitalist restructuring, globalism and the third way: lessons from the Swedish model. London, Routledge.
Saxton, K., A. Falconi, S. Goldman-Mellor and R. Catalano (2013). "No evidence of programmed late-life mortality in the Finnish famine cohort." Journal of Developmental Origins of Health and Disease **4**(1): 30–34.
Schmallegger, D. and D. Carson (2010). "Whose tourism city is it? The role of government in tourism in Darwin, Northern Territory." Tourism and Hospitality: Planning and Development **7**(2): 111–129.
Sköld, P. (1997). "Escape from catastrophe: The Saami's experience with smallpox in eighteenth-and early-nineteenth-century Sweden." Social Science History: 1–25.
Sköld, P. and P. Axelsson (2008). "The Northern Population Development. Colonization, and mortality in Swedish Sápmi 1786–1895." International Journal of Circumpolar Health **67**(1): 29–44.
Taylor, A. (2011). The forecasting of remote area populations: numbers aren't everything. Demography at the Edge: Remote human populations in developed nations. D. Carson, R. Rasmussen, P. Ensign, L. Huskey and A. Taylor. Farnham, United Kingdom, Ashgate Publishing Ltd.: 39–52.
Taylor, A. (2014). "Population projections for sparsely populated areas: reconciling 'error' and context." International Journal of Population Research **early view**.
Taylor, A. and D. Carson (2009). "Indigenous mobility and the Northern Territory Emergency Response." People and Place **17**(1): 29–38.
Taylor, A., D. B. Carson, D. A. Carson and H. Brokensha (2015). "'Walkabout' tourism: The Indigenous tourism market for Outback Australia." Journal of Hospitality and Tourism Management **24**: 9–17.
Taylor, J. (1989). "Public policy and aboriginal population mobility: insights from the Katherine region, Northern Territory." Australian Geographer **20**(1): 47–53.
Taylor, M. A. P. and S. K. Freeman (2010). "A review of planning and operational models used for emergency evacuation situations in Australia." Procedia Engineering **3**: 3–14.

Chapter 6
Land Use Planning for Demographic Change After Disasters in New Orleans, Christchurch and Innisfail

David King and Yetta Gurtner

Abstract Land use planning is dominated by the growth paradigm—planning and development strategies of cities and regions to encompass increased demand for housing and infrastructure. Urban and Regional planning strategies are focused on enhancing development and growth to counter decline. In contrast, an emerging literature is concerned with planning for decline—managing population and infrastructure loss, decommissioning settlements and planning for reduced population and economy. The advent of a disaster is frequently a catalyst for local decline, but such loss is often connected to longer term issues and trends of population decline. New Orleans, Christchurch and Innisfail are examined in this chapter, to illustrate issues of population loss and demographic change against the impacts of specific disasters. The case studies exhibit multiple patterns of migration both spatially and temporally. Net migration has reflected population loss, but is not homogenous across the community. Specific demographic, cultural and socio-economic groups exhibited different patterns of migration and mobility. Reconstruction of such settlements faces changed demography with a shift in service and infrastructure needs. A reduced population requires land use rezoning, new strategic plans, land use change, removal of structures and re-siting of infrastructure while climate change related adaptation strategies identify protect, accommodate or retreat. Case studies illustrate various approaches to these issues.

Keywords Population loss · Land use · Planning · Urban planning · Regional planning · Disaster

D. King (✉) · Y. Gurtner
College of Science and Engineering, James Cook University, Townsville, Australia
e-mail: david.king@jcu.edu.au

Y. Gurtner
e-mail: yetta.gurtner@jcu.edu.au

D. Karácsonyi et al. (eds.), *The Demography of Disasters*,
https://doi.org/10.1007/978-3-030-49920-4_6

6.1 Introduction

Urban planning and regional planning are dominated by the growth paradigm, and while the growth of towns and cities have been driven by urbanisation from industrialisation alongside rapid population increase, there has also a demand for growth from businesses and industry. The influx of migrants into cities creates a demand for housing, services and infrastructure supplied by private and public sectors, and in turn creates jobs and economic activity. An increasing population boosts the market and creates a further opportunity for business expansion and profit. The private sector drives and invests in growth and thus demands policies and development strategies from all levels of government that enhance growth. Planning tools and concepts such as strategic planning schemes are frequently geared to growth. The positivist approach to city scaling (Bettencourt et al. 2007; Youn et al. 2016) shows that urban growth generates a value adding factor of about 15%, so that the effects of growth compound, creating both greater efficiencies and further growth.

While there is a "growth is good" paradigm "the key obstacle here is the notion … that a healthy city always grows in population and that only unhealthy ones shrink" (Hollander et al. 2009, p. 27). However, there are many conflicting messages in the literature regarding increasing urbanisation, including the speed with which cities have to absorb new migrants from rural areas, issues of city size, high density living, ecological impact, urban angst and alienation, inequality, social and economic divisions, social change, crime and pollution. Planners are faced with dealing with the negative aspects of cities, but work towards optimistic outcomes that rely on the drivers of growth to generate jobs and resources that contribute to solutions. Optimistic growth and modernism have been expressed most powerfully in the developing world where most urbanisation is presently occurring. In contrast, the developed world planning profession is increasingly faced with emerging trends of slow growth, environmental sustainability and population decline. Planning for shrinkage or population loss requires a significant re-thinking of urban and regional planning (Hollander et al. 2009). Newman (2006) critiques planning approaches based on population impact and ecological footprint, favouring a sustainability approach. A planning approach based on adverse local population impact is best tackled by a reduction in population and migration, which is not popular with governments or business as it runs directly counter to growth. Sustainability on the other hand encompasses both the positive impacts and opportunities of the city. Newman (2006) stresses the importance of sustainability in planning both for city growth and the scenario of stagnation or decline.

The reality of declining and disadvantaged regions is increasingly observed in towns and cities. Population migration and settlement relocation are part of adapting to a changing environment and economic opportunities. Alongside rapid urbanisation, there is a worldwide trend of shrinkage of cities, especially in the developed world (Oswalt et al. 2006; Hollander et al. 2009), where urban decline, inequity and social disadvantage are linked issues. Hollander et al. (2009, p. 223) claim that "developed, modern cities throughout the world are facing declines at an unprecedented scale". Many medium- and small-sized cities below 500,000 population are shrinking while larger than these cities (in particularly global and mega cities) are rapidly growing (United Nations, Department of Economic and Social Affairs, Population Division 2018).

To an even greater extent than in city planning, regional planning has been concerned with economically disadvantaged or declining regions for many decades. Such areas are often characteristic of endemic rural poverty, remoteness, paucity of services and infrastructure, and areas of structural economic change as a consequence of post-industrial and resource change. A primary aim of regional economic development has been to arrest regional decline, initiate economic growth and in more recent decades, to adapt to changed social and economic realities through exploration of the strengths and opportunities of places to plan for new alternative economies and communities (Ehrenfeucht and Nelson 2011).

Hollander et al. (2009, p. 228) identify a broad range of processes which have led to demographic decline and change in both cities and regions. These include: suburbanisation and out migration, change in economic activity, post-industrial shift from manufacturing to service industries, war, fire, disease, agricultural crisis, an ageing population or low fertility rate, political changes that have shifted development and economic priorities and long-term subsidies, and perhaps most markedly impacts associated with significant natural or human induced disasters. A complicating factor of growth or decline is the failure of urban boundaries to shift as population moves out of city centres. The population of a city may show apparent decline while the broader peri-urban region may even have grown, or at least the decline may be much less overall. As populations exit the city for peripheral or peri-urban locations, they may also increase their vulnerability as they move into unfamiliar areas subject to different hazards.

In this chapter, we specifically address population decline of towns, cities and regions as a consequence of disasters. Population decline is not equal or evenly spread, but is expressed through different demographic indicators—ethnicity, socio-economic status, age and gender for example. Evacuation is an immediate response to a hazard, with a gradual return of some of the population during the recovery period, possibly over many years. This period of dislocation and population loss is an opportunity to re-appraise planning priorities—to plan and manage for a different community (see also Chapter 2). We explore these planning and demographic concepts and issues through an examination of three places that have lost population following recent disasters; New Orleans in the USA, Christchurch in New Zealand and Innisfail in Australia. The chapter examines theoretical aspects of planning for decline, the demographic impacts of disasters with emphasis on the three case studies and planning strategies and approaches.

6.2 Planned Decline, Planning for Decline and Disaster Recovery

The concept of shrinkage, shrinking cities (Martinez-Fernandez et al. 2016) and decline with consequent loss of population and services is an emerging trend in planning. As planning tools are geared to growth, shrinkage requires a re-thinking of planning. Hollander et al. (2009) recommend a paradigm shift to proactively plan for shrinkage—where shrinking cities sounds better than urban decline. An alternative expression might be "rightsizing"—planning for a different size and composition of community. Hollander et al. (2009) advocate that the appropriate planning response is to re visualise the city—to bring about smaller, less dense, redesigned cities that continue to be productive and sustainable. Shrinkage brings a range of issues including; land use and vacant land, reduction and/or decommissioning of infrastructure, cleaning and greening for maintenance of land values, redevelopment and revised landscape options, persistence of cultural attractions, historic preservation and social equity issues.

Shrinking cities, however, are not necessarily in absolute decline. Shrinkage creates a redefined density with the opportunity for nodes of redevelopment. Such strategies may include: environmental mitigation and ecological restoration, revegetation, stormwater and hazard management through vacant land, community gardens ("the new ruralism", Ellis and Fanning 1977), urban agriculture, mixed used redevelopment, tourism, temporary use initiatives such as retail exhibitions, urban installations and street markets, media/industry investments and recreation. Planners are in a unique position to reframe decline as an opportunity, a chance to re-envision cities and to explore non-traditional approaches to growth, liveability and safety (Hollander et al. 2009). A positive "sense of place" and resilience for a locality are especially significant after a major disaster.

Planning and management in the post-disaster recovery context represents another emerging area of planning research and policy (Olshansky and Chang 2009; Blanco and Alberti 2009). Planners play a critical role in the redevelopment of an area after an event with unique time-sensitive challenges regarding reconstruction and recovery of housing and infrastructure and social, economic and environmental systems. Specific issues include resource availability, public interest (with particular concern regarding speed versus quality) and opportunities for community betterment.

Olshansky and Chang (2006, p. 201) assert, "However defined, it is clear that post-disaster recovery demands the skills of planners. Recovery is a microcosm of all the challenges of urban planning—developing land use and economic development strategies to improve lives, acting in the absence of sufficient information, making trade-offs between deliberation and expediency, navigating local politics, engaging the public and identifying funding sources to supplement inadequate local resources." The reimagining of the reduced city enables planners to re-address the physical vulnerabilities of cities, polarisation, re-urbanisation, sense of place, mixed use redevelopment, social equity and quality of life—to incorporate new urbanist principles, to create liveable cities, resilient cities and climate change adaptation through new approaches to land use change, zoning and integrated strategic planning (Hollander et al. 2009).

For environmental systems, large-scale depopulation with reduction and decommissioning of infrastructure allows for the removal of buildings and paved surfaces from floodplains, with enhanced opportunities for green urbanism, environmental mitigation and ecological restoration, green spaces for reforestation with environmental functions, watersheds (storm water management), wildlife habitats and the establishment of concentrated areas of vegetation to improve air quality and reduce urban heat-island effects. Socially and economically, a decreased population size and density post-disaster may facilitate a revised community focus, redistribution of resources and access, new social and community networking opportunities, diversified employment prospects and the development of enhanced resilience, safety and well-being.

The resized city requires strategic sustainable planning through positive de-densification (purchase of empty spaces and maintenance of adjoining blocks), demolition or re-purpose demolition of benign neglect, stabilisation of transitional cities and neighbourhoods for a sustainable future. Planning concepts require sustainable redevelopment, planning for a radically altered future alongside the uncertainty of climate change scenarios impacts. Enablers are participatory processes, local leadership, cooperation all levels of government, external funds and resources, pre-existing planning documents and institutions, with engagement and consensus on policies regarding future development (Hollander et al. 2009).

6.3 Shrinkage and Out Migration

While the growth paradigm remains popular, it is widely acknowledged by planners that cities, towns and regions lose population and may experience a consequent decline in economic opportunities, service provision and community resilience. Migration is a fundamental human characteristic. It is a rational response to changing resources, threats and opportunities. People are mobile over time and space. For the individual, out migrant relocation may enhance resilience and quality of life, although it may not necessarily reduce vulnerability factors. For the community that loses population to emigration, the process may reduce resilience and adaptive capacity and may increase overall vulnerability to economic and social change, including future disasters brought on by natural hazards and climate change. Emigration is highly significant for disaster risk reduction, although there is limited social science research literature or systematic data which estimates the impact of disasters upon population flows post-disaster (Love 2011).

Sociological disaster research predominantly investigates the recovery of social units such as individuals, family and households, organisations and community within the context of social networks, systems and institutions (Boon et al. 2012, 2013, 2016). Economic impacts further encompass businesses, economic costs (short term rather than long term), economic security and capacity, insurance coverage, finance, employment and livelihoods. Physical destruction and damage to the built environment consequent to a disaster generally connects facets such as logistics, resources, lifelines, critical infrastructure (see Chap. 10), housing and basic service provision, environment and revised policy. Despite the significant implications for changed communities and regions, there is limited literature on planning for recovery and reconstruction, being focused more on mitigation of risks for an existing population, rather than being concerned with the absent or diaspora population.

Additionally, disaster driven out migration takes place within a context of the pre-existing processes of net migration. Disasters increase out migration, but in the long term, it occurs within the pre-disaster economic trend (see also Chap. 5, the case of Katherine flooding). There are five types of out migration processes that are enhanced by a disaster.

(1) Fear of risk and a consequent preparedness to leave the place to seek residence elsewhere (King et al. 2014). This is based on hazard vulnerability and perception of vulnerability by the individual or household, or business operation. Prior experience, or prior knowledge of the hazard in this place or others, influences an attitude of intent to move if threatened further. Age, local ownership of property, and economic capacity influence relocation capacity. Older people with local roots and property are much more constrained than younger economically active families, who cite safety of children as well as having a greater capacity to seek opportunities elsewhere (King et al. 2014). Disaster risk reduction for this population may put out migration as the most effective strategy and relocation may occur at any time prior to a hazard event occurring.

(2) Evacuation prior to a disaster where early warning is possible, such as for floods, cyclones and bushfires, is organised by authorities and emergency services and may be forced or advisory. This takes people out of harm's way and reduces the death rate, although the experience of severe bushfires has shown that voluntary evacuation which is left too late may significantly add to deaths. People may be accommodated temporarily and permitted to return soon after the danger has passed, but within the constraints of the extent of damage to the home community.

(3) Evacuation after an event has severely damaged a community, organised by authorities and emergency services and may be compulsory. Evacuation is usually a response to sudden onset disasters, such as earthquakes, where damage is widespread, secondary hazards pose a risk, and the community is unable to be supported through loss of housing, basic services and lifelines.

(4) Further evacuation from the impacted community or region progressively occurs as a consequence of the loss of structures, economy, services and infrastructure, making post-disaster recovery slow and inhibiting early return. Temporary accommodation nearby in anticipation of a rapid return becomes increasingly unviable, forcing people to move further away for work and accommodation.

(5) The longer term loss of economy and dwellings prompt emigration to places of greater opportunity where people may remain as they settle into a new community and lifestyle. Many of those initially displaced by advisory evacuations, accommodation loss and sustained damage choose this option.

In reviewing the demographic impacts of disasters, Love (2011) refers to populations in transition. While the long-term decline of a city is not necessarily a direct outcome of a significant high impact disaster, current research and literature indicate a period of population transition reflecting evacuations (voluntary and/or mandatory), residential damage, and displacement. Decisions to leave or return are highly dependent on the level of residential damage at localised geographical levels. Love (2011) identified that ninety per cent of residents return within 6 months, but if return does not occur within 2 years, it is likely to be permanent (see also Chaps. 2 and 11 on permanent resettlement). This is especially evident where there has been displacement of vulnerable populations from areas where they do not have the resources to recover. A concentration of vulnerable populations also occurs in areas where the better resourced were able to leave. Displacement as opposed to intentional out migration alters population dynamics, demographic composition and trends (Love 2011). Consequently, the people who leave and return to a disaster-impacted town or city are not a representative cross section of the population. In disrupting the functions and dynamics of pre-existing social systems, disasters distort social and demographic patterns.

As disaster rebuilding takes place in depopulated and changed physical and community landscapes, the critical issue emerges for planners of which infrastructure to rebuild, standards of improvement and best practice (Hollander et al. 2009, Ehrenfeucht and Nelson 2011). Comprehensive and strategic planning are crucially important in recovery as it is through land use planning and flexible rebuilding strategies that it is feasible to "build back better". Land is the basis of residential communities and economy, and thereby provides community security (Lundin 2011). There is an opportunity for change, urban transition, reconstruction and redevelopment—moving people away from direct physical exposure, thereby increasing mitigation and resilience rebuilding. These are priorities of the Sendai Framework—Disaster Risk Reduction (UNISDR 2015) and build back better (see Chap. 12)—and determined by how we conceive recovery and its narratives in recovery management decision making. Opportunities exist for planners to integrate New Urbanist principles (Congress for the New Urbanism 1999) into a resized urban landscape, driving land use change, development strategies encompassing considerations of economics, energy, transportation, sustainable development, well-being, social capital, resilience and liveability, as well as paying attention to climate change mitigation and adaptation in a greater future hazard risk environment.

6.4 Recovery and Loss Following Major Disasters: Case Studies

Since the new millennium, numerous towns and cities around the world have demonstrated population loss following a major disaster. This chapter proceeds to examine the experience of three disparate places from the perspectives of planning responses and disaster recovery. Population loss from each place is examined in the context of longer term demographic movements, with the disaster events extending these established trends.

Each of the three case studies is quite different, and as such, the aim of this review is not to make comparisons or contrasts, but rather to examine the direct experiences of urban places in the developed world that have had a notable decline in population following the impact of relatively recent disasters. New Orleans in Louisiana, USA, had an overall population of 1.34 million (Sastry 2009) before it was devastated by Hurricane Katrina in 2005. Christchurch City in the South Island of New Zealand with a pre-disaster population of 367,720 was severely damaged by two earthquakes in 2010 and 2011, with the latter quake causing more damage and significant loss of life (Love 2011). Innisfail and its Local Government Area of the Cassowary Coast in Far North Queensland, Australia, had a population of around 30,000 in 2011 (ABS 2016), and was hit by two category four cyclones—Cyclone Larry in 2006 and Cyclone Yasi in 2011. Although there was no loss of human life from either of these cyclone events, there was a compounding effect upon economic recovery.

Love (2011) stresses that the outmigration and movement of people after disaster events in Christchurch and New Orleans cannot be compared, proposing that Christchurch's experience is closer to the impact of Hurricane Andrew on Miami (1992), and the Kobe earthquake (1995). Likeswise Zaninetti and Colten (2012) compare New Orleans to the experience of Galveston, Texas after the category 4 hurricane[1] in 1900, following which economic and population growth shifted to Houston. Galveston was already in decline pre-1900. Both New Orleans and Innisfail experienced population stagnation and loss of economic importance over an extended period before the recent disasters, while Innisfail and Christchurch serve rural hinterlands, but these case studies are not otherwise comparable. They are presented as examples of different processes and experiences following disasters.

6.4.1 Christchurch

Christchurch city, the largest city on the New Zealand South Island, is renowned as a popular heritage tourism destination, surrounded by a peri-urban rural residential area (Swaffield 2012). Love (2011) refers to the widely cited figure of 70,000 overall population loss from Christchurch after the 2010–2011 earthquakes. An estimated

[1]According to the Saffir-Simpson scale hurricanes are scaled from 1 (min.) to 5 (max.). Category four usually means a maximum wind speed between 209–251 km/h.

Table 6.1 Population change in christchurch pre- and post-2010/2011 earthquakes (Newell et al. 2012)

		Waimakariri district	Christchurch city	Selwyn district	Greater christchurch
Estimated population	2006	44,060	361,820	35,000	440,860
	2011	48,600	367,720	41,100	457,420
Population change	2008–2009	800	3,700	1,100	5,600
	2009–2010	800	4,000	1,000	5,700
	2010–2011	900	−8,900	1,600	−6,500
Population change range scenarios 2011–2012*	Lowest outlier	400	−10,300	600	−9,300
	Low mid range	400	−7,900	900	−6,500
	Mid range	700	−4,700	1,200	−2,900
	High mid range	700	−2,800	1,200	−900
	Highest outlier	800	−700	1,300	1,400

*Estimates modelled by Newell et al. 2012

regional population of 460,000 in the Greater Christchurch area was affected, with 150,000 homes damaged to varying degrees. There was actually little population change after the 2010 quake, but damage from the 2011 quake resulted in a significant outflow and relocation (Parker and Steenkamp 2012). Newell et al. (2012) show in Table 6.1 that the population loss after this event was particularly concentrated in the city while less damaged peri-urban areas appeared to increase in population.

Newell et al. (2012) estimate around 15%, i.e., 55,000 of the Christchurch population left the city after the 2011 earthquake. Many Maori residents returned to traditional Iwi (Gawith 2011), but many who relocated a short distance subsequently returned in the short term. School enrolments in Christchurch (Newell et al. 2012) suggest less than 10% out migration occurred immediately post-quake with a steady population return. Gawith (2011) reviews a compilation of media reports on the Christchurch earthquakes, giving the media estimates of 26,000 people departing, and school rolls showing 4496 pupils having moved to new schools—partly driven by damage and closure of educational facilities. While the 2013 census shows that the population in Christchurch city was down 2% from 2006 estimates, the greater Christchurch region was estimated to have increased in population by 2.6% over this period (Statistics New Zealand 2014a). These census data also suggest that the highest rates of decline (35–36%) occurred in the most damaged areas with two-thirds of those displaced by the damage moving to another dwelling within the same territorial authority. Census data are not directly comparable to other population estimates, especially those of local governments and city councils. The 2018 New Zealand census estimated a 10% undercount, especially of Pacific Islanders, but only a 2.4% undercount in 2013 (BERL 2018; Statistics New Zealand 2019; Statistics New Zealand 2014b).

Before the earthquakes of 2010 and 2011, the trend in Christchurch's population growth had been a continuous increase, predominantly through in migration, although it also had an ageing population (Love 2011). Newell (2012) places the observed out migration from Christchurch within a broader context of migration patterns of in and out migration and international arrivals, which show a marked decrease because of a loss of capacity for the tourism industry and related short-term visitor sectors, such as international students. A reduction of births was also observed. Building approvals, on the other hand, were highly robust from 2012, with the construction industry being a dominant driver of growth. The number of technicians and trade workers in the region increased by 6.9% from 2006 to 2013 in contrast to a national trend of decline in this labour force sector (Statistics New Zealand 2014a). Although the estimates of population out migration and return after the quakes appear highly variable, it is evident that the population loss from Christchurch was closely correlated with the extent of damage and loss of housing, services and infrastructure. Although the population dynamics have changed, the 2018 census indicates that the population has grown and by 2018, the census estimates that the population is around 404,600 (Statistics New Zealand 2019).

6.4.2 New Orleans

By comparison to Christchurch, the major US port city of New Orleans was experiencing long-term population and economic decline pre-Katrina (Love 2011). Love (2011) claims that New Orleans City had been noticeably declining since 1970, while Zaninetti and Colten (2012) assert that the history of New Orleans, known as The Crescent City or the Big Easy, has been one of declining importance since the American Civil War (early 1860s). The core city area reached a population peak of 627,525 in 1960 (Plyer 2011), followed by absolute population decline of this core city area and progressive expansion into the outer suburbs, encroaching on the surrounding protective wetlands and subsequently increasing the physical vulnerability to flooding. The population of greater New Orleans (metropolitan area) pre-event in 2005 was estimated at 1.34 million, of whom 69% identified as African-American (Sastry 2009). In the decades prior, the population declined 18% between 1970 and 2000 followed by a further 6% over the next 5 years. Unemployment rates were above the national average, over 20% were living below the poverty line in 2000, with high levels of socio-economic disadvantage (United States Census 2019; Simo 2008). In this context, Sastry (2009) suggests that many people who may already have been planning to leave the city for economic reasons may have been prompted by the Katrina evacuations. Katrina consequently accelerated demographic shifts.

With the impact of Hurricane Katrina in 2005 and the associated levee system failure, the entire New Orleans city centre population of 455,000 had to evacuate and resettle for over a month. There were at least 1500 recorded fatalities and more than

Table 6.2 Return rates to New Orleans post-Katrina (Fussel 2015; Love 2011)

Housing status	One year return percentage
Undamaged	96
Damaged but habitable	81
Uninhabitable	54
Destroyed	30

a million people were displaced within the broader region of impact. Approximately 71.5% of the 188,251 housing units in New Orleans were damaged, with 55.9% having major or severe damage (Fussell et al. 2010). Given the magnitude of this damage and the substantial drainage requirements, many evacuated residents ended up displaced for long periods of time. Four months after Katrina, the population of New Orleans was estimated at 158,353, a 64% loss on pre-Katrina numbers (Kates et al. 2006). Of 66,000 school students in New Orleans parish pre-Katrina, there were only 5000 by the end of the year. Most displaced students re-enrolled in other states (Sastry 2009).

With reconstruction and rebuilding slowed significantly by legal and property rights, provision of temporary accommodation, insurance, ecological restoration plans, race and other political issues all of which impacted people's return, the Census Bureau estimated the population of New Orleans at approximately 223,000 in July 2006—less than 50% of the pre-Katrina level (Zaninetti and Colten 2012). Fussell et al. (2010) calculated return rates against housing status one year after Hurricane Katrina (Table 6.2) suggesting an overall return rate of 70%, although other estimates were closer to 60%. Recovery of urban infrastructure, housing, businesses and services continued to be slow and highly variable. In 2010, a number of neighbourhoods that did not flood were near to 90—100% of their pre-Katrina population, if not exceeding. In contrast, vacancy rates across the entire city were approximately 25%, reflected in empty lots and abandoned homes, commercial and institutional buildings (Plyer 2011). The Data Center (Plyer 2015) states that ten years after Hurricane Katrina, New Orleans (city core area) had an estimated population of approximately 385,000, a net loss of 15%. Further analysis demonstrates the differential social, spatial and temporal population dimensions of recovery and the continued socio-demographic composition of vulnerability (Fussel 2015).

6.4.3 Innisfail

Innisfail, a small rural town in North Queensland, Australia, serves a surrounding agricultural population predominantly producing sugar, bananas and other tropical fruit crops. Historically the township has been significantly impacted by a number of cyclones. In 1918, a cyclone destroyed 1200 houses with 90 people killed. Of these fatalities, 37 people were specifically from Innisfail and 40–60 indigenous people were killed nearby (Boon et al. 2013). In 1986, Cyclone Winifred hit Innisfail resulting in 3 deaths, 12 injuries and 200 people displaced (Boon et al. 2013). In this event, fifty homes were destroyed and a further 1500 buildings (95% of the town) were damaged resulting in insurance reparations totalling $65 million. Local media were running a historical remembrance of Winifred at the time Cyclone Larry hit in 2006.

Of the two category 4 cyclones[2] that recently devastated the area within a five year period, Cyclone Larry passed directly over Innisfail in March 2006 and Cyclone Yasi crossed near the small town of Mission Beach (50 km south of Innisfail) in February 2011, accompanied by a 5 m storm surge. Cyclone Larry caused the most physical destruction in Innisfail with over half the dwellings and infrastructure damaged. These damage rates were higher in older dwellings and in many of the surrounding townships. Cyclone Yasi destroyed approximately 150 homes, made 650 uninhabitable, with a further 2275 sustaining moderate damage across the entire impact area. No one died (directly from cyclone impact) in either event, and there were very few injuries. In both events, the direct impact on primary industry was devastating, losing hundreds of millions of dollars in damaged crops with consequent loss of employment and economic livelihoods. Insured damage from the whole region impacted by Cyclone Larry was estimated to be over $500 million dollars, while the damage from Cyclone Yasi five years later was estimated to be $3.5 billion (Boon et al. 2013).

Given a long period of ineffective local governance with economic weakness and instability, the rural shire of Innisfail reflected an extended population decline with an ageing population and out migration of younger residents. Stagnation of the tourism industry also led to a steady population decline in the wider Cassowary Coast region (Table 6.3). Both Cyclone Larry and Yasi occurred half a year before the 2006 and 2011 censuses. Many people were displaced with a significant number of agricultural employees without work by the time the formal censuses were undertaken. Despite the economic activity generated by cyclone recovery, the population of Cassowary Coast declined by 1.9% between 2001 and 2006 and a further 0.3% from 2006 to 2011. The trend of population decrease has continued in Innisfail since 2011 with both the township and Cassowary Coast declining by an overall rate of approximately 1.7% between 2001 and 2015.

6.5 Socio-Demographic Impacts of Disasters and Planning Strategies

Consistent with the finding of Love's review (2011), each of these case studies reveal populations in transition, characterised by obvious accelerated decline post-disaster event. Patterns of residents staying, leaving or returning have shown a strong correlation with the level of residential damage at the localised geographical level.

[2]According to the Saffir-Simpson scale. However Cyclone Yasi reached Category 5 according to the Australia-South Pacific tropical cyclone classification but was downgraded to Category 4 after landfall.

Table 6.3 Population change in Innisfail pre- and post-2006/2011 cyclones (Adapted from ABS 2016)

	As at 30 June	Innisfail cassowary coast (statistical area 3)	Innisfail (statistical area 2)
Estimated population	2001 (census)	35,408	9,719
	2005	35,203	9,822
	2006 (census)	34,711	9,664
	2010	34,921	9,627
	2011 (census)	34,718	9,576
	2015	34,820	9,556
Population change	2001–2006 (census)	−697 (−1.9%)	−55 (−0.6%)
	2005–2006 (TC Larry)	−492 (−1.4%)	−158 (−1.6%)
	2010–2011 (TC Yasi)	−203 (−0.6%)	−51 (−0.5%)
	2006–2011(census)	−101 (−0.3%)	−88 (−0.9%)
Total change	2001–2015	−588 (−1.7%)	−163 (−1.7%)

While the historical context and demographic circumstances were variable for each case study pre-event, there have been a number of socio-demographic factors and issues evident within the recovery process. Population movement during recovery and reconstruction has not been homogenous spatially, temporally or socially. Changes in gender profiles, age structure, race and ethnicity, employment, income, livelihoods, insurance coverage, housing affordability, rebuilding and redevelopment, service provision and rehabilitation of the surrounding environment each have implications on measures of vulnerability and resilience to future events that should be taken into account in planning decision making.

Sub-national population estimates for Christchurch city in 2012 revealed a population decline of about 13,500 (3.6%) compared to pre-earthquake estimates, although the greater Christchurch region experienced a growth of 2.6% (Statistics New Zealand 2012; Statistics New Zealand 2014a, b). The net migration loss from the city was partly offset by natural increase yet there were some observed differences in both gender and age groups (Statistics New Zealand 2012; Newell et al. 2012). In the two years after the earthquake, there was a significant loss of women from the workforce, fewer young adults and a net outflow of children and their parents. The population continued to age with an increase in the number of people over 50, also indicating this cohort was less likely to have left Christchurch than people of other age groups. A significant number of those displaced by damage and destruction to the housing stock relocated in proximate suburban regions with the 2013 census showing a large increase in the percentage of workers now commuting to the city for employment. There was however an 81.1% increase in the number of unoccupied dwellings as a consequence of the earthquake, many of these "red-stickered" as unsafe for occupation (Statistics New Zealand 2014a).

In Greater Christchurch construction replaced manufacturing as the industry sector with the highest proportion of employment after the quakes, as the inflow of workers and extensive rebuilding activity helped the economy remain reasonably resilient. Retail and tourism were hit hardest, but insurance helped to buffer the economic impact (Parker and Steenkamp 2012). Reflecting the extensive level of destruction, there was a marked decrease in the number of workers in central Christchurch City. However with offsets in the south-west and near the airport there was no long-term change in overall employment rates between 2006 and 2013 censuses (Statistics New Zealand 2014a). Gawith (2011) lists many social, emotional, psychological, traumatic, economic and financial impacts, as well as a loss of places and community. Relocation was thus not just physical necessity but about moving forward. Greater Christchurch strategic planning was in place pre-earthquake, based on performance criteria that stressed sustainability (Swaffield 2012). Although recovery and rebuilding have been slow there is an optimism about the future of Christchurch. There is not an alternative central place in that region of New Zealand.

Even prior to the impact and devastation caused by Hurricane Katrina, the core city of New Orleans was a poor community associated with high levels of poverty, crime, illiteracy, inadequate basic services including health care and education, substandard housing stock and a lack economic opportunities. Sastry (2009) identified that 23% of the New Orleans population was below the poverty line, with a 35% poverty rate amongst African-Americans. Rental rates were high, with lower than national average rates of home ownership. The distribution of predominantly African-American and more socially and economically disadvantaged people were concentrated in low-lying areas that took the brunt of floods. There were consequently high levels of permanent displacement of people and many uninsured losses following Hurricane Katrina. The principally poor population lacked the finances or resources to rebuild (Zaninetti and Colten 2012). Raising houses on stilts was seen by planners as too expensive for a poor population.

The recovery process in New Orleans has further embedded distinct geographical patterns of social vulnerability. Zaninetti and Colten (2012) highlight a change in the city's ethnic landscape with differentiation in population distribution on class and racial lines. The city and metro area became more ethnically diverse; as African-American and white non-Hispanic populations declined in overall number, there was an increase in Hispanic and Asian residents (Plyer 2015). This was also an ageing population with a noticeable growth in the proportion of residents aged 65 and older. With the business and tourist centres relatively undamaged by the flooding, there has been a significant increase in property values and consequently the historic downtown has revealed gentrification and relative affluence. Recovery in other areas has reflected a temporal redistribution of population in clusters of settlements associated with less flood-prone neighbourhoods, the level of damage and extent of rebuilding.

The worst flooded areas have been characterised by depopulation with abandoned properties and blight rather than spatial contraction of the city and infrastructure. In 2012, vacancy still represented a loss of over 11% of the city, particularly a shrinkage of the metropolitan area (Zaninetti and Colten 2012). Population growth and relocation in New Orleans has moved from below sea level towards higher ground (vertical migration). Exposure to flood has been reduced since Katrina through population redistribution, but population loss has also reduced the tax base and capacity of the city to provide services and infrastructure. High demand and high rental costs for viable dwellings have made much of post-Katrina housing unaffordable. The reality of post-Katrina New Orleans is the emergence of unequal and disparate cities—the downtown cultural and business areas which have recovered, middle-class neighbourhoods which reflect variable degrees of redevelopment and recovery, and the disadvantaged areas (Olshansky 2006).

While continued population stagnation and decline in both Innisfail and Cassowary Coast reflect an established demographic trend, patterns over the last ten years have been underscored by significant reduction in standards of living. During the period 2001 to 2006, incomes grew faster than rents and mortgage repayments, but in the second half of the decade post-Cyclone Larry and accompanying the global recession and Cyclone Yasi, the reverse has been the case with housing costs increasing at much higher rates than household income, generating concerns regarding affordability (Boon et al. 2013). As redevelopment and rebuilding post-disaster resulted in an increase in housing approvals and the overall number of available dwellings, vacancy rates have also escalated.

Despite the influx of trades, technicians and the temporary construction boom associated with cyclone recovery, young people and families have continued to move to larger cities seeking education and employment opportunities. Numbers of youth and residents aged 25–44 declined by 1.8% with a distinct decline (10%) in couples with children and family household compositions. Reflecting an aging population the percentage of residents aged over 55 is increasing (see Chap. 2 for comparison on disaster-induced aging). The median age in Innisfail has subsequently increased by 7.2 years to 42.4 between the 2005 and 2011 censuses (QGSO 2017). With net population loss, there has been a reduction in both business and job prospects in the Innisfail township.

Growth and decline in coastal Queensland, including the Cassowary Shire remain heavily influenced by resource exploitation. Tropical fruits and sugar cane have declined in terms of relative commodity prices. Farmers are ageing, and the families of many have left the area to seek qualifications and off-farm careers. On top of these longer term trends, hazard events are major drivers of community change. The Cassowary Coast has 9.4% of its population Indigenous and experiences high population mobility, with 44% of the population of the local government area in the lowest *Socio-Economic Indexes for Areas* (SEIFA) quintile. Similar to the Maori of Christchurch (Gawith 2011), aboriginal farm workers and banana packers in Cassowary Shire found themselves put out of work following Larry and Yasi, returning to their home communities on Cape York Peninsula between 400 and 1000 km to the north.

Many other seasonal farm workers in the region were backpacker tourists who also lost opportunities in the shire and ceased to travel in this area. Tourism was hard hit generally, with resorts going out of business or into "moth balls", or operating part time, seasonally or for booked events. For many potential tourists, the stigma and perception of a disaster afflicted environment, and infrastructure preserved well beyond its useful life—the tourism industry has been slow to recover. Within this context of recurrent cyclones, flooding and coastal hazards, population loss and economic insecurity the Cassowary Coast Regional Council has continued to pursue a strategic planning direction of growth based on recognition of Innisfail as the major regional activity centre, supported by value added agriculture and tourism (AECOM 2012).

6.5.1 Planning Approaches

While population adjustment is fundamental to human response to disasters, population redistribution is a part of adaptation. In the aftermath of a disaster event, there are significant concerns regarding the prospect of sustained population exodus from a town or city. Task forces consisting of government representatives, planners, non-government organisations, industry and community members are often established to develop strategic approaches to arrest further population flight and facilitate reconstruction, resettlement and recovery. Strategies may be protective, defensive, offensive, opportunistic, or landscape and urban design oriented with different resource orientations (Lima and Eischeid 2017).

Planning for effective post-disaster recovery requires all scales and levels of government and a vision and openness to imagine a radically different, new community or city. Similarly, it is necessary to recognise patterns of demographic change and transition that may represent short, medium or even longer term population loss (King et al. 2016). There is a need for data and resource sharing and extensive communication. Recovery has to involve the diaspora population and planning for recovery needs funding (Olshansky 2006). Rather than traditional planning premised on anticipated future growth and development, post-disaster recovery should be prepared to plan-to-scale or right sizing for greater resilience and sustainability. Hollander et al. (2009) identify a number of strategies for "shrinking cities" or depopulated areas including de-densification, use of vacant land and underused property, green urbanism and environmental improvement, historic preservation and redistribution of access and resources aimed at enhancing equity, liveability, safety and sustainability. A number of these strategies are evidenced in the case studies.

Primary responsibility for planning for the recovery and revitalisation of Christchurch was given to the Canterbury Earthquake Recovery Authority (CERA). The main objective of this group was to strategically manage issues of centralisation, land use and infrastructure in the recovery process with an appreciation of potential future seismic activity and climate change considerations including floods, storms and sea-level rise (Miles et al. 2014). Given the high levels of damage and destruction in the CBD precinct, the vision was based on decentralisation and changes in the peri-urban landscape (CERA 2015). This process imagined a greener, more compact, more accessible and safer central business district dominated by low-rise buildings.

Further to this was the plan for a green frame or buffer around the CBD that would blend in with the Avon River, to be developed as a corridor of parkland through the city—emphasising ecological restoration and environmentally sensitive transport, including a new light rail network, and connectivity of city through recreational pedestrian boardwalks and cycle lanes.

Priority activities within the initial recovery framework included restoration of critical infrastructure (particularly water and sewerage) and the residential land use hazard assessments for resettlement, relocation and rebuilding (Miles et al. 2014). Specific designations were informed by the changes in building codes and regulations, where code changes required different structural requirements for houses, especially for foundations. Recommendations also advocated lower urban density with dispersed and mixed use environments in neighbourhood centres (Chang et al. 2014; Swaffield 2012). Subsequent to the demolition and removal of damaged buildings in the CBD, many vacant gaps and gravel quadrants were creatively inhabited by both temporary and permanent art installations, cultural activities, entertainment, open space, recreation and public areas including weekly street markets and even a retail mall made from shipping containers.

Five years after the 2011 quake, formal planning and ideas progressively transitioned from recovery to regeneration with the new Master Plan focused on urban renewal and development (CERA 2015). This plan promotes a vibrant, attractive, resilient city, with abundant open space and themed districts supporting restaurants, small shops and cafes, music, sports and recreation to stimulate business growth and economic activity. Restoration and maintenance of iconic historic buildings has been similarly supported to reinvigorate tourism and help re-establish a sense of place for residents. Despite the extensive rhetoric of building back residential dwellings and commercial premises more sustainably utilising green options, there was no legislation to enforce building "green" and insurance payments limited owners to like-for-like (with due consideration to new codes) (Miles et al. 2014). Businesses, services and populations have progressively returned to a revitalised more resilient region but the new strategic direction still appears to be dominated by top-down leadership and planning directives rather than a participatory process with consideration of resident aspirations (Chang et al. 2014). A subsequent earthquake affecting Christchurch in February 2016 recorded limited physical damage in the city, but the extent of psychological impact on an already traumatised and recovering community has yet to be fully appreciated.

Even prior to the impact of Hurricane Katrina, the parish of New Orleans lacked a strong history of traditional urban planning practice (Collins 2015). In the destruction and confusion post-event residents and displaced populations were confronted with a diversity of highly conflicting proposals ranging from campaigns to build back "bigger and better" to complete abandonment of the city (Olshansky 2006). Given the historical, cultural and symbolic significance of New Orleans, it was imperative for the community to rebuild and recover. The simultaneous emergence of differing city wide and neighbourhood plans from organisations such as the Federal Emergency Management Agency (FEMA), the Louisiana Recovery Authority (LRA), and various not for profit community groups did little to address the immediate demand for permanent dwellings, basic services and critical infrastructure (Collins 2015). Kates et al. (2006) observed very slow and inadequate reconstruction and recovery processes that were influenced by race, class and government incompetence.

Integral to any effective plan for population return and resettlement was the resolution of the complexities and conflicts regarding a comprehensive land use and zoning process. Ehrenfeucht and Nelson (2011) identified a range of strategies relating to targeted investment and consolidation; restoration of the natural protective wetlands and environmental quality, alternatives for underused areas; mechanisms to maintain or reintegrate abandoned parcels; plans for infrastructure and service provision, and interventions to address issues of social inequality. However most lacked wholesale community and political support, sufficient funding and resources and activities were highly sporadic and ad hoc.

The recovery of New Orleans was primarily resident driven, but the rebuilding of houses and reconstruction of levees achieved replacement rather than building better structures that reduce future disaster risks. Increased physical vulnerability of New Orleans to flood risk is seen by Zaninetti and Colten (2012) as "maladaptive evolution caused by planning" (p. 680). Collins (2015) suggests that the first five years post-Katrina were focused almost entirely on recovery, the next five years were taken up with resolving the complex zoning processes with the eventual development and adoption of The *Plan for the twenty-first Century,* (commonly referred to as the Master Plan); a City Charter-mandated planning framework that shapes New Orleans' physical, social, environmental and economic future. This master plan reflects the values and priorities of liveability, opportunity and sustainability that emerged through a participatory community decision making process (Collins 2015). However, it fails to adequately address vulnerability and hazard resilience.

In terms of population size and the scale of destruction to buildings, infrastructure and services, recovery for Innisfail and the Cassowary Coast post-cyclone events was significantly smaller and less resource intensive than either the Christchurch and New Orleans case studies. However for the community and residents of this region, effective planning strategies were equally as significant to lifestyle, livelihoods, economic viability and decisions to persevere or migrate to other locations. Following Cyclone Larry, the Operation Recovery Task Force was established through the Queensland Government to coordinate basic needs, rebuilding and planning priorities (Queensland Government 2007). After Cyclone Yasi, this responsibility was delegated to the Queensland Reconstruction Authority (QRA).

In both cases, regional planning strategies were aimed at protecting the character of local townships, limiting exposure to natural hazards, establishing long-term economic stability and maximising infrastructure and transport provision efficiencies (AECOM 2012). Equally, provision was made to protect, maintain and sustain the region's unique natural assets and environment through biodiversity conservation and coastal protection. Specific initiatives identified in the master plan for Innisfail included the intensification and renewal of its CBD within the existing footprint, the development of greenfield sites within the urban footprint, the inclusion of open space for public, cultural and community amenity, in fill development and increased density for established industrial zones, and industry sector reform based on diversification opportunities, value added agriculture and technological innovation (AECOM 2012). While statistical population trends suggest small yet continued net out migration the Cassowary Coast Planning Scheme 2015 anticipates population growth, tourism and expanded industry development within the region.

6.6 Conclusion

The analysis of planning responses to disasters in these three case studies illustrate consistent issues such as the priority restoration of critical infrastructure: water, waste, utilities, transport (roads, bridges, rail), as well as challenges and contention concerning dwellings, business, industry, support services, schools, public transport, hospitals, heritage, and environment, and sustainability provisions such as the enhancement of active transport of walking and cycling (although these were a lower priority). Recovery was complicated by residential zoning, complex and challenging regulations, competing visions and directions, a lack of public participation and engagement, the issue of contested demolition and what to do with unused and vacant lands, and a plethora of complex issues around insurance and the lack of insurance amongst the poor, resources and funds. In the midst of these challenges, controversy surrounded priorities of quick reconstruction and return to some sort of normalcy as opposed to effective long-term processes and transparency in all recovery and reconstruction activities.

We can summarise some basic findings and implications.

- Disaster is often inevitably a component of long-term population trends.
- Planning for community recovery and stability may be aimed at smaller rather than larger settlements in the future.
- Adaptation may be geared towards smaller populations and an altered demographic and socio-economic structure.
- Planning policy of retreat—decommissioning structures, settlements and infrastructure.
- Changed land use patterns—consolidation and principles of New Urbanism.
- Building back better.
- Rezoning hazard-prone areas.
- Sustainable planning and resource use.

King et al. (2016) analysed strategies and policies in the UNISDR Global Assessment Review of the Hyogo framework and subsequent Sendai framework (Boon et al. 2016). Many of these strategies have been adopted at local government levels, and some are implemented as policies, but for local governments, especially the three case studies examined in this chapter, it is a work that is in progress. The urgency of the pressure to recover and rebuild pushes longer term strategic views to one side, but we see these emerging after the immediate recovery phase.

There is inevitably short- to medium-term population loss as many people temporarily move away from danger and the loss of services, infrastructure and economic support. The return of this population has to be phased over a long period as the city is rebuilt, but a portion of this population may never return, or a different demographic or social group replaces some of those who have left. The challenge of recovery and adaptation is correctly identifying and anticipating this demographic change in order to adopt approaches that suit the altered settlement. Within this change are activities and strategies to build back better, and to enhance both resilience and sustainability. Planning responds to a primary paradigm of growth, but growth is not always ideal or desirable. Loss of population and corresponding services and infrastructure is not necessarily detrimental. It is clearly an opportunity for planners in envisioning new and altered places. The emerging concept is planning to scale, or right sizing, where there is a clear and accurate awareness and recognition of demographic change and transition that enables an appropriate and better place to emerge from disaster. The latest trend in urban design is placemaking, which within the last two years has been defined as a movement on a scale with, and complementary to New Urbanism (Kent 2019). Planners are responding enthusiastically to the ideas in placemaking as the movement incorporates many of the issues that we have noted in relation to the recovery of the three case study settlements, but with a stronger emphasis on people, place and ethical integration as the core of urban design and planning (Kent 2019; Eckenwiler 2016). Placemaking transcends recovery and demographic planning. If we approach the planning issues, identified in these case studies, of adaptation, retreat, the decommissioning of structures settlements and infrastructure, changed land use patterns, consolidation and principles of New Urbanism, building back better, rezoning hazard-prone areas and sustainable planning and resource use from a placemaking perspective where we put people at the centre of well designed good places, it will matter less whether the place is smaller or whether it is growing. The quality and good design of the post-disaster community is far more important than its demographic impacts or recovery.

References

ABS. (2016). Regional population growth, Australia, 2014–15 ABS 3218.0. *Australia Bureau of Statistics*, https://www.abs.gov.au/AUSSTATS/abs@.nsf/DetailsPage/3218.02014-15?OpenDocument. Sighted 20/01/17.

AECOM. (2012). *Innisfail urban growth strategy prepared for the Cassowary Coast Regional Council*. AECOM Australia Pty Ltd: Fortitude Valley.

BERL. (2018). *Census 2018 undercount*. https://berl.co.nz/economic-insights/government-and-fiscal-policy/census-2018-undercount, Sighted 21/12/19.

Bettencourt, L. M. A., Lobo, J., Helbing, D., Kühnert, C., & West, G. B. (2007). Growth, innovation, scaling, and the pace of life in cities. *Proceedings of the National Academy of Sciences of the United States of America*, 104(17):7301–7306.

Blanco, H., & Alberti, M. (2009). Introduction. Blanco & Alberti (eds) Shaken, shrinking, hot, impoverished and informal: Emerging research agendas in planning. Special issue of *Progress in Planning*, 72(4):195–250.

Boon, H. J., Cottrell A., King D., Stevenson R. B. & Millar J. (2012). Bronfenbrenner's bioecological theory for modelling community resilience to natural disasters. *Natural Hazards* 60381–408.

Boon, H.J., Millar, J, Lake, D., Cottrell, A., King, D. (2013). *Recovery from Disaster: Resilience, Adaptability and Perceptions of Climate Change.* National Climate Change Adaptation Research Facility, Gold Coast.

Boon, H., Cottrell, A. & King, D. (2016). *Disasters and social resilience.* Routledge explorations in environmental studies, Taylor & Francis, London.

Cassowary Coast Planning Scheme. (2015). *Cassowary coast regional council planning scheme.* https://www.cassowarycoast.qld.gov.au/documents/1422210/42234576/CCRC%20Planning%20Scheme%202015%20-%20V4, Sighted 12/1/20.

CERA. Canterbury Earthquake Recovery Authority. (2015). *Greater Christchurch Earthquake recovery: Transition to regeneration.* Christchurch: Canterbury Earthquake Recovery Authority.

Chang, S. E., Taylor, J. E., Elwood, K. J., Seville, E., Brunsdon, D., & Gartner, M. (2014). Urban disaster recovery in Christchurch: The central business district cordon and other critical decisions. *Earthquake Spectra*, 30(1):513–532.

Collins, R. (2015). No more "planning by surprise": City planning in New Orleans ten years after Katrina. *The Data Center*. The Data Center.org, Sighted 12/1/20.

Congress for the New Urbanism. (1999). Leccese, Michael; McCormick, Kathleen (eds.). *Charter of the new urbanism.* McGraw-Hill Professional.

Eckenwiler, L. (2016). Defining ethical placemaking for place-based interventions. *American Journal of Public Health*, 106(11):1944–1946.

Ehrenfeucht, R. & Nelson, M. (2011). Planning, population loss and equity in New Orleans after Hurricane Katrina. *Planning Practice & Research*, 26(2):129–146.

Ellis, W.N. & Fanning, O. (1977). The new ruralism. *Habitat International,* 2(1–2):235–245.

Fussell, E., Sastry, N., & VanLandingham, M. (2010). Race, socioeconomic status, and return migration to New Orleans after Hurricane Katrina. *Population and Environment*, 31(1–3):20–42.

Fussel, E. (2015). The long-term recovery of New Orleans' population after Hurricane Katrina. *American Behavioral Scientist*, 59(10):1231–1245.

Gawith, L. (2011). How communities in Christchurch have been coping with their earthquake. *New Zealand Journal of Psychology,* 40(4):121–130.

Hollander, J. B., Pallagst, K., Schwarz, T., & Popper, F. J. (2009). Planning shrinking cities. *Progress in Planning,* 72(4):223–232.

Kates, R. W., Colten, C. E. Laska, S. &. Leatherman, S. P. (2006). Reconstruction of New Orleans after Hurricane Katrina: A research perspective. *Proceedings of the National Academy of Sciences of the United States of America*, 103(40):14653–14660.

Kent, E. (2019) "Leading urban change with people powered public spaces. The history, and new directions, of the Placemaking movement", *The Journal of Public Space*, 4(1):127–134.

King, D., Bird, D., Haynes, K., Boon, H., Cottrell, A., Millar, J., Okada, T., Box, P., Keogh, D. & Thomas, M. (2014). Natural disaster mitigation through relocation and migration: Household Adaptation Strategies and Policy in the Face of Natural Disasters. *International Journal of Disaster Risk Reduction,* 8(June):83–90.

King, D., Gurtner, Y., Firdaus, A., Harwood, S. & Cottrell, A. (2016). Land use planning for disaster risk reduction and climate change adaptation: Operationalizing policy and legislation at local levels. *International Journal of Disaster Resilience in the Built Environment*, Special Issue, 7(2):158–172.

Lima, M.F. & Eischeid, M.R. (2017). Editorial: Issue 7: Shrinking cities: Rethinking landscape in depopulating urban contexts. *Journal of Landscape Research*, 42(7):691–698.

Love, T. (2011). Population movement after natural disasters: a literature review and assessment of Christchurch data. Sapere Research Group, Wellington.

Lundin, W. (2011). *Land use planning after a natural fisaster*. University of New Orleans Theses and Dissertations. Paper 1386.

Martinez-Fernandez, C., Weyman, T., Fol, S., Audirac, I., Cunningham-Sabot, E., Wiechmann, T. & Yahagi, H. (2016). Shrinking cities in Australia, Japan, Europe and the USA: From a global process to local policy responses. *Progress in Planning*, 105(April):1–48.

Miles, S., Brechwald, D., Davidson, R., Demeter, K., Johnston, D., Pampanin, S. & Wilkinson, S. (2014). *Building Back Better Case Study of the 2010–2011 Canterbury, New Zealand Earthquake Sequence*. A report prepared by the Earthquake Engineering Research Institute in collaboration with the New Zealand Society for Earthquake Engineering and the Natural Hazards Platform for the Global Facility for Disaster Reduction and Recovery of The World Bank.

Newell, J., Beaven, S., Johnston, D.M. (2012). Population movements following the 2010–2011 Canterbury Earthquakes: Summary of research workshops November 2011 and current evidence, *GNS Miscellaneous Series* 44. 23 p + Appendix C.

Newell, J. (2012). *Indicative population estimates for "Greater Christchurch" post June 2011*. Monitoring and Evaluation Research Associates Limited, Wellington (NZ).

Newman, P. (2006). The environmental impact of cities. *Environment and Urbanization,* 18(2):275–295.

Olshansky, R. (2006). Planning after Hurricane Katrina. *Journal of the American Planning Association*, 72(2):147–153.

Olshansky, R. & Chang S. (2009). Chapter 2. Planning for disaster recovery: emerging research needs and challenges. In Blanco & Alberti (eds) Shaken, shrinking, hot, impoverished and informal: Emerging research agendas in planning. *Progress in Planning*, 72(4):195–250.

Oswalt, P., Rieniets, T. & Schirmel, H. (2006). *Atlas of shrinking cities*. Publisher: Ostfildern Hatje Cantz, Maidstone.

Parker, M. & Steenkamp, D. (2012). The economic impact of the Canterbury earthquakes. *Reserve Bank of New Zealand: Bulletin*, 75(3)13–25.

Plyer, A. (2011). Population loss and vacant housing in New Orleans neighborhoods. *The Data Center*. The Data Center.org, https://www.datacenterresearch.org/reports_analysis/population-loss-and-vacant-housing/ Sighted January 18 2017.

Plyer, A. (2015). Fact for features: Katrina recovery. *The Data Center*. The Data Center.org, https://www.datacenterresearch.org/data-resources/katrina/facts-for-features-katrina-recovery/ Sighted January 18 2017.

QGSO. Queensland Government Statistician's Office. (2017). *Regional Profiles: Time Series Profile for Innisfail - Cassowary Coast Statistical Area Level 3*, Queensland Treasury, Queensland, https://statistics.qgso.qld.gov.au/qld-regional-profiles Sighted 13/1/20.

Queensland Government. (2007). *The final report of the operation recovery task force: Severe Tropical Cyclone Larry*. Department of Premier and Cabinet. Queensland Government: Brisbane.

Sastry, N. (2009). Tracing the effects of Hurricane Katrina on the population of New Orleans: The displaced New Orleans residents pilot study. *Sociological Methods & Research*, 38(1):171–196.

Simo, G. (2008). Poverty in New Orleans: Before and after Katrina. *Vincentian Heritage Journal,* 28(2):309–320.

Statistics New Zealand. (2012). *Subnational Population Estimates: At 30 June 2012 – Media Release*, ihttps://www.stats.govt.nz/browse_for_stats/population/estimates_and_projections/SubnationalPopulationEstimates_MRYe30Jun12.aspx Sighted January 24 2017.

Statistics New Zealand. (2014a). https://archive.stats.govt.nz/browse_for_stats/population/census_counts/PostEnumerationSurvey_HOTP13.aspx, Sighted 21/12/19.

Statistics New Zealand. (2014b). *2013 Census QuickStats about greater Christchurch*. www.stats.govt.nz Sighted on January 21 2017.

Statistics New Zealand. (2019). *Christchurch estimated resident population for urban areas, at 30 June (1996+)*. https://www.stats.govt.nz/topics/population-estimates-and-projections Sighted 23/12/19.

Swaffield, S. (2012). Reinventing spatial planning at the urban rural interface: A Christchurch, New Zealand case study. *Planning Practice & Res*earch, 27(4):405–422.

United Nations, Department of Economic and Social Affairs, Population Division. (2018). *The World's Cities in 2018—Data Booklet* (ST/ESA/ SER.A/417).

UNISDR. (2015). *Sendai Framework for Disaster Risk Reduction 2015–2030*. UNISDR, Geneva.

United States Census. (2019). *QuickFacts, New Orleans city, Louisiana*. https://www.census.gov/quickfacts/fact/table/neworleanscitylouisiana/PST120218 Sighted 23/12/19.

Youn, H., Bettencourt, L., Lobo, J., Strumsky, D., Samaniego, H., & West, G. (2016). Scaling and universality in urban economic diversification. *Journal of the Royal Society Interface*, 13(114):1–7.

Zaninetti, J & Colten, C. (2012). Shrinking New Orleans: Post Katrina population adjustments. *Urban Geography*, 33(5):675–699.

Chapter 7
Disasters and Demographic Change of 'Single-Industry' Towns—Decline and Resilience in Morwell, Australia

Deanne Bird and Andrew Taylor

Abstract In 2014, an open-cut coal mine fire burned for 45 days in the small single-industry town of Hazelwood in Victoria (Australia) spreading smoke and ash across the adjacent community of Morwell. This chapter examines the extent to which the mine fire acted as a catalyst for demographic and socio-economic change and considers how, if at all, it impacted Morwell's resilience to disasters. We report on a range of secondary data analyses augmented with qualitative insights captured in government reports (namely, the Hazelwood Mine Fire Inquiry reports), as well as from related research papers and media articles. We suggest that a succession of structural and demographic changes meant that the town and its residents were accustomed and resilient to relatively large shocks. In this sense, the Morwell and broader Latrobe Valley population banded together around various community-led initiatives to fight for a better future for their community.

Keywords Disaster · Hazelwood · Mine fire · Resilience · Social capital · Demography

Deanne Bird is employed by the Department of Health and Human Services, Victoria and is an Associate of Monash University. The views presented in this chapter reflect those of the authors and do not necessarily represent the views of the Department of Health and Human Services or Government of Victoria.

D. Bird (✉)
Latrobe Health Innovation Zone, Department of Health and Human Services, Government of Victoria, Melbourne, Australia
e-mail: deanne.bird@gmail.com

Monash University Disaster Resilience Initiative (MUDRI), Monash University, Clayton, Australia

A. Taylor
Northern Institute, Charles Darwin University, Brinkin, Australia
e-mail: andrew.taylor@cdu.edu.au

D. Karácsonyi et al. (eds.), *The Demography of Disasters*,
https://doi.org/10.1007/978-3-030-49920-4_7

7.1 Introduction

Throughout Australia's history, disasters have challenged the existence of rural and remote towns. At the extreme end, for example, the rural and remote towns of Gundagai in New South Wales and, Clermont and Grantham in Queensland were relocated following fatal floods in 1852, 1916 and 2011, respectively. For single-industry towns (see Text Box 7.1), disasters involving the pre-dominant industry present added threats since even temporary closures or wind-backs in economic production are likely to significantly erode the job base, cut money circulation in the town and lead to permanent out-migration of some residents. Through these processes, closures also thin-out the locally based pool of experienced workers, encouraging companies who own or run the (single) industry to 'import' temporary workers in the form of fly-in-fly-out (FIFO) or drive-in-drive-out (DIDO) workers. In turn, these factors tend to erode the social fabric and cohesiveness of communities, making them less resilient to shocks and stressors (Mitchell and O'Neill 2016). The shock and stressor considered in this chapter are from the Hazelwood mine fire disaster that impacted the adjacent, single-industry town of Morwell, Australia in 2014.

Before describing the case study location of Morwell, the Hazelwood power station, open-cut coal mine and mine fire disaster in Sect. 7.2, we first consider the terms *disaster* and *resilience*. There has been much debate around what constitutes a disaster (Quarantelli 1998; Perry and Quarantelli 2005). Based on that debate, Perry (2005) concluded that disasters are disruptive, social occasions that are related to social change and that disaster research generally focuses on 'some change in circumstances' (p. 316).

The definition of resilience in relation to disasters has also been the subject of much debate in academic discourse (e.g. Klein et al. 2003; Manyena 2006; Norris et al. 2008; Manyena et al. 2011; Alexander 2013; Cutter et al. 2014; Weichselgartner and Kelman 2015). For this chapter, we consider resilience as a '... measure of how well people and societies can adapt to a changed reality and capitalise on the new possibilities offered' (Paton 2006, p. 8).

In light of the above, the aims of this chapter are to examine the extent to which the Hazelwood mine fire acted as a catalyst for demographic and socio-economic change in the single-industry town of Morwell, and consider how, if at all, the Hazelwood mine fire impacted Morwell's resilience to disasters, shocks and stressors.

Text Box 7.1: Rural, Remote and Single-Industry Towns in Australia

Rural and remote towns are hallmarked in the national psyche of many developed nations for their purported resilience in the face of adversity. Perhaps nowhere has this been more the case than in Australia. Although 68% of Australians live in major cities and with cities recently accounting for almost 80% of national population growth (ABS 2019), there remains an affinity for rural and remote towns. Despite the passing centuries, the exponential growth of cities and a boon in overseas migration numbers to cities of late, Australian's affinity with 'the bush' remains. Rural and remote settlements are lauded for their historical and contemporary contributions to the nation and their economic and demographic resilience. This reflects, at least in part, remembrance of the significance of the livestock industry, and particularly wool, for transforming the economy from a colonially dependent backwater to a 'modern' thriving one with comparatively high living standards (ABS 2003). The nation is said to have 'ridden to success' on the sheep's back for more than a century from the mid-eighteenth century to mid-nineteenth century as a result of wool exports. Even today, agricultural or resource products round-out eight of the top ten Australian export commodities (Australian Government 2018).

Over time, the importance of inland Australia and its towns was re-enforced by other major historical developments like the Overland Telegraph Line, which ran through the 'middle dirt' of the landmass, and for some time was the only means of direct communication between Australia and elsewhere (notably Great Britain). Carrying Morse code signals, it stretched 3,200 km from Port Augusta in South Australia to Darwin in the north of the country. Likewise, the national airline Queensland and Northern Territory Aerial Services (QANTAS) was established in 1920 in the Queensland desert town of Winton. Some settlements were established and then either thrived or survived from the extraction of resources. In particular, gold and later a suite of resource-based settlements were established across the continent.

Many of Australia's rural and remote towns were therefore settled and established because of a single industry. The nation's economy has at various times been driven by agricultural and later resource-led economic growth. In many places, commodifiable resource(s) including gold, coal and iron ore were either the reason for rural town establishments or the reason for its continuation and growth after initial settlement. Often called 'single-industry' towns, these have been written about for their resilience in the face of volatile population change and susceptibility to 'boom and bust' economic times (for example, Carson and Carson 2014). While some have disappeared and become 'ghost towns', most have either continued to exist, have grown or transformed their economic basis despite declines in the main industry because of rationalisation, technological changes and market forces. Nevertheless, for some single-industry towns there remains a legacy and mythology whereby the perception of the town being single industry persists among outsiders. This is in part because of the past pre-eminence of the industry in question for jobs, business and maintaining population. Examples include Burnie in Tasmania whose industry profile was once dominated by a papermill and associated forestry activity (see The Advertiser 2013), as well as Nhulunbuy in the Northern Territory which was established specifically to house workers for the nearby bauxite mine and alumina smelter that closed in 2013 (see Carson and Carson 2014; Collin 2017).

7.2 Morwell and the Hazelwood Mine and Power Station

Morwell, centrally located in the State of Victoria's Latrobe Valley 150 km east of Melbourne (Fig. 7.1), is often recognised as a single-industry town. While the region is located in an important agricultural area, Morwell's history is intricately linked to the production of electricity produced from the extensive brown coal reserves in the

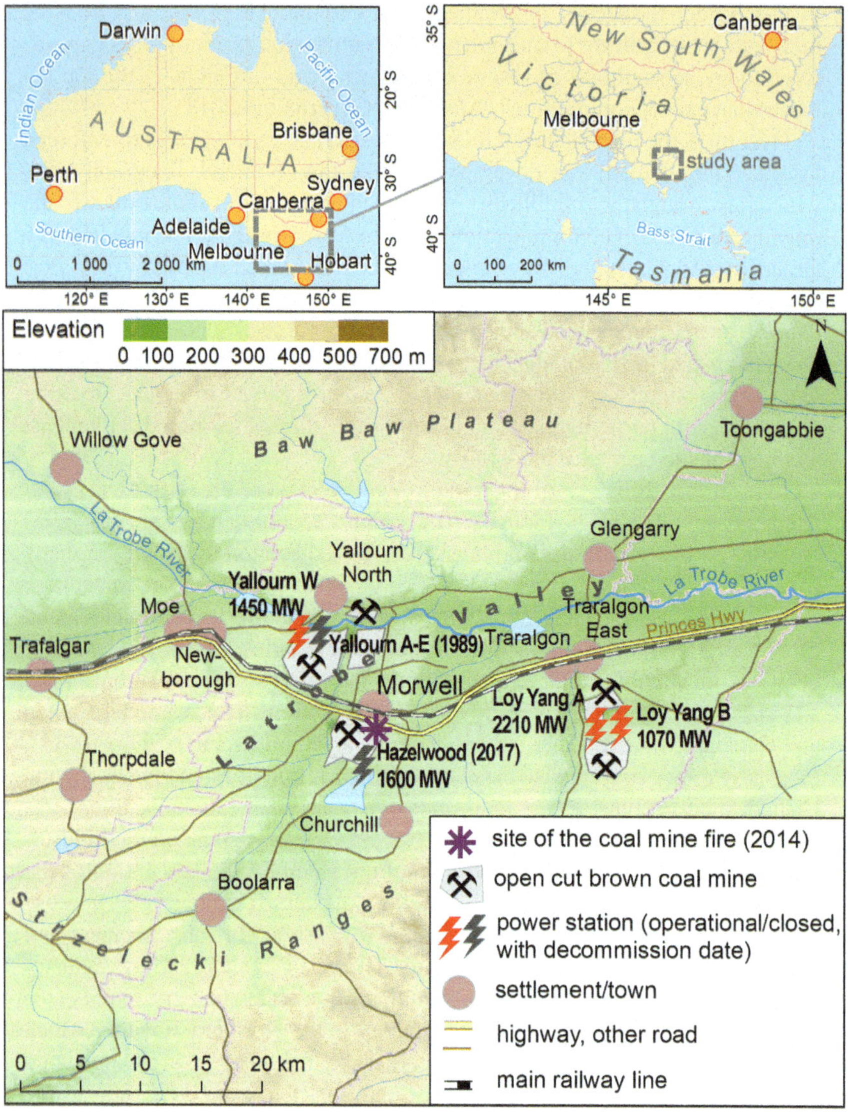

Fig. 7.1 Location of the town of Morwell (Authors: Bird, Taylor, cartography by Karácsonyi)

region. Teague and Catford (2016) report that around 95% of Victoria's base load electricity was produced from Latrobe Valley mines (Hazelwood, Yallourn and Loy Yang) and collectively they constituted the largest brown coal mining operation in the Southern Hemisphere (Davison 2015).

The Hazelwood mine and power station hailed from the 'Morwell project' which consisted of an open-cut coal mine and briquette works. It was initiated by the State Electricity Commission (SEC) of Victoria in the late 1940s (Morwell Advertiser 1949). A decade later, approval was given for the development of the Hazelwood Power Station to service Morwell's open-cut mine (later known as the Hazelwood open-cut mine). At its peak, the Hazelwood Power Station met 25% of the State and over 5% of the nation's electricity demands alone and its extensive dimensions reached a maximum depth of 120 m from ground level with an 18 km perimeter (Teague and Catford 2016).

From these beginnings, and given the coal mine pit begins its descent a few hundred metres from Morwell's residential area (Fig. 7.2), the town's identity had been interwoven with, up until recently, the Hazelwood Power Station and open-cut mine:

> …the development and expansion of coal mining in the area over time has had a direct impact on the people of Morwell due to the town overlaying a significant coal deposit. In the context of the Hazelwood mine and power station being built to the south of Morwell, the town has expanded to the east and to the north. Despite such expansion away from the mine, the southern perimeter of Morwell is still remarkably close to the mine site. (Teague et al. 2014, p. 51)

At times, more than a third of jobs in the region were with the SEC alone, not taking into account associated industry and businesses associated directly with the coal mine and power station (Duffy and Whyte 2017). However, it was far from smooth sailing

Fig. 7.2 Looking approximately northwards, this aerial photograph shows the proximity of Morwell (residential area outlined in green) to the Hazelwood Power Station (circled in red) and open-cut coal mine. (Teague et al. 2014, p. 410)

for Morwell residents and the plant. Most notably, the state government privatised the SEC in the mid-1990s and with it the Hazelwood Power Station and open-cut mine. More recently, Hazelwood's open-cut coal mine was at the centre of one of Australia's worst environmental and public health disasters (Doig 2015) when it burned for 45 days after embers from a nearby bushfire set the coal alight in February 2014. Three years later, the Hazelwood Power Station and coal mine were decommissioned.

In this chapter, we plot demographic change in Morwell through periods of coal mining development, expansion, decline and the mine fire disaster, right through to post-decommissioning. The mine fire in 2014 '… disrupted the community to a significant extent and was beyond the capacity of the community and support agencies to cope' (Walker et al. 2016, p. 16). However, using a range of data from the Australian Bureau of Statistics (ABS) alongside community consultations and research conducted after the 2014 Hazelwood Mine Fire, we examine population resilience through a long history of disaster familiarity (after previous major fires in 1977, 2006 and 2008). Our main contention is that, rather than accounting for a big change in population numbers and characteristics, the mine fire disaster added to pre-existing challenges faced by a community which itself had learned to adapt to and survive, in demographic and economic terms, to shocks, disasters and other challenges.

7.3 Mapping Resilience Through Demographic Change

In response to academic debate, Cutter (2016) critically examined methods for understanding and quantifying resilience in relation to disaster. In doing so, she identified the most commonly used attributes, assets (economic, social, environmental, infrastructure) and capacities (social capital, community functions, connectivity, and planning) for measuring community resilience. This chapter focuses on demographic and socio-economic indicators as measures of resilience. Our reason for doing so is based on the premise that much can be revealed about the resilience or otherwise of single-industry towns through the study of change among pre- and post-event demographic indicators, including socio-economic profiling. Such research and analysis are useful because they can:

- Provide a baseline for evaluating demographic and economic impacts from individual disasters (where suitable data is available);
- Plot interrelationships between economic and demographic transformations pre- and post-disaster;
- Be applied in a range of ways to ascertain likely demographic and economic futures given the post-disaster population structure and size; and
- Be studied for applied lessons which might help other towns increase their resilience and map out their futures.

Importantly, socio-economic profiling can also be validated, compared and contrasted to qualitative research 'on the ground' to enrich the research application.

In this chapter, we examine a range of publicly available demographic and socio-economic data for Morwell and compare to the wider Latrobe Valley region and the State of Victoria. We first augment our analysis with qualitative insights captured in government reports (namely, the Hazelwood Mine Fire Inquiry reports which provide detail on community consultations, health improvement forums and public submissions), as well as from related research papers and media articles. This latter analysis allows us to consider the social fabric and cohesiveness of Morwell with a focus on social capital and connectivity.

In a presentation given at a Municipal Association of Victoria forum on resilient cities and communities, Duckworth (2015, p. 6) highlighted 'resilience is not possible without the networks and links between individuals, communities, organisations, businesses and government'. Underlying these networks and links is social capital, which appears as a key feature in the broad range of tools, indices and scorecards that have been developed to measure community resilience (Cutter 2016). However, social capital is more than just connectivity through these networks and links. It is about cooperation among different groups of people and collective action to produce a mutual benefit, and it is reliant on trust (Bridger and Luloff 2001). Putnam et al. (1993) described the kind of trust required as social trust, which emanates from norms of reciprocity and networks of engagement.

In this chapter, three types of social capital are considered for the Morwell population, reflecting the networks of engagement and the situations where norms of reciprocity evolve.

1. *Bonding social capital* describes emotionally close, strong connections arising from family and friend networks.
2. *Bridging social capital* describes loose connections between acquaintances and individuals of diverse social groups, often stemming from membership or involvement in organisations, clubs and associations.
3. *Linking social capital* describes network connections between regular people (such as residents) and officials (such as government representatives) (Aldrich and Meyer 2015).

Before Morwell's pre- and post-fire demographic trajectories are examined, we first consider the Hazelwood mine fire alongside the health and economic impacts of that event and government response.

7.4 The 2014 Hazelwood Mine Fire Disaster

7.4.1 The Mine Fire

In February 2014, during a record-breaking hot and dry summer, a number of large bushfires burned across Victoria. Two of these were near Morwell and the adjoining Hazelwood coal mine. Embers from these fires set the northern, eastern and south-eastern batters and floor of the Hazelwood coal mine alight on 9 February 2014. The fires spread quickly and proved extremely difficult to extinguish due to the highly combustible nature of brown coal and the relatively thin layer of soil and clay covering the massive coal seams in the Latrobe Valley. Moreover, the mine operator was not prepared to manage such an event. In this instance, the mine's firefighting infrastructure was either not present or had not been maintained in the areas of the mine that were alight. Furthermore, once called in to assist, firefighting personnel experienced difficulties accessing and navigating the mine (Doig 2015; Teague et al. 2014).

More than 7,000 fire services personnel, a number equivalent to half the size of the town's population, from across Australia fought the fire for 45 days until finally, on 25 March 2014 the fire was officially declared extinguished (Teague et al. 2014). While it burnt out of control, the mine fire produced a significant amount of ash causing physical and mental health issues for nearby communities, especially the population of Morwell. For example, on 16 February 2014, the daily average of $PM_{2.5}$[1] reached approximately 28 times the advisory level (Teague et al. 2014). On the same day, carbon monoxide levels were almost four times the compliance standard (Teague et al. 2014). In relation to the particulate matter, an older respondent in the Walker et al. study commented:

> It was like sand on a windy day. Sand on the beach on a windy day hitting your face, that's what it was like yet you couldn't see anything with the naked eye. (Walker et al. 2016, p. 42)

7.4.2 Health Impacts and Government Responses

Commonly reported short-term physical health impacts during the disaster included skin and eye irritations, blood noses and headaches. Despite the apparent adverse conditions in the early stages of the mine fire, the government was slow to issue public health alerts due to bureaucratic protocols around decision-making and over reliance on validated air quality data when indicative data would have sufficed (Teague et al. 2014).

> It was the same old information we were getting day after day after day, "There's nothing wrong, nothing to worry about, nothing wrong, nothing to worry about, nothing wrong, nothing to worry about" and that's from people either in Traralgon or in Melbourne. They weren't in Morwell, trying to breathe this rubbish. (Walker et al. 2016, p. 56)

[1] Particulate matter equivalent to or less than 2.5 microns and known to cause adverse health effects.

> The general conversation was the same... they knew something wasn't right, and the smoke was different, and it was different, I've never smelt anything like it, or tasted anything like it. (Walker et al. 2016, p. 42)

On 25 February 2014, in recognition of the need for better protocols, the Environment Protection Authority and Department of Health[2] developed specific guidelines around health actions relating to $PM_{2.5}$ levels. On 28 February, when $PM_{2.5}$ levels had increased again, the Chief Health Officer advised temporary relocation for vulnerable groups, including preschool aged children, pregnant women, people with pre-existing cardiovascular and respiratory conditions and people over 65 years (Teague et al. 2014). However, this advice was deemed 'too late'. Furthermore, as the advice singled out particular demographic groups, the community found it to be illogical and divisive (Teague et al. 2014). Overall, the community did not feel that their views were being heard.

> While distributing considerable amounts of information to the community, government departments and agencies did not engage to any significant extent in listening to, or partnering with local residents and community groups, which are identified as important strategies in best practice risk and crisis communication. (Teague et al. 2014, p. 400)

A few weeks into the mine fire disaster, and in response to community outcry, Monash University's School of Public Health and Preventative Medicine were commissioned to conduct a Rapid Health Risk Assessment. Based on the results of that assessment, the Board of Inquiry concluded that:

> ... the level of exposure to smoke and ash experienced by the community in Morwell would not be expected to cause any deaths if the level of exposure remained at that level for six weeks. However, the study was based on a standard Victorian population and was not adjusted for the poorer health status prevailing in Morwell. (Teague et al. 2014, p. 24)

The Board highlighted that several vulnerable groups characterised the population with an ageing population, higher incidence of cardiovascular and respiratory disease, a high percentage of low-income households and a higher percentage of residents with a disability (Teague et al. 2014). With this in mind, the Board reflected that the mine fire 'added further insult to an already vulnerable community' (Teague et al. 2014, p. 24). Nevertheless, Walker et al. (2016) report stoicism and resilience, particularly among older adults within the community. One service provider claimed their older clients were 'pretty resilient' relying on lived experience to deal with the situation at hand. Moreover, people relied on family and friends for essential assistance rather than the authorities, stating that they had it 'under control, we're at our families' and 'I've just gone to my friend in Traralgon' (Walker et al. 2016, p. 62).

While the short-term health impacts were of concern, the community were also gravely concerned about the potential for medium to long-term health impacts.

---

[2]The Department of Health became the Department of Health and Human Services on 1 January 2015.

> People are still waiting and people have moved on but I think the impact is longer lasting because the response post mine fire wasn't swift. [...] They denied that there was health issues, there was some cover-ups or perceived cover-ups, they could have done a lot more and they didn't. I think that affects the community pride, community connectedness and realizing that we're not as valued as somewhere else in the state. (Jones et al. 2018, p. 539)

The community petitioned the government to take further action, with over 21,000 signatures of support (Duffy and Whyte 2017). In response, the Department of Health and Human Services commissioned the Hazelwood Health Study, which commenced in November 2014. The study showed an increase in the occurrence of gestational diabetes among pregnant women exposed to mine fire-related air pollution than those who were not exposed during the disaster, with the greatest risk associated with exposure during the second trimester (Johnston et al. 2019). Furthermore, participants of the Health Study that were exposed to mine fire-related air pollution were more likely (compared to those not exposed) to report respiratory symptoms (wheeze, night-time and resting shortness of breath, chronic cough and phlegm, chest tightness and nasal symptoms) along with psychological distress in adults (Hazelwood Health Study 2019) and school children (Allen et al. 2019). Yell et al. (2019) also note that community wellbeing was greatly affected, with a distinct loss of trust in government authorities responsible for dealing with the disaster. The impact on community wellbeing is clearly articulated by the following comments captured in the Jones et al. study:

> Bushfires were [...] only short lived with buildings and people losing properties and everything like that, whereas here the destruction was, I suppose, the Valley itself [...]. (Jones et al. 2018, p. 539)

> I am very proud of my town [...] and my town was copping not a very good rap [...] It wasn't doing our town's image any good at all. We got known as a smoke town. Morwell has copped enough over the years without having to have that added to it. (Jones et al. 2018, p. 539)

> I think that my experience here in the Valley's changed. So, up to the mine fire I think that life was a certain way, post that, it's almost like my attachment to the area is not as much. My feeling for the area, has waned – my commitment to the region – to where I live. My interest is no longer there. (Jones et al. 2018, p. 540)

7.4.3 Economic Impacts

On top of dealing with the long-term physical and mental health impacts of the mine fire, the community were faced with economic challenges including fears of falling house prices and local businesses struggling due to significantly reduced trading (Walker et al. 2016). The neighbourhood houses[3] were also feeling a pinch. Whyte

[3]Neighbourhood Houses are described by Neighbourhood Houses Victoria as places that "bring people together to connect, learn and contribute in their local community through social, educational, recreational and support activities, using a unique community development approach" (https://www.nhvic.org.au/neighbourhoodhouses/what-is-a-neighbourhood-house).

(2017) provides an insight from a 2014 interview with the Morwell Neighbourhood House coordinator:

> We've been impacted financially here because we've closed all our classes, people aren't engaged, that component of the community that were coming in to do courses have either left or have bunkered down because they've got their own social and emotional issues and they can't engage on that level. (Whyte 2017, p. 14)

Then, in November 2016, the owner at the time, Engie, announced the closure of the Hazelwood Power Station and open-cut coal mine. A submission from Latrobe City Council aptly noted:

> While residents can learn to co-exist with [mining operations] the more that those operations intrude on the lives of those residents, the more resentful residents become. This in turn can lead to a diminishing of a community's ability to feel empowered and resilient. (Teague et al. 2016, p. 70)

The site was officially decommissioned in March 2017 at which time, 750 personnel were in direct employment at Hazelwood (Engie 2017). Later that same year, Carter Holt Harvey closed its Morwell sawmill on 28 September leading to the loss of another 160 jobs in the region.

> What has emerged from [the Hazelwood Health Study interviews] (and ongoing conversations with community members) is that recovery is no longer only framed in relation to the mine fire event. It is talked about as also needing to address the ongoing impacts of the privatisation of the power industry in the late 1980s and early 1990s and the more recent closure of the Morwell open-cut mine and Hazelwood power station. (Hazelwood Health Study 2017, p. 49)

Having captured the essence of some of the impacts from the Morwell disaster on town residents and others, we now map out the transition of the population from pre- to post-disaster to evaluate the extent to which the fire may have altered the town's demographic and economic trajectories.

7.5 Morwell's Pre and Post-fire Demographic Trajectory

7.5.1 Plotting Structural and Demographic Change

Twenty-seven years of population data demonstrate the impacts of several structural upheavals on Morwell's resident population (Fig. 7.3). While population growth for Morwell and the Latrobe Valley has fluctuated quite widely, and at times been negative, State growth since 1991/92 was positive and comparatively consistent.

Demographic fluctuations for Morwell can be traced to a range of factors. During the 1990s, the town experienced population decline from restructuring and privatisation of the power industry by the State Electricity Commission of Victoria (SEC). This led to jobs and population losses in the region with Morwell most affected in proportional terms, particularly in 1993/94. Following privatisation in 1996, the then labelled 'Hazelwood Power Corporation' was sold to a consortium led by the

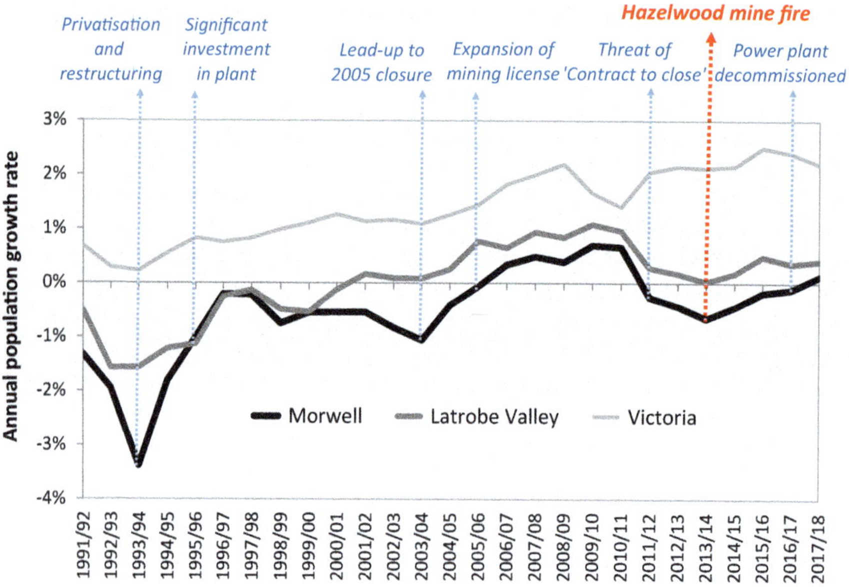

Fig. 7.3 Annual population growth rates and major shocks, 1991/92 to 2017/18. (Calculation by Bird and Taylor using data from ABS.Stat)

British organisation National Power. However, population growth rates did not return to positive until the 2004 announcement of the proposed expansion of the mine to incorporate Phase 2 of the Hazelwood Mine West Field Development. Before this, the plant was due to be decommissioned in 2005 by the SEC, placing employment in the region under threat. This was evident in the negative population growth rates during the first half of 2000s.

In 2011/12, the Australian Government considered a Contract for Closure program as part of its Clean Energy Act, causing uncertainty in the region. However, that program was soon scrapped and no plants were closed. Nevertheless, as reported earlier, Engie closed Hazelwood at the end of March 2017 giving workers and local communities just five months' notice (ABC 2017). However, since closure of the plant in early 2017, there has been no evidence of a significant decline in growth rates for Morwell. Although growth rates were seemingly better than for most years since from the early 1990s onwards, from 2011/12 to 2016/17 rates were slightly negative.

The impacts of the events described above can be seen on the town's estimated resident population (Fig. 7.4) which fell by more than 300 during the lead up to the proposed 2005 closure (which did not take place). It recovered subsequently before falling again after the threat of the 'contract to close' became apparent in 2011/12.

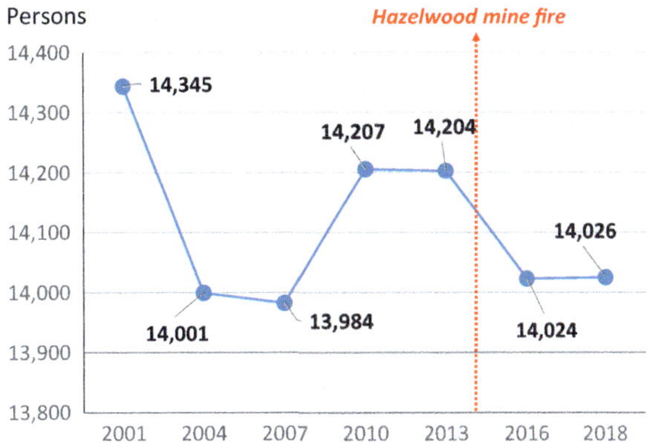

Fig. 7.4 Morwell's Estimated Resident Population, 2001–2018. (Calculation by Bird and Taylor using data from ABS.Stat)

Nevertheless, in 2018, the population was estimated at 14,026, just 2% lower than its peak of 14,345 during the past eighteen years.

Changes in Morwell's Age and Gender Composition

While the population size has not fallen significantly since the mine closure, there have been progressive changes in the composition of the population. The population pyramid for the town (Fig. 7.5) for 2006 (black bars) and 2016 (for males blue with orange and females khaki with orange outline bars) shows there was a reduction in

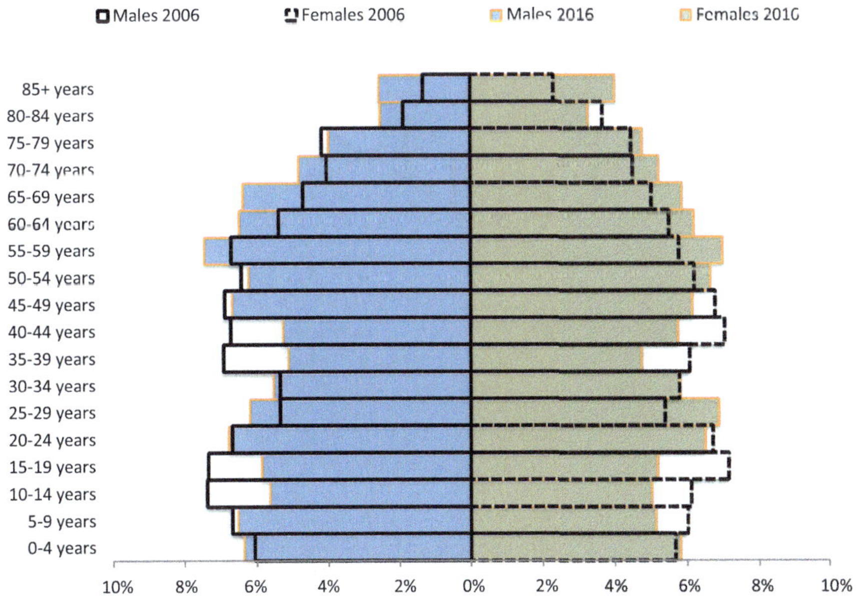

Fig. 7.5 Population by age and sex, 2006 and 2016. (Calculation by Bird and Taylor using ABS Census of Population and Housing data obtained from Table Builder)

the proportion of mid-career working-age residents (aged 30–44 years) and children aged 10–19 years. The latter is likely associated in part with the former as well as young people leaving for post-school education. In its place is a relatively prominent population ageing trend with an increase in the proportion of males and females aged 55 and above. Of note, the proportion of young children has remained stable, while those aged 25–29 years made up a greater part of the overall population in 2016 compared to 2001.

The changes described above are emphasised when comparing age-based estimates for the total population from 2007 to 2017 (Fig. 7.6). Changes in age composition for Morwell are shown by the red and black bars with the red indicating a reduced proportion of the population in 2017 at that age group and the black indicating an increased proportion. Hence, while the total population size has not altered significantly, its composition has.

Migration
During 2011 to 2016 the town experienced a net loss of 149 males and 162 females (a total of 311) residents through migration. Analysis of this by life-stages (Fig. 7.7) shows net losses of early-career residents and children were substantial, while there were net gains in late-career males and retiree-aged males (those 65 years and over).

Other Demographic Indicators
Comparing a range of other demographic indicators for Morwell to the region and to the State of Victoria highlights some significant differences and changes in the town

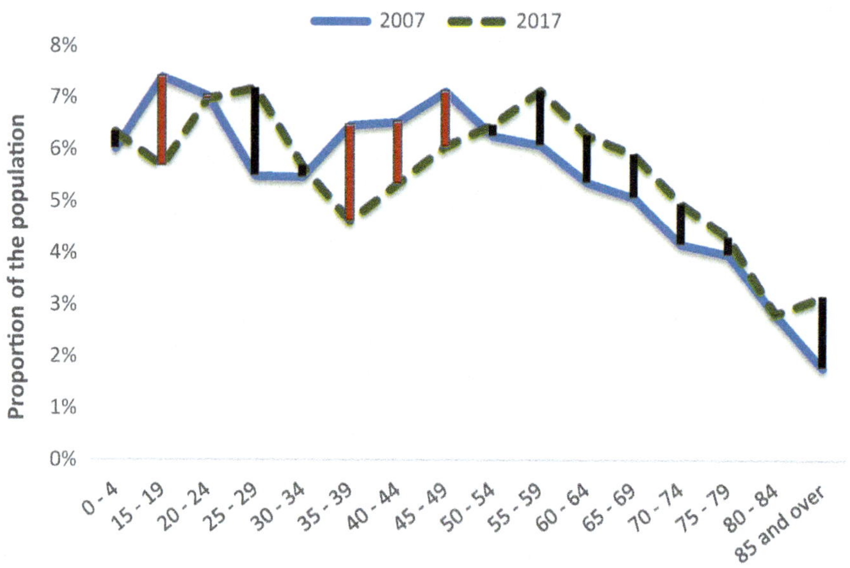

Fig. 7.6 Changes in Morwell's population distribution by age, 2007 and 2017. (Calculation by Bird and Taylor using data from ABS.Stat)

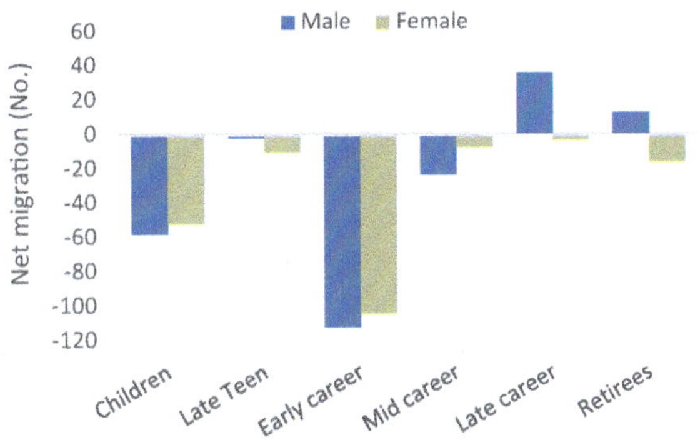

Fig. 7.7 Net migration by age for Morwell during 2011 to 2016. (Calculation by Bird and Taylor using ABS Table Builder data)

for the 15 years to 2016. With a median age of 43 in 2016, Morwell's population in 2016 was older than for the region and Victoria by quite some margin, having risen considerably over the decade (Table 7.1). This indicates population ageing in the town through either resident's ageing-in-place or the out-migration of younger people (or likely a combination of both). Morwell was relatively less 'multicultural' than Victoria by 2016 and its proportion born overseas had diminished, unlike for the State as a whole. Morwell, however, had a higher proportion of Indigenous residents (2.5%) compared to both the region (1.5%) and State (0.8%).

Also of note were lower median rents and house mortgage repayments in 2016 for Morwell when compared to the region and State. This is in part due to the higher proportion of householders who were renting from State housing organisations, at 6% in 2016 compared to 3% for Victoria (Table 7.2). In addition, a greater proportion rented from a real estate agent in 2016 (21%) than for Victoria (15%), having risen from 17% of households in 2006. These data indicate socio-economic status continues to be relatively lower in Morwell. However, there does not appear to be any collapse in these housing indicators subsequent to the fire and in the lead up to the closure of Hazelwood mine suggesting that the housing market may not have declined significantly in the post-disaster period; although certainly not increasing to the extent of mortgages and rents for the State as a whole.

7.5.2 *Employment, Income, Industry and Housing Profile Changes*

For towns such as Morwell, industrial composition and the distribution of jobs across industries are generally dominated by the main operation, usually a single large resourced-based extractive and/or processing industrial unit. Although the label might be appropriate as a descriptor, in reality, no town is truly 'single industry' with jobs in other sectors invariably prevalent and important including in retail, health and government administration and services; as well as a range of professions across various industries (such as medical or technical services). Economists point out there

Table 7.1 Selected demographic indicators, 2006, 2011 and 2016

Indicator	Morwell			Latrobe valley			Victoria		
	2006	2011	2016	2006	2011	2016	2006	2011	2016
Estimated resident population	13,578	14,004	13,808	68,859	72,216	73,099	4,932,422	5,354,039	5,926,624
Median age	39	40	43	37	39	41	37	37	37
Proportion born overseas (%)	16	17	15	13	14	12	24	26	28
Per cent Indigenous (%)	2.2	2.8	2.5	1.3	1.5	1.6	0.6	0.7	0.8
Average household size	2.3	2.2	2.1	2.4	2.4	2.3	2.6	2.6	2.6
Median mortgage repayment (monthly)	737	975	953	867	1,200	1,200	1,252	1,700	1,728
Median rent (weekly)	115	150	180	120	160	200	185	277	325

Table constructed by Bird and Taylor using ABS, 2003.0 2016 Census of Population and Housing Time Series Profile (Morwell SA2, Latrobe Valley SA3 and Victoria State). Estimated Resident Population sourced from data extracted ABS.Stat

Table 7.2 Tenure and landlord type for Morwell and Victoria, 2006, 2011 and 2016

	Morwell			Victoria		
	2006 (%)	2011 (%)	2016 (%)	2006 (%)	2011 (%)	2016 (%)
Owned or being purchased	68	66	64	74	72	73
Rented real estate	17	19	21	15	17	16
Rented state or territory housing	8	7	6	3	3	3
Other	7	8	9	9	9	9
Total	100	100	100	100	100	100

Table constructed by Bird and Taylor using ABS, 2003.0 2016 Census of Population and Housing Time Series Profile (Morwell SA2 and Victoria State)

are a range of benefits and challenges for towns like Morwell which were 'specialised' (also termed 'agglomerated' or 'concentrated') and for towns with employment diversified across industries (for a good discussion on these issues see ABS 2014). In this section, we profile employment and industry indicators for Morwell over time to examine the extent of industry concentration and identify whether and how the Hazelwood mine fire and power station closure may have altered the economic make-up of the town.

Labour Market and Income Indicators

In 2011, the last Census before the closure of the power station, and at the end of a period of relatively high population growth from investment in Hazelwood, Morwell had a relatively high unemployment rate (12%) compared to its surrounding region (8%) and to the State of Victoria (5%). Participation rates in the workforce were also much lower (Table 7.3). In general, the range of indicators would suggest that the comparative socio-economic status of Morwell's residents was low before the fire. Despite the scale of the fire and its impacts, the suite of indicators in Table 7.3 does not suggest a dramatic worsening of socio-economic conditions. Although the town did experience an increase in unemployment and decrease in participation rates during 2011 to 2016, these can reasonably be described as a continuation of pre-existing trends or as minor changes to date.

Industry and Employment Profiles.

All else being equal, observed structural adjustments might be anticipated in the profile of jobs and industry compositions for small towns like Morwell after a disaster and when the major employer ceases to operate. Interestingly for Morwell, employment in the 'Mining' and 'Electricity, Gas, Water and Waste Services' industry sectors have never dominated overall employment patterns in the town. This is likely due to variable distributions of jobs in the mine and plant across a range of industry sectors including two which show in Census data to have been prominent: 'Manufacturing' and 'Construction' (Table 7.4). The former has significantly declined in Australia, and this is most evident for the State of Victoria where the sector has fallen

Table 7.3 Comparative employment and income indicators, 2006, 2011 and 2016

Indicator	Morwell			Latrobe valley			Victoria		
	2006	2011	2016	2006	2011	2016	2006	2011	2016
Unemployment rate (%)	12	12	14	8	8	10	5	5	7
Participation rate (%)	49	48	46	56	56	54	60	61	61
Median personal income	$326	$391	$470	$376	$468	$544	$456	$561	$644
Median family income (weekly)	$809	$930	$1,092	$1,053	$1,236	$1,414	$1,189	$1,460	$1,715

Table constructed by Bird and Taylor using ABS, 2003.0 2016 Census of Population and Housing Time Series Profile (Morwell SA2, Latrobe Valley SA3 and Victoria State)

from being the largest employer in the State in 2011 to the sixth by 2016. Similarly, in Morwell, manufacturing no longer featured in the top four industries by 2016. Construction jobs in Morwell likely reflects stages of development in the plant and infrastructure associated with the mine and power station while the prominence of health care and social assistance by 2016 reflects a national population ageing trend and data in demographic analysis earlier in this chapter.

It is also pertinent, when considering possible local economic impacts from the disaster, to examine baseline industry concentrations and differential employment profiles for men and women, particularly considering the male-dominated profile of resource-based industrial operations. In terms of industry concentrations, the share of employment in the top four employing industries (known as the C4 Index) was the same in Morwell and the Latrobe Valley region in 2006 and 2011 (47% and 46%) but fell to 42% in 2016; the same as for the State of Victoria. Put another way, jobs in the town became more widely distributed across industries after the fire. Of concern, however, have been differential gendered outcomes for employment seen in the employment to population ratio. For males, there was a significant decline in the proportion of those 15 years and over who were employed, down from over half in 2006 to 43% in 2016. While this also declined in the Latrobe Valley and Victoria overall, the fall was larger in Morwell and is likely indicative of a range of long-term (for example, technology changes and population ageing) and event-related impacts including the mine fire and structural workforce adjustments in other industries. Meanwhile, the employment to population ratio for women was more stable for all three jurisdictions although much lower than for males (Table 7.5).

7.6 Discussion and Conclusion

The small Victorian town of Morwell, with a 2018 population of 14,000, has long been subject to shocks and stressors. In this chapter, we postulated that Morwell might previously have been considered a 'single-industry' town, given the historical importance of the coal mine and power station to employment and wealth generation. As a single-industry town, and in consideration of the apparent vulnerabilities that exist in relation to the poor health and wellbeing statistics of Morwell (see Teague et al. 2014), we anticipated that the 2014 mine fire, located adjacent to the town, might have led to substantial impacts on the economy and population. We know that health and wellbeing alongside socio-economic status impact the resilience of a population (100 Resilient Cities 2019; Latrobe City Council 2017; Wisner et al. 2004). However, our analysis of population and economic data over time suggests, aside from serious impacts during privatisation in the 1990s when growth rates were negative, the economy and population of the town have never substantively collapsed in the face of such shocks and stressors. Even subsequent to the fire and after the closure of the power station in 2017, the population size did not drop significantly when we might have anticipated otherwise.

Table 7.4 Top four industries and trends in jobs for Morwell and beyond

Top 4 industries by employment size	Rank	2006		2011		2016	
		Industry	Employment	Industry	Employment	Industry	Employment
	Morwell						
	1	Retail trade	682	Retail trade	668	Health care and social assistance	630
	2	Health care and social assistance	534	Health care and social assistance	612	Retail trade	626
	3	Manufacturing	533	Manufacturing	487	Accommodation and food services	382
	4	Construction	422	Construction	400	Construction	327
	Latrobe valley						
	1	Retail trade	3,923	Health care and social assistance	3,890	Health care and social assistance	4,279
	2	Health care and social assistance	3,173	Retail trade	3,794	Retail trade	3,427
	3	Manufacturing	3,082	Construction	2,972	Construction	2,670
	4	Construction	2,800	Manufacturing	2,910	Public administration and safety	2,361
	Victoria						
	1	Manufacturing	287,108	Health care and social assistance	292,417	Health care and social assistance	341,999
	2	Retail trade	263,447	Retail trade	273,715	Retail trade	279,636

(continued)

Table 7.4 (continued)

Top 4 industries by employment size	Rank	2006		2011		2016	
		Industry	Employment	Industry	Employment	Industry	Employment
	3	Health care and social assistance	236,552	Manufacturing	271,051	Education and training	236,276
	4	Education and training	174,423	Construction	210,972	Construction	228,149

Table constructed by Bird and Taylor using ABS, 2003.) 2016 Census of Population and Housing Time Series Profile (Morwell SA2, Latrobe Valley SA3 and Victoria State)

Table 7.5 Indicators for industry diversity and gendered employment outcomes

Indicator	2006 (%)	2011 (%)	2016 (%)
Morwell			
C4 index	47	46	42
Employment to population ratio—males	51	48	43
Employment to population ratio—females	36	37	36
Latrobe valley			
C4 index	47	46	45
Employment to population ratio—males	58	57	53
Employment to population ratio—females	45	48	46
Victoria			
C4 index	43	42	42
Employment to population ratio—males	63	64	61
Employment to population ratio—females	51	53	52

Table calculated by Bird and Taylor from ABS Table Builder data

Part of the explanation for lower than anticipated impacts on the economy and population after these shocks and stressors may lay in the historical diversity of industrial units comprising the mine and power station operations, as well as the employment profile it helped generate in the town through successive periods of renewal and growth (see Fig. 7.3). While we might have anticipated mining and power generation to be dominant industries in the town, particularly when the mine and power station expanded, the employment profile instead suggests retail and health or aged care to have been more prominent. Nevertheless, employment in manufacturing, likely the response option which many power plant employees selected in the Census, has suffered tremendously over time; but this is common across Australia and especially in the State of Victoria (see Table 7.3). However, declines in construction employment leading up to the closure may signify pending economic issues for the town in the form of downstream impacts on the whole economy and possibly on population size. Morwell is by no means alone in this respect with the construction industry historically subject to high variations across rural and remote Australia through boom and bust cycles.

The employment profile in Morwell may bring into question the voracity of the application of the label 'single-industry town', and this is further supported by industrial concentration measures derived for analysis in this chapter. Jobs in Morwell have never been significantly more concentrated in the top four industries than for the surrounding region and over time have become less concentrated with the C4 index (see Table 7.4) in 2016 the same as for the State of Victoria as a whole (42%). However, in spite of what appears to be stoicism in the face of major shocks, there is evidence that the fire and (what was then) the impending closure of the mine and power station worsening gendered employment outcomes with a lower proportion of working-age males employed in 2016 (see Table 7.4). This might also explain the increased unemployment rate and reduced participation rate in the five years to 2016 when the State experienced the opposite with a buoyant economy and rapid population growth driven primarily by the city of Melbourne. Migration data supports this supposition with a noticeable net loss of population aged 35–54 years to other parts of Australia during 2007 to 2016 (see Fig. 7.6), and particularly for males (see Fig. 7.5).

Nevertheless, our demographic and socio-economic data suggests a certain stoicism and resilience inherent within the Morwell population. If we reconsider our working definition of resilience, it is pertinent to question how Morwell is adapting to the changed realities and capitalising on any new possibilities offered?

Walker et al. (2016) noted a certain level of stoicism and resilience in the community post-mine fire disaster with people relying on lived experience to deal with the situation at hand and their social connections for support. It has already been noted that Morwell has lived experience resulting from a long history of disaster familiarity (previous major fires in 1977, 2006 and 2008). In terms of social connections for support, however, we now consider the community's capacities in relation to the three different types of social capital—bonding social capital, bridging social capital and linking social capital.

If we look at the various community-led action groups established to secure a better future for Morwell and the Latrobe City area, it is evident that bonding and bridging social capital were prevalent. Most notably, strong and loose connections exist among the generations of families that have worked together in the mining industry and through attachment to place—i.e. people are connected because they are proud of the region where they live. Work place connections and attachment to place connections are clearly articulated by Doig (2019). Through these networks, within which a certain level of social trust evolves, the population collectively banded together around various community-led initiatives to produce a mutual benefit—a better future for their community. One such initiative was the formation of the community action group, Voices of the Valley in response to the Hazelwood Mine Fire. Voices of the Valley was established, through community fundraising, to ensure public concerns regarding health issues resulting from the mine fire were heard by government and the broader Victorian population. Another key network of collective action emanated from Morwell Neighbourhood House and together, they were instrumental in petitioning the government to take further action (Doig 2015, 2019). This action resulted in the establishment of the Hazelwood Health Study and reopening of the Hazelwood Mine Fire Inquiry. Regarding linking social capital, however, the Hazelwood Mine Fire Inquiry reports clearly articulate the fact that network connections and social trust between residents and the government was missing.

The Hazelwood Mine Fire Inquiry was originally completed in 2014 with 12 recommendations. In the second iteration, however, the Board of Inquiry made 246 recommendations for government, which included many bold initiatives aimed at re-establishing the lost social trust between residents and government officials, while at the same time, improving health and wellbeing across Latrobe City. Three of the key items included the designation of Latrobe City as a Health Innovation Zone, the establishment of the Latrobe Health Assembly and appointment of the Latrobe Health Advocate (Teague et al. 2016), which have been achieved.

The purpose of the Latrobe Health Innovation Zone, within which the Latrobe Health Assembly and Advocate sit, is to give voice to community aspirations in the planning and delivering of better health and wellbeing outcomes. With a membership of about 45 members, the majority of whom are Latrobe City residents, the Latrobe Health Assembly is a mechanism for increased community engagement leading to health improvement and integration of services. Based on community feedback and concerns, the Latrobe Health Advocate is providing independent and direct advice to the Minister for Health on system and policy issues affecting public health and wellbeing.

While it is too early to determine whether these initiatives are improving health and wellbeing and enhancing social trust between residents and the government (at the time of writing, they had been operating as a collective for 1 year), we are witnessing some positive demographic outcomes with Morwell's population growth rates tracking close to zero. We consider this a positive outcome in demographic terms in light of the closure of the main employer in the town. One key factor here is the ability of former Hazelwood employees to source work elsewhere in the region while remaining as residents in the town. Much of the transitioning of workers has been achieved by the Latrobe Valley Authority, which was established by the Victorian Government in November 2016 on the back of Engie's announcement to close the Hazelwood Power Station and open-cut coal mine.

The Latrobe Valley Authority is charged with the Worker Transition Service (providing upskilling, training and support to all ex-employees of Hazelwood including contractors, supply chain employees and their families) and the Back to Work Scheme (providing support to employers for hiring and training unemployed Latrobe Valley residents). This is part of the Victorian Government's 'Economic Growth Zone' which is injecting $266 million into the Latrobe Valley for the creation of local jobs and businesses.

The town's relative proximity to Melbourne, a fast-growing and large city, may also position it as a future commuter town where affordable housing and lifestyle amenity secure its demand as a place to live and for businesses and the economy to continue to shift focus to servicing an emerging commuter population. Nevertheless, the memories of the fire disaster are likely to linger for remaining residents in perpetuity with the presence of the open-cut coal pit both a reminder of the town's past events, as well as potential ongoing environmental issues.

While the data and analysis conducted here provides consistent evidence of the resilience of the Morwell population, in reality, it may be too early to discern the full impacts from the fire and the de-coupling of the town from the power plant. A better understanding will come from repeating the sorts of analysis undertaken and reported on here with 2021 Census data and other data sources. This will provide a sturdier assessment on whether this seemingly resilient town was able to withstand the major shock of the power plant closure. Similarly, future studies should consider a more robust analysis of the strengths inherent within this community, particularly in relation to social capital and other social indicators of resilience highlighted by Cutter (2016). This will be critical to identify whether or not the various forms of social capital are negatively impacted by the possible diversification of the population as new people move in in search of the new jobs and business opportunities being created.

It must also be noted, in relation to studying demographic indicators at these low levels of geography, there is some possibility that, for several reasons, they may not proximate actual 'on the ground' changes and trends in the community. Some of these issues are explained by Carson and colleagues in Chap. 5 and include data accuracy, changes in methods for data collection and ensuring changes are looked at using the most appropriate level or unit of geography (as far as possible).

In the face of the inevitable shift from fossil fuel to other sources, towns like Morwell were always going to experience a big shock at some point in time. Morwell, whose population and economy, although transitioning, has more than survived, may be a valuable case study for other towns undergoing such transitions from changing global resource consumption patterns. In reality, shock after shock may have helped 'prepare' the town to transition. The enforced transition with the closure of Hazelwood can be viewed as a litmus test for other rural and remote towns in Australia (and elsewhere) whose economic basis has been strongly tied to a main source (whether or not this is labelled 'single industry'). Climate change will no doubt lead to other towns facing similar challenges, testing their capacity to adapt and thrive.

References

100 Resilient Cities. 2019. What is Urban Resilience? Available: https://www.100resilientcities.org/resources/#section-2 [Accessed 21 July 2019].

ABC (2017). *Hazelwood power station closure: What does it mean for electricity bills, the environment and the Latrobe Valley?* Accessed 1 May 2019 at https://www.abc.net.au/news/2017-03-30/hazelwood-power-plant-shutdown-explained/8379756

ABS (2003). 1301.0 - Year Book Australia, 2003. Accessed 30 July 2019 at https://www.abs.gov.au/ausstats/abs@.nsf/featurearticlesbyCatalogue/1476D522EBE22464CA256CAE0015BAD4?OpenDocument

ABS (2014). 1381.0.55.001 - Research Paper: A Review of Selected Regional Industrial Diversity Indexes, 2011. Accessed 30 July 2019 at https://www.abs.gov.au/ausstats/abs@.nsf/0/4276628C76F84A10CA257DAF0018B599?Opendocument

ABS (2018). Regional Statistics, ASGS 2016, 2011–2018, Annual 2011 to 2018. Accessed 28 May 2019 at https://itt.abs.gov.au/itt/r.jsp?databyregion

ABS (2019). 3218.0 - Regional Population Growth, Australia, 2017–18. Accessed 29 May 2019 at https://www.abs.gov.au/AUSSTATS/abs@.nsf/Lookup/3218.0Main+Features12017-18?OpenDocument

Aldrich, D. P. & Meyer, M. A. 2015. Social Capital and Community Resilience. *American Behavioral Scientist,* 59, 254-269.

Alexander, D. E. 2013. Resilience and disaster risk reduction: an etymological journey. *Nat. Hazards Earth Syst. Sci.,* 13, 2707-2716.

Allen, S., Carroll, M., Berger, E., Maybery, D. & Campbell, T. 2019. Research Summary: The ongoing experiences of students following the Hazelwood mine fire. Available: https://www.monash.edu/__data/assets/pdf_file/0007/1766104/Research-Summary-Schools-Study-Round-2-Interviews.pdf [Accessed 2 July 2019].

Australian Government (2018). Australia's trade statistics at a glance. Accessed on 31 January 2019 at https://dfat.gov.au/trade/resources/trade-at-a-glance/Pages/top-goods-services.aspx

Bridger, J. C. & Luloff, A. E. 2001. Building the Sustainable Community: Is Social Capital the Answer? *Sociological Inquiry,* 71, 458-472.

Carson, D.A., & Carson, D.B. (2014). Mobilities and Path Dependence: Challenges for tourism and "attractive" industry development in a remote company town. *Scandinavian Journal of Hospitality & Tourism, 14*(4), 460-479.

Collin, N. 2017. *Business Reinvention And The Town That Changed Itself*. Available https://www.nigelcollin.com.au/blog/town-of-reinvention-and-change/ [Accessed 5 October 2019].

Cutter, S. L. 2016. The landscape of disaster resilience indicators in the USA. *Natural Hazards,* 80, 741-758.

Cutter, S. L., Ash, K. D. & Emrich, C. T. 2014. The geographies of community disaster resilience. *Global Environmental Change,* 29, 65-77.

Davison, M. 2015. Minerals Council of Australia, Victorian Division's submission to Terms of Reference 8–10 (Mine Rehabilitation) of the Board of Inquiry into the Hazelwood Mine Fire. Available: https://hazelwoodinquiry.vic.gov.au/wp-content/uploads/2015/09/Minerals-Council-of-Australia-Victorian-Division-Submission.pdf [Accessed 14 July 2019].

Doig, T. 2015. *The Coal Face,* Australia, Penguin.

Doig, T. 2019. *Hazelwood*, Penguin Random House Australia.

Duckworth, M. 2015. Community cohesion and resilience. Paper presented at the 'Resilient cities and communities: the new global imperative' forum, 1 December 2015. Municipal Association of Victoria.

Duffy, M. & Whyte, S. 2017. The Latrobe Valley: The politics of loss and hope in a region of transition. *Australasian Journal of Regional Studies,* 23, 421-446.

Engie. 2017. Hazelwood - The life of Hazelwood and its people 1964 - 2017. Available: https://engie.com.au/wp-content/uploads/flipbook.pdf [Accessed 10 July 2019].

Hazelwood Health Study 2017. Hazelwood Health Study Annual Report 3–16 November 2017. Monash University. Medicine, Nursing and Health Services.

Hazelwood Health Study. 2019. Research Summary: Adult Survey - Mine fire smoke exposure and health. Available: https://www.monash.edu/__data/assets/pdf_file/0006/1766094/20190123-Adult-Survey-Volume-2-Research-Summary.pdf [Accessed 2 July 2019].

Johnston, F., Melody, S., Venn, A., Wills, K. & Ford, J. 2019. Research Summary: The Latrobe ELF Study | Exposure to mine fire smoke and the risk of pregnancy-related health problems. Available: https://www.monash.edu/__data/assets/pdf_file/0006/1795830/Research-Summary-ELF-Exposure-to-mine-fire-smoke-and-the-risk-of-pregnancy-related-health-problems.pdf [Accessed 2 July 2019].

Jones, R., Lee, S., Maybery, D. & McFarlane, A. 2018. Experiences of a prolonged coal-mine fire. *Disaster Prevention and Management: An International Journal, 27*, 534-545.

Klein, R. J. T., Nicholls, R. J. & Thomalla, F. 2003. Resilience to natural hazards: How useful is this concept? *Global Environmental Change Part B: Environmental Hazards,* 5, 35-45.

Latrobe City Council. 2017. Living Well Latrobe Health and Wellbeing Plan. Available: https://www.latrobe.vic.gov.au/Our_Community/Living_Well_Latrobe/Living_Well_Latrobe_Health_and_Wellbeing_Plan [Accessed 21 July 2019].

Manyena, S. B. 2006. The concept of resilience revisited. *Disasters,* 30, 434-450.

Manyena, S. B., O'Brien, G., O'Keefe, P. & Rose, J. 2011. Disaster resilience: a bounce back or bounce forward ability? *Local Environment,* 16, 417-424.

Mitchell, C. J. & O'Neill, K. 2016. Tracing economic transition in the mine towns of northern Ontario: An application of the 'resource-dependency model'. The Canadian Geographer/Le Géographe canadien, 60, 91-106.

Morwell Advertiser. 1949. Development of Morwell Project. Progress to Date. Available: https://trove.nla.gov.au/newspaper/article/69133750 [Accessed 11 April 2019].

Norris, F. H., Stevens, S. P., Pfefferbaum, B., Wyche, K. F. & Pfefferbaum, R. L. 2008. Community Resilience as a Metaphor, Theory, Set of Capacities, and Strategy for Disaster Readiness. *American Journal of Community Psychology,* 41, 127-150.

Paton, D. 2006. Disaster Resilience: Building Capacity to Co-Exist with Natural Hazards and their Consequences. *In:* Paton, D. & Johnston, D. (eds.) *Disaster Resilience An Integrated Approach.* Illinois: Charles C Thomas Publisher Ltd.

Perry, R. W. 2005. Disasters, Definitions And Theory Construction. *In:* Perry, R. W. & Quarantelli, E. L. (eds.) *WHAT IS A DISASTER? New Answers to Old Questions.* United States of America: Xlibris Corporation.

Perry, R. W. & Quarantelli, E. L. 2005. *What is a disaster? New answers to old questions,* United States of America, Xlibris Corporation.

Putnam, R. D., Leonardi, R. & Nonetti, R. Y. 1993. *Making Democracy Work Civic Traditions in Modern Italy,* Princeton University Press.

Quarantelli, E. L. 1998. *What is a Disaster? Perspectives on the Question,* London, Routledge.

Teague, B. & Catford, J. 2016. Hazelwood Mine Fire Inquiry Report 2015/2016 Volume IV—Mine Rehabilitation. Melbourne: Victorian Government.

Teague, B., Catford, J. & Petering, S. 2014. Hazelwood Mine Fire Inquiry Report. Melbourne: Victorian Government.

Teague, B., Catford, J. & Roper, A. 2016. Hazelwood Mine Fire Inquiry Report 2015/2016 Volume III—Health Improvement. Melbourne: Victorian Goverment.

The Advertiser. 2013. *Battered industrial city of Burnie is looking to reinvent itself*. Available https://www.adelaidenow.com.au/battered-industrial-city-of-burnie-is-looking-to-reinvent-itself/news-story/7f63552fdeb28217261a314ea2595db6 [Accessed 6 October 2019].

Walker, J., Carroll, M. & Chisholm, M. 2016. Policy review of the impact of the Hazelwood mine fire on older people: Final Report—Version 1.0–30 November 2016 Monash University, Medicine, Nursing and Health Sciences: Hazelwood Health Study.

Weichselgartner, J. & Kelman, I. 2015. Geographies of resilience: Challenges and opportunities of a descriptive concept. *Progress in Human Geography,* 39, 249-267.

Whyte, S. 2017. 'They go into bat for me' Morwell Neighbourhood House, the Hazelwood Mine Fire and Recovery. Churchill, Australia: Centre of Research for Resilient Communities, Federation University.

Wisner, B., Blaikie, P., Cannon, T. & Davis, I. 2004. *At risk: natural hazards, people's vulnerability, and disasters,* New York, Routledge.

Yell, S., Duffy, M., Whyte, S., Walker, L., Carroll, M. & Walker, J. 2019. Research Summary: Community perceptions of the impact of the Hazelwood mine fire on community wellbeing, and of the effectiveness of communication during and after the fire. Available: https://www.monash.edu/__data/assets/pdf_file/0006/1766103/community-perceptions-of-the-impact.pdf [Accessed 2 July 2019].

Chapter 8
Migration as a Potential Heat Stress Adaptation Strategy in Australia

Kerstin K. Zander, Carmen Richerzhagen, and Stephen T. Garnett

Abstract As the climate changes, natural disasters are becoming more frequent and severe. Some disasters are sudden and briefly devastating. Research shows that, in response, many people emigrate temporarily but return when the danger is past. The effect of slow-onset disasters can be equally disruptive but the economic and social impacts can last much longer. In Australia, extreme heat and the rising frequency of heat waves is a slow-onset disaster even if individual periods of hot weather are brief. This chapter investigates the impact of increasing heat stress on the intention of people living in Australia to migrate to cooler places as an adaptation strategy using an online survey of 1344 people. About 73% felt stressed by increasing heat of which 11% expressed an intention to move to cooler places in response. The more affected people had been by the heat, the more likely they were to intend to move. Tasmania was a preferred destination (20% of those intending to move), although many people (38%) were unsure where they would go. As Australia becomes hotter, heat can be expected to play a greater role in people's mobility decisions. Knowing the source and destination of this flow of internal migrants will be critical to planning and policy-making.

Keywords Climate change · Extreme heat · Mobility · Online survey · Relocation · Planned behaviour

K. K. Zander (✉)
Northern Institute, Charles Darwin University, Darwin, Australia
e-mail: Kerstin.Zander@cdu.edu.au

C. Richerzhagen
The German Development Institute, Bonn, Germany
e-mail: Carmen.Richerzhagen@die-gdi.de

S. T. Garnett
Research Institute for the Environment and Livelihoods, Charles Darwin University, Darwin, Australia
e-mail: Stephen.Garnett@cdu.edu.au

D. Karácsonyi et al. (eds.), *The Demography of Disasters*,
https://doi.org/10.1007/978-3-030-49920-4_8

8.1 Introduction

Climate change is expected to increase the frequency and severity of many weather-related disasters, including floods, storms, droughts and heat waves. Heat waves[1] are the most dangerous natural hazard worldwide (Fang et al. 2015). The 2003 heat wave in Europe, for example, killed more than 70,000 people (Robine et al. 2008), particularly older people and those with illnesses (see Chap. 10 as well on 2009 south-eastern Australia heatwave). As climate change increases, the frequencies of heat waves (Perkins et al. 2012; Perkins and Alexander 2013) mortality and morbidity rates will also rise from associated disasters such as bushfires and droughts. Temperatures are also increasing in general due to climate change (Coumou et al. 2013), and this increased exposure to heat, even if not on consecutive days, are also of public health concern. Being exposed to too much heat can cause heat stress and associated illness with symptoms, ranging from being uncomfortable to hot flushes, headaches, fatigue, dizziness, nausea, cramps and, eventually, heat strokes and collapse (Parsons 2003). Any of these symptoms can compromise people's wellbeing and productivity (Zander et al. 2015).

Short-term in-situ adaptation to heat includes hydration (drinking), cooling (including air-conditioning) and resting. In the long-term, however, adaptation plans to heat and heat waves must include alterations to infrastructure such as insulation, ventilation, air-conditioning (Barnett et al. 2015; Hatvani-Kovacs et al. 2016) as well as landscape design. The latter is likely to be particularly important in cities suffering from higher temperatures as a consequence of human activity, known as the heat island effect (Solecki et al. 2005; Chen et al. 2014).

Migration is the most extreme form of adaptation to heat stress. Migration as an adaptation to climate change has been recognised since the first report of the Intergovernmental Panel on Climate Change (IPCC) as well as debated extensively in academic literature focused on developing countries (Hugo 2011; de Sherbinin et al. 2011). Migration as a response can be seen as either a failure of in-situ adaptation or as part of a portfolio of adaptation measures (Bardsley and Hugo 2010; de Sherbinin et al. 2011) which includes taking advantage of opportunities offered by cooler places (Tacoli 2009; Scheffran et al. 2012). Those who stay may be well-adapted and resilient. However, of particular concern to developing countries are those people who lack the resources to move away, that is that they are *trapped* (Black et al. 2011b). However, even successful in-situ adaptation to increasing temperatures might not prevent people from moving away (Sakdapolrak et al. 2014) because migration is rarely caused by a single factor alone but by many factors that collectively contribute to an individual's decision to move (McLeman and Smit 2006; Black et al. 2011a). Interaction effects and the wider range of climatic effects on migration are not yet well understood (Carleton and Hsiang 2016).

---

[1] Several definitions of the term heat wave exist within the international meteorological community. The IPCC defines a heat wave as a 'period of abnormally and uncomfortably hot weather' (IPCC 2014).

In developing countries, the fact that people move, or are pushed, away because of climate change impacts is increasingly the subject of research (Bardsley and Hugo 2010; Massey et al. 2010; Gray and Mueller 2012a, 2012b; Warner and Afifi 2014; Ocello et al. 2015), with a few recent examples concentrating on heat stress (Mueller et al. 2014) and increasing temperatures (Bohra-Mishra et al. 2014; Gray and Wise 2016). There is much less research in this field from developed countries and is mostly limited to Indigenous people (Zander et al. 2013; King et al. 2014). For most people in developing countries, responding to climate change by migrating is impeded by shortages of economic resources and human capital (Black et al. 2011a) or inhibited for cultural reasons such as a strong attachment to traditional lands (Mortreux and Barnett 2009; Adger et al. 2013). Most people in developed economies such as Australia, in contrast, have few constraints on decisions about when and where to relocate to improve economic prosperity and wellbeing. As heat increases, it is therefore increasingly likely that people in developed economies will move to avoid it, regardless of their stage in life.

This chapter extends the work by Zander et al. (2016) on this topic using the same dataset: a cross-sectoral sample of people living in Australia between 18 and 65 years. This chapter explores peoples' intention to move from their current place of residence to cooler places because of heat stress. It investigates determinants of the intention to move, the timeframe of moving and the geographical frame of their mobility.

Understanding how people adapt to heat is particularly important to Australia, a generally hot continent (see also Chap. 10). Australia's climate has already warmed by 0.9 °C since 1910 and temperatures are projected to rise by 0.6–1.5 °C by 2030 compared with the climate of 1980–1999 (BoM 2014). As has happened globally (WMO and WHO 2015), both the duration and frequency of heatwaves had increased over the period 1971–2008 with the hottest days during heatwaves across most of Australia becoming even hotter (Perkins and Alexander 2013) and the chances of standalone hot days also being higher (Min et al. 2013; Perkins and Alexander 2013).

8.2 Data and Methods

8.2.1 Theory of Planned Behaviour

This chapter focuses on people's intentions to move away from their current place of residence voluntarily because of heat. This intention reflects a willingness to change place of residence (de Jong 1999). Although this is not the same as actual moving (Fishbein and Ajzen 1975; Lu 1998; van Dalen and Henkens 2008; de Groot et al. 2011), it is a moderate to strong indicator of future movements, based on Ajzen's theory of planned behaviour (1991), that is particularly valued in the fields of human geography and psychology (Manski 1990; Sandu and de Jong 1996; van Dalen and Henkens 2008). Previous studies have shown a significant positive relationship between intentions and actual behaviour (van Dalen and Henkens 2008; Thissen et al. 2010).

8.2.2 *Data Collection and Sampling*

Data were collected through an online survey conducted in the last two weeks of May and the first two weeks of October 2014.[2] To carry out the survey, a research company (MyOpinions) we commissioned that has a continuously updated online panel of more than 300,000 verified respondents within Australia. Online surveys have many benefits over mail-out surveys and in-person interviews and are more cost-effective (Berrens et al. 2003; Dillman 2007; Fleming and Bowden 2009). Some studies have shown that results do not differ across survey modes (Lindhjem and Navrud 2011; Windle and Rolfe 2011). To avoid self-selection bias, not uncommon in online surveys, the topic of the survey was not made known when members of the panel were invited to participate. Respondents were paid $2 upon completion of the survey, which took between 13 and 15 min of their time.

In total, 9406 people from the MyOpinions panel were sampled (4913 in the first wave and 4493 in the second wave). The overall response rate was 20.5%, including a 3.3% drop-out rate. There were 1925 responses received in total, consisting of 847 and 1078 from each wave, respectively.

8.2.3 *Questionnaire*

The questionnaire comprised of three parts: (1) questions on general mobility including frequencies and reasons for past movements, (2) questions on intentions for future movements and whether heat would influence this movement, and questions about the timeframe of the movements and the intended destination, and (3) questions on respondents' socio-demographic background, attitudes and climate change beliefs.

The first question within the second section was a general question about whether or not people had been feeling heat stressed in the previous 12 months. Those who denied this were assumed not to think about heat at all when deciding whether to move and were not included in the analysis on intentions to move because of heat.

---

[2]The surveys were staggered into two waves to reduce the chance of surveying during (or soon after) particularly hot periods, choosing late May and early October as being times when exceptional heat would be unlikely. The other reason for conducting the survey in two sessions was to ensure that to the demographics of both samples were representative of Australian society and that results could be replicated.

8.2.4 Data Analysis

The data were tested for different factors likely to affect respondents' probability of moving to cooler places because of heat using ANOVA. If the data required it, Tukey tests were then used for multiple comparisons of means.

8.3 Results and Discussion

Eighty-six of the 1925 responses obtained were largely incomplete and therefore omitted from further analysis. Out of the remaining 1839 respondents, 27% did not feel heat stressed in the year preceding the survey and were also not included in the analysis of movements because of heat stress. The final data set explored here, therefore, contains information of 1344 respondents.

8.3.1 Demographic Sample Characteristics

Slightly less than half of the respondents (48%) were female. The average age was 41 years (SD: 12.2), with a median of 41, which is slightly higher than the 37 median age at a national level (ABS 2015). One of the reasons for a higher median is that the survey was targeted to people over 18. Fifty-six per cent of respondents said they had children. Most respondents (72%) had tertiary education (Diploma: 35%, University: 37%) and most (96%) were in paid employment (Full-time: 57%; Part-time: 34%; Casual: 5%; Not in paid job: 4%). The average annual personal income was AUD 58,000 (SD: 76,000) with a median of AUD 50,000 which is very similar to the national median of working people between 18 and 65 (AUD 46,000; ABS 2012). In line with the national population distribution (ABS 2012), about 64% of the respondents were from the three most populated states (Victoria (VIC): 26%, New South Wales (NSW): 24% and Queensland (QLD): 16%) and proportionally fewer from the other states/territories (Western Australia (WA): 13%, South Australia (SA): 9%, Australian Capital Territory (ACT): 5%, Tasmania (TAS): 4%, Northern Territory (NT): 3%).

8.3.2 Past Movements and Their Reasons

About 18% of respondents were highly mobile, in that they usually moved once a year (3%) or every 2–3 years (15%) in the past. Almost half were moderately mobile (46%), moving once every five (20%) or ten years (26%) and about a third were relatively sedentary, having never moved either in their lives (11%) or within the last 15 years (25%).

Respondents' mobility was correlated with age and family status. Younger people were more likely to be highly mobile ($P < 0.001$) which is consistent with the life cycle theory of migration which postulates that young people are most mobile, usually as they search for education and employment (Polachek and Horvath 1977; Coulter and Scott 2015). The average age of highly mobile people was 33.8 years, that of moderately mobile people 40.8 years and that of people with low mobility 44.5 years (Fig. 8.1). Those people with low and medium mobility were significantly more likely ($P < 0.001$) to have children (58% and 61% respectively) than those who were highly mobile (43%). Again, this can be attributed to respondents' life cycles, stipulating that families with children are less mobile than singles or couples without children (Long 1972), particularly regarding long-distance movements (which is expected if one moves to escape the heat). The reasons behind this could be that the costs of moving increase as the number of persons living in a family rises and that the presence of additional members in the family means that more ties must be broken at the place of origin and established at the destination (Long 1972). Gender, income and work situation (workload and employee) did not have a significant impact on the degree of general mobility. The state from which people lived also had no significant impact on people's general mobility.

In line with migration/mobility theory (Clark and Withers 2007), employment was the most important reason for moving among respondents who had moved in the past (Fig. 8.2). Although the weather was the least important of the reasons for past movements, 41% of respondents listed it as having an important or very important influence in their decision to move in the past (this can include people moving to warmer as well as cooler climates).

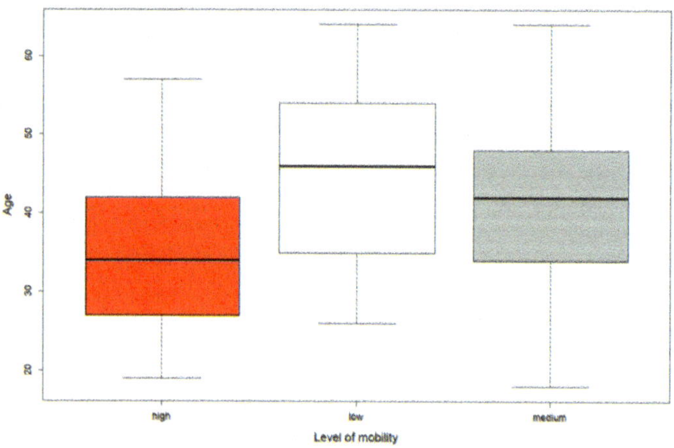

Fig. 8.1 Boxplot showing the significant relationship between age of respondents and their level of mobility, showing that highly mobile respondents were significantly younger than those whose mobility was low or medium and those with medium mobility younger than those with low mobility

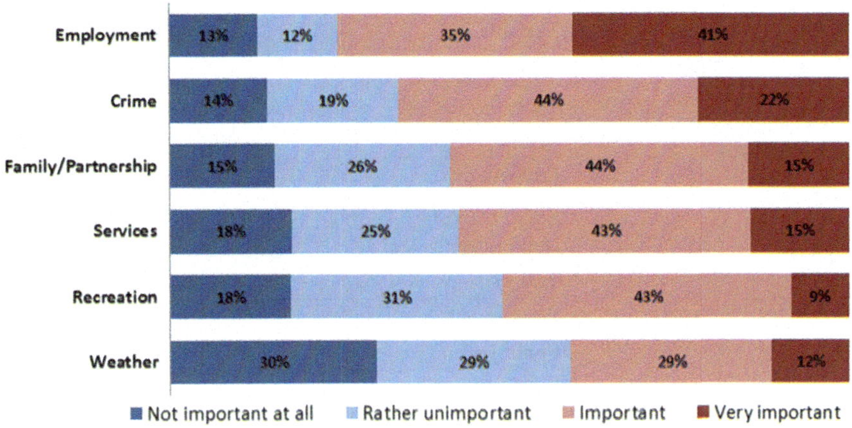

Fig. 8.2 The importance of different factors in past movements (% respondents, $N = 1839$)

8.3.3 Intention to Move Because of Heat

Almost 11% ($N = 133$) of respondents who said that they were stressed by heat intended to move because of this heat stress; the large remainder (89%; 1133) would not move for this reason. One result that stood out was that respondents living in the NT were about three times more likely to intend to move because of heat (36% versus the average of 11%; $P < 0.001$). This was not surprising given that most respondents from the NT live in Darwin, the capital, situated in the tropical Top End of Australia. Because of the high humidity, the perceived heat is particularly high for at least half the year (Goldi et al. 2015).

The intention to move was also affected by gender ($P < 0.005$), degree of perceived heat stress ($P < 0.001$) and general mobility ($P < 0.001$) (Table 8.1). A higher percentage of male respondents said they would move because of heat than female respondents (13% versus 8%; $P < 0.01$). Not surprising, the more heat stressed respondents said they felt, the more likely were they to intend to move because of heat stress with almost a third (28%) of those feeling very heat stressed intending to move.

Also not surprising was the result that highly mobile people (see previous section) were more likely to intend to move because of heat stress (21%). This is consistent with migration intention studies (de Jong et al. 1985). This can also mean that people see heat only as a factor in the mix of reasons why they would move, and they would probably move anyway after a certain time in the current location. On the other hand, it can mean that those who have never moved, are 'trapped' where they are, even if it becomes very hot, to the extent that it affects health. Often these people wish to stay because of tight connections (family, culture) to where they live, or because they have employment that they do not want to leave. In developing countries 'trapped' people do not have the resources to move away (Black et al. 2011b). In Australia, this might

Table 8.1 Percentage of respondents ($N = 1344$) intending to move away because of heat stress—differences by socio-economic characteristics

Variable	Percentage
Gender:	
Male	12.7[a]
Female	8.2[b]
Education:	
University degree	13.5[a]
Diploma/certificate	8.8[a]
Year 12	8.0[a]
Years 11 or below	9.3[a]
Employee:	
Self-employed	15.4[a]
Public sector	13.7[a]
Private sector	8.4[b]
Not in paid job	8.9[b]
Workload:	
Full-time	11.5[a]
Part-time	9.1[a]
Casual	10.2[a]
Not in paid job	8.9[a]
Location:	
NT	35.9[b]
QLD	13.1[a]
NSW	11.4[a]
WA	9.6[a]
SA	8.6[a]
VIC	7.9[a]
ACT	7.2[a]
TAS	5.2[a]
Heat stress:	
Rarely	5.7[a]
Sometimes	7.1[a]
Often	19.9[b]
Very Often	28.1[b]
Mobility:	
High	20.6[a]
Medium	8.9[b]

(continued)

Table 8.1 (continued)

Variable	Percentage
Low	7.0[b]

Note Differences in superscript letters signify statistically significant differences in the intention to move within the categories. They show, for example, that men are significantly more likely to intend to move than women (12.7% versus 8.2%) and that different workloads have no impact on the intention to move

also be true for the very poor, noting that 13.3% of the population in Australia live below the poverty line (ACOSS 2016). This survey did not confirm that income was of concern when deciding to migrate because of heat stress. However, even though the income distribution of respondents resembled that of the general population, the sampling method may have excluded the very poorest people.

Other core demographic factors such as age, income, having children and workload did not have significant impacts on the intention to move because of heat. The conclusion that age did not affect moving intention was unexpected. Previous studies have shown that movement intentions (mainly international migration) were positively correlated with age (de Jong et al. 1985; de Groot et al. 2011). This means that heat can influence intentions to move regardless of a person's stage in life. This can have consequences for service provision by altering normal movement patterns. For instance, if more middle-aged people in established careers move than would previously have been expected, this could leave a shortage of skills and labour in areas that experience more frequent uncomfortably hot weather.

8.3.4 *Moving When?*

Of those people who intend to move because of heat, most thought they might move in the distant future (33%), 17% in two to three years, 20% in about a year and 14% within the next three months (Table 8.2). A substantial proportion (13.5%) was in the process of moving at the time of the survey. Men stated significantly more often that they are already in the process of moving because of heat than women (19.3% versus 4%; $P = 0.01$). Women were significantly more likely to intend to move in the distant future (Table 8.2). Those of either gender who said they would move in the distant future were also older ($P < 0.01$) than those thinking of moving within the next three months (mean age 42.6 versus 34.3). Respondents with children were more likely to intend to move at later stages than those without children ($P < 0.05$).

Those already in the process of moving were more often stressed by heat than those intending to move later than within the next three months ($P < 0.005$). Those intending to move in the distant future exhibited lower general mobility than those who would move earlier ($P < 0.005$). Education, gender, income and location (their state) did not significantly explain the timeframe of potential heat related movements.

Table 8.2 Time horizon, by gender, age and family status, of peoples' intention to move away from where they currently live because of heat ($N = 133$)

	All (%)	Female (%)	Male (%)	Mean age (SD)	Having children (%)
Already in the process of moving	13.5	4.0^{a}	19.3^{b}	35.9^{ab} (9.8)	6.8^{a}
Within the next 3 months	15.8	10.0^{a}	19.3^{a}	34.3^{a} (9.8)	17.8^{b}
In about a year	20.3	22.0^{a}	19.3^{a}	40.5^{ab} (13.4)	20.5^{b}
In about 2–3 years	16.5	22.0^{a}	13.3^{a}	43.3^{ab} (10.7)	23.3^{b}
In the distant future	33.8	42.0^{a}	28.9^{b}	42.6^{b} (11.0)	31.5^{b}

Note Differences in superscript letters signify statistically significant differences in the intention to move across gender, age and family status

Those intending to move in the short-term (within a year, categories grouped) were significantly younger ($P < 0.005$), on average 37.2 years, than those planning movement in the long-term (42.8 years). This is again the highly transient population of young people. Of all men intending to move, more than half (58%) would do so in the short-term, while only 36% of those women intending to move would do so in the short-term ($P < 0.01$).

8.3.5 *Moving from Where to Where?*

More than a third (38%) of those wanting to move because of heat stress did not know where they would move to. Most people (91%) would move within Australia with the remaining 9% intending to move overseas. Significantly, more women intended to move overseas than men (16% versus 5%; $P < 0.05$).

Only 9% would move within their state of origin, most (91%) would change states. Almost 20% would move to TAS (Table 8.3), particularly from NSW (31%) and VIC (19%). TAS was the only state that would receive more people than it would lose, which was expected given its cold climate.

Nobody would move to the NT or SA (Table 8.3), with a particularly high proportion of those from the NT (36%) intending to move because of heat stress (Table 8.1); this proportion was only about 9% in SA. Besides the NT, many of the respondents intending to move were from QLD (ratio: origin (%)/sample distribution % = 1.24; Table 8.3) and NSW (ratio = 1.06). This was not surprising, given that parts of the NT and QLD lie in the tropical humid north, and given that a large proportion of respondents from the NT stated an intention to move. It was surprising, however, that so few people from SA intended to move because of heat, given that Adelaide and surroundings saw record temperatures and increasing frequencies of heat waves in the last few years (Steffen et al. 2014) with many people suffering (Xiang et al. 2014; Hatvani-Kovacs et al. 2016; Zander et al. 2017).

Table 8.3 Origin and destination of respondents intending to move because of heat stress (%; $N = 133$)

Location	Destination	Origin	Origin (%)/sample distribution (%)	Sample distribution ($N = 1344$)
Tasmania	19.5	2.3	0.51	4.5
New South Wales	9.8	24.8	1.06	23.5
Victoria	9.0	19.5	0.76	25.6
Queensland	8.3	19.5	1.24	15.7
Western Australia	4.5	12.0	0.91	13.2
Australian Capital Territory	1.5	3.8	0.73	5.2
Northern Territory	0.0	10.5	3.28	3.2
South Australia	0.0	7.5	0.82	9.1
Overseas	9.0	NA	NA	NA
Unknown	38.3	NA	NA	NA

Those currently in the NT would move mainly to QLD (29%), WA (21%) or NSW (14%). Those originating in SA mostly did not know their potential destination (40%) or stated TAS (30%), QLD (20%) or VIC (10%) as destinations.

8.4 Conclusion

Our results provide one, out of several accounts, of the general population in a developed economy intending to relocate because of climatic heat stress. So far most research on migration as a response to climate change is from developing countries, or, if in developed countries, then in the context of Indigenous people. Because people in Australia are free to move if financial resources permit, this research applied Ajzen's theory of planned behaviour to investigate peoples' intention to move away to cooler places. Results of an Australian-wide online survey with about 1900 respondents showed that 11% of those respondents who said that they were stressed by heat intended to move away because they felt heat stressed in their current place of residence. As expected, people who have been highly mobile in the past were more likely to intend to move because of heat, while age did not affect the intention to move because of heat stress. Men were also more likely to intend to move because of heat stress. Most heat-stressed men intended to move in within the next three months while most women (42% of those intending to move) would do so in the distant future. Younger respondents are more likely to move in a short-time frame, many of those already in the process of moving. People living in the NT, notably from the tropical (hot and humid) Top End of Australia, were three times more likely to move because of heat stress. Although slightly more than a third of respondents intending to move did not know where to, among those who knew, the preferred destination was Tasmania which has one of the lowest mean temperatures in Australia. Many people (38%) were unsure about a potential destination and Tasmania was a preferred destination for those who were not.

References

ABS (Australian Bureau of Statistics) (2012) Census of population and housing - 2011. Data generated using ABS Table Builder. Commonwealth of Australia, Canberra.

ABS (Australian Bureau of Statistics) (2015). 3101.0 - Australian Demographic Statistics, Jun 2015. Commonwealth of Australia, Canberra. https://www.abs.gov.au/ausstats/abs@.nsf/featurearticlesbyCatalogue/7A40A407211F35F4CA257A2200120EAA?OpenDocument [accessed 5 December 2016].

ACOSS (Australian Council of Social Service) (2016) Poverty in Australia 2016. Social Policy Research Centre, University of New South Wales, Sydney. https://www.acoss.org.au/wp-content/uploads/2016/10/Poverty-in-Australia-2016.pdf [accessed 5 December 2016].

Adger WN, Barnett J, Brown K, Marshall N, O'Brien K (2013) Cultural dimensions of climate change impacts and adaptation. *Nature Climate Change* 3: 112–117.

Ajzen I (1991) The theory of planned behaviour. *Organizational* Behavior *and Human Decision Processes* 50:179–211.

Bardsley DK, Hugo GJ (2010) Migration and climate change: examining thresholds of change to guide effective adaptation decision-making. *Population and Environment* 32: 238–262.

Barnett G, Beaty RM, Meyers J, Chen D, McFallan S (2015) Pathways for adaptation of low-income housing to extreme heat. In: Palutikof JP, Boulter SL, Barnett J, Rissik D (eds) *Applied studies in climate adaptation*. Wiley-Blackwell, Oxford, pp. 364–371.

Berrens R, Bohara A, Jenkins-Smith H, Silva C, Weimer D (2003) The advent of internet surveys for political research: A comparison of telephone and internet samples. *Political Analysis* 11: 1–22.

Black R, Adger WN, Arnell NW, Dercon S, Geddes A, Thomas D (2011a) The effect of environmental change on human migration. *Global Environmental Change* 21: S3–S11.

Black R, Bennett SRG, Thomas SM, Beddington JR (2011b) Migration as adaptation. *Nature* 478: 447–449.

Bohra-Mishra P, Oppenheimer M, Hsiang SM (2014) Nonlinear permanent migration response to climatic variations but minimal response to disasters. *Proceedings of the National Academy of Sciences of the United States of America* 111: 9780–9785.

Bureau of Meteorology (BoM) (2014) *State of the climate 2014*. BoM and CSIRO, Canberra.

Carleton TA, Hsiang SM (2016) Social and economic impacts of climate. *Science* 353: aad9837.

Chen D, Wang X, Thatcher M, Barnett G, Kachenko A, Prince R (2014) Urban vegetation for reducing heat related mortality. *Environmental Pollution* 192: 275–284.

Clark WAV, Withers SD (2007) Family migration and mobility sequences in the United States: spatial mobility in the context of the life course. *Demographic Research* 17: 591–622.

Coumou D, Robinson A, Rahmstorf S (2013) Global increase in record-breaking monthly mean temperatures. *Climatic Change* 118: 771–782.

Coulter R, Scott J (2015) What motivates residential mobility? Re-examining self-reported reasons for desiring and making residential moves. *Population Space and Place* 21: 354–371.

de Groot C, Mulder CH, Das M, Manting D (2011) Life events and the gap between intention to move and actual mobility. *Environmental Planning A* 43: 48–66.

de Jong G, Root BD, Gardner RW, Fawcett JT, Abad RG (1985) Migration intentions and behaviour: decision making in a rural Philippine province. *Population and Environment* 8: 41–62.

de Jong GF (1999) Choice process in migration behaviour. In: Pandit K, Withers S (eds) *Migration and restructuring in the United Sates*. Rowman and Littlefield, Lanham, pp. 273–293.

de Sherbinin A, Castro M, Gemenne F, Cernea MM, Adamo S, Fearnside PM, et al. (2011) Preparing for resettlement associated with climate change. *Science* 28: 456–457.
Dillman DA (2007) Mail and internet surveys: the tailored design method - 2007 update with new internet, visual, and mixed-mode guide. Wiley, New York.
Fang J, Li M, Shi P (2015) Mapping heat wave risk of the world. In: Shi P, Kasperson R (Eds.), *World atlas of natural disaster risk.* Springer, Heidelberg, pp. 69–102.
Fishbein M, Ajzen I (1975) *Belief, attitude, intention, and behavior: an introduction to theory and research.* Addison-Wesley, Reading.
Fleming C, Bowden M (2009) Web-based surveys as an alternative to traditional mail methods. *Journal of Environmental Management* 90: 284–292.
Goldi J, Sherwood SC, Green D, Alexander L (2015) Temperature and humidity effects on hospital morbidity in Darwin, Australia. *Annals of Global Health* 81: 333–341.
Gray C, Mueller V (2012a) Drought and population mobility in rural Ethiopia. *World Development* 40: 134–145.
Gray CL, Mueller V (2012b) Natural disasters and population mobility in Bangladesh. *Proceedings of the National Academy of Sciences of the United States of America* 109: 6000–6005.
Gray C, Wise E (2016) Country-specific effects of climate variability on human migration. *Climatic Change* 135: 555–568.
Hatvani-Kovacs G, Belusko M, Skinner N, Pockett J, Boland J (2016) Heat stress risk and resilience in the urban environment. *Sustainable Cities and Society* 26: 278–288.
Hugo G (2011) Future demographic change and its interactions with migration and climate change. *Global Environmental Change* 21: S21–S33.
IPCC (2014) Annex II: Glossary. In: Mach KJ, Planton S, Stechow C von (Eds.), *Climate Change 2014: Synthesis Report. Contribution of Working Groups I, II and III to the Fifth Assessment Report of the Intergovernmental Panel on Climate Change.* IPCC, Geneva, pp. 117–130.
King D, Bird D, Haynes K, Boon H, Cottrell A, Millar J, Okada T, Box P, Keogh D, Thomas M (2014) Voluntary relocation as an adaptation strategy to extreme weather events. *International Journal of Disaster Risk Reduction* 8: 83–90.
Lindhjem H, Navrud S (2011) Are Internet surveys an alternative to face-to-face interviews in contingent valuation? *Ecological Economics* 70: 1628–1637.
Long LH (1972) The influence of number and ages of children on residential mobility. *Demography* 9: 371–382.
Lu M (1998) Analyzing migration decision making: relationships between residential satisfaction, mobility intentions, and moving behavior. *Environmental Planning* A 30: 1473–1495.
Manski CF (1990) The use of intentions data to predict behavior: a best-case analysis. *Journal of the American Statistical* Association 85: 934–940.
Massey DS, Axinn WG, Ghimire DJ (2010) Environmental change and out-migration: evidence from Nepal. *Population and Environment* 32: 109–136.
McLeman R, Smit B (2006) Migration as an adaptation to climate change. *Climatic Change* 76: 31–53.
Min S-K, Cai W, Whetton P (2013) Influence of climate variability on seasonal extremes over Australia. *Journal of Geophysical Research: Atmospheres* 118: 643–654.
Mortreux C, Barnett J (2009) Climate change, migration and adaptation in Funafuti, Tuvalu. *Global Environmental Change* 19: 105–112.
Mueller V, Gray C, Kosec K (2014) Heat stress increases long-term human migration in rural Pakistan. *Nature Climate Change* 4: 182–185.
Ocello C, Petrucci A, Testa M, Vignoli D (2015) Environmental aspects of internal migration in Tanzania. *Population and Environment* 37: 99–108.
Parsons K (2003) *Human thermal environment. the effects of hot, moderate and cold temperatures on human health, comfort and performance.* 2nd Ed. CRC Press, New York.
Perkins SE, Alexander LV, Nairn JR (2012) Increasing frequency, intensity and duration of observed global heatwaves and warm spells. *Geophysical Research Letters* 39: L20714.

Perkins S, Alexander L (2013) On the measurement of heat waves. *Journal of Climate* 26: 4500–4517.

Polachek S, Horvath F (1977) A life cycle approach to migration: analysis of the perspicacious peregrinator. In: Ehrenberg R (Ed), Research in labor economics. Greenwich, CT: JAI Press.

Robine J-M, Cheung SLK, Le Roy S, Van Oyen H, Griffiths C, Michel J-P, et al. (2008) Death toll exceeded 70,000 in Europe during the summer of 2003. *Comptes Rendus Biologies* 331: 171–178.

Sakdapolrak P, Promburom P, Reif A (2014) Why successful in-situ adaptation with environmental stress does not prevent people from migrating? Empirical evidence from Northern Thailand. *Climate and Development* 6: 38–45.

Sandu D, de Jong GF (1996) Migration in market and democracy transition: migration intentions and behaviour in Romania. *Population Research and Policy Review* 15: 437–457.

Scheffran J, Marmer E, Sow P (2012) Migration as a contribution to resilience and innovation in climate adaptation: Social networks and co-development in Northwest Africa. *Applied Geography* 33: 119–127.

Solecki WD, Rosenzweig C, Parshall L, Pope G, Clark M, Cox J, Wiencke M (2005) Mitigation of the heat island effect in urban New Jersey. *Environmental Hazards* 6: 39–49..

Steffen W, Hughes L, Pearce A (2014) Heatwaves: hotter, longer, more often. Public Climate Council of Australia Limited. https://www.climatecouncil.org.au/uploads/7be174fe8c32ee1f36 32d44e2cef501a.pdf [accessed 5 December 2016].

Tacoli C (2009) Crisis or adaptation? Migration and climate change in a context of high mobility. *Environment and Urbanization* 21: 513–525.

Thissen F, Fortuijn JD, Strijker D, Haartsen T (2010) Migration intentions of rural youth in the Westhoek, Flanders, Belgium and the Veenkoloniën, The Netherlands. *Journal of Rural Studies* 26: 428–436.

van Dalen HP, Henkens K (2008) Emigration intentions: mere words or true plans? explaining international migration intentions and behavior. *CentER Discussion Paper* 2008–60, Center for Economic Research, Tilburg University.

Warner K, Afifi T (2014) Where the rain falls: evidence from 8 countries on how vulnerable households use migration to manage the risk of rainfall variability and food insecurity. *Climate and Development* 6: 1–17.

Windle J, Rolfe J (2011) Comparing responses from internet and paper-based collection methods in more complex stated preference environmental valuation surveys. Economic Analysis and Policy 41: 83–97.

WOE, WHO (2015) Heatwaves and health: guidance on warning-system development. World Meteorological Organization and World Health Organization. https://www.who.int/globalchange/publications/heatwaves-health-guidance/en [accessed 5 December 2016].

Xiang J, Bi P, Pisaniello D, Hansen A (2014) The impact of heatwaves on workers' health and safety in Adelaide, South Australia. *Environmental Research* 133: 90–95.

Zander KK, Petheram L, Garnett ST (2013) Stay or leave? - Potential climate change adaptation strategies among Aboriginal people in coastal communities in northern Australia. *Natural Hazards* 67: 591–609.

Zander KK, Botzen WJW, Oppermann E, Kjellstrom T, Garnett ST (2015) Heat stress causes substantial labour productivity loss in Australia. *Nature Climate Change* 5: 647–651.

Zander KK, Surjan A, Garnett ST (2016) Exploring the effect of heat on stated intentions to move. *Climatic Change* 138: 297–308.

Zander KK, Moss SA, Garnett ST (2017) Drivers of heat stress susceptibility in the Australian labour force. *Environmental Research* 152: 272–279.

Chapter 9
Designing Resilient Cities that Work for Women, Too

Jessica L. Barnes

Abstract Urban landscapes can—and do—influence multiple aspects of our lives, including our overall quality of life and disaster resilience. Research has confirmed that some populations experience negative outcomes in disasters at least partially attributed to poorly designed urban environments; and women's and girls' resilience in particular can be impacted by their experience of the urban landscape. In response, urban designers have an opportunity and an obligation to incorporate gender-sensitive design approaches in all of their projects to ensure the whole community has access to the benefits of urban landscapes. This chapter examines current evidence and strategies for successful urban design that supports resilience in women and the cities they occupy.

Keywords Gender mainstreaming · Urban design · Landscape architecture · Inclusive design · Resilient cities

9.1 Introduction

Urban landscapes can—and do—influence multiple aspects of our lives, including our overall quality of life (Rondeau et al. 2005; Urban Development Vienna 2013). This observation is what draws me to landscape architecture and urban design—the potential to have large-scale, positive impacts on the people who create our communities and cities. It is only recently, though, that urban designers are starting to understand that these landscapes do not affect everyone equally. Indeed, researchers have confirmed that social, economic, and political factors influence our perceptions of and interactions with the spaces we occupy (Garcia-Ramon et al. 2004), and in some instances, these varied experiences can impact individuals' and communities' disaster resilience (Tidball and Krasny 2013).

I believe that urban designers have an obligation to reduce disaster risk and support resilience in their communities (see also Chap. 12). This is not only because the

J. L. Barnes (✉)
Disaster Resilience Leadership Academy, Tulane University, New Orleans, LA, USA
e-mail: jessi@jessibarnes.com

D. Karácsonyi et al. (eds.), *The Demography of Disasters*,
https://doi.org/10.1007/978-3-030-49920-4_9

incidence of disasters is increasing (UNDRR 2019), but also because, as Wisner et al. (2003) put it: "… disasters can be perceived within the broader patterns of society, and … analyzing them in this way may provide a much more fruitful way of building policies that can help to reduce disasters and mitigate hazards, while at the same time improving living standards and opportunities more generally" (p. 4). In other words, disaster resilience and quality of life are interlinked, and improving one will have a positive impact on the other. This—to me—is the essence of what it means to be an urban designer: improving the quality of life for those in the communities we serve by taking a multifaceted approach to community development that recognizes the role that economic, social, and gender inequalities have in creating truly resilient communities.

When considering disaster resilience and the urban landscape, much has been said about the importance of engineering, infrastructure, and natural systems in protecting communities from physical damage; however, there is less discussion about the importance of connecting the urban landscape to the economic, social, institutional, and community capacities that also make up disaster resilience (Cutter et al. 2010; Tidball and Krasny 2013). Confining the field of urban design to infrastructure and natural systems misses opportunities to build capacities in these interrelated areas.

To expand and build upon the scope of urban design's contribution to disaster resilience, I suggest that designers focus on a near-universally marginalized group of people who routinely experience substandard outcomes in disasters and who account for half of the global population: women and girls. This chapter connects women's disaster resilience to their experiences of and participation in the built environment. To my surprise, while the field of gender and disaster has been well established for some decades (Ashraf and Azad 2015; Enarson and Chakrabarti 2009; Enarson and Morrow 1998a), few studies have specifically examined the intersection between gender, disaster, and the built environment. Likewise, researchers have long-understood that our built environments have various and measurable impacts on our health and well-being (Ekkel and de Vries 2017; Markevych et al. 2017; WHO Regional Office for Europe 2016), but few of these studies have gender dis-aggregated their data or connected it to resilience. With this in mind, I'd like to spark a conversation between researchers, design professionals, and women and their communities to consider the myriad of potential benefits of combining efforts. As I will demonstrate, bringing these stakeholders together to design and build our cities has the potential to proactively support resilience among over half of our world's human population and will likely have positive outcomes for the other half as well.

Why focus on women's disaster resilience? As Wisner et al. (2003) observe in *At Risk: Natural Hazards, People's Vulnerability, and Disasters*, "gender is a pervasive division affecting all societies, and it channels access to social and economic resources away from women and towards men" (p. 48). Enarson and Chakrabarti (2009) assert that gender must be a mandatory, critical dimension of all initiatives in order to create more sustainable societies. To understand how urban landscapes affect communities differently, city officials must identify which stakeholders have the most limited voices when it comes to making decisions about the built environment. Since men overwhelmingly dominate the field of urban planning and design

(Rustin 2014; The American Institute of Architects 2015), women represent a prime target demographic for this purpose. In short, our urban landscapes have a male bias—a bias that must be addressed and corrected to improve a community's overall disaster resilience.

In this chapter, I review both published articles and gray literature that document gender-specific outcomes for women and girls in disasters. I compare these studies to research concerning the measurable impacts that the built environment can have on individual and community capacities while looking for overlap and commonalities between the two fields. Finally, I speculate on how these connections between urban design and disaster resilience might influence the future of the design of the built environment and how the lessons gleaned so far might have a broader impact on resilience.

9.2 Definitions

Discussing women and girls as a homogeneous demographic group disguises the rich diversity among them. Women as individuals will all have unique and valuable skills, knowledge, and experiences that can contribute to building more resilient cities (Hankivsky 2005). While I note specific demographics of women when such information is available, generally I use the term with broad strokes. In this paper, "women" refers to people of all ages who were born female, people who identify as women, and/or people who perform labor traditionally assigned to women, like caregiving and domestic work. Each broad group of women may overlap, and some might include men. Perhaps because of this diversity, cities designed with and for women have the potential to benefit the community across many demographics (Micklow et al. 2015; Women in Cities International 2010).

When discussing women's disaster resilience, I'm referring to resources that support an individual's and a community's capacity to respond to and recover from a hazard. Capacity relates to five major areas of resilience: social resilience, economic resilience, infrastructure resilience, institutional resilience, and social capital (Cutter et al. 2010). Vulnerability, on the other hand, describes those barriers that limit an individual's or a community's ability to respond to disasters. Importantly, all communities have both vulnerability and resilience.

Gender mainstreaming, also called gender-based analysis, means normalizing decision making to consider how any planned decision, policy, or program might impact both women and men (Bellitto 2015). It is considered the most favored strategy for gender inclusion among urban design professionals around the world (Hankivsky 2005). Though often associated more strongly with women, gender mainstreaming emphasizes that women's *and* men's needs are equally important and must equally be accounted for when making decisions (Bellitto 2015).

As already discussed, considering gender alone has the potential to be too broad, missing valuable opportunities and feedback. Hankivsky (2005) argues that one of the main weaknesses of gender mainstreaming is that it creates a dichotomy of male versus female at the expense of embracing intersectionality. She advocates for highlighting the relationships among gender, class, sexuality, race, ethnicity, and power (Hankivsky 2005). I acknowledge that gender mainstreaming, as compared to diversity management, has the potential to exclude marginalized populations. However, scholars like Bellitto (2015) believe that gender mainstreaming might be the most acceptable entry point to challenging the status quo in countries, such as the USA, that lag behind in achieving gender equality, since it can be implemented from both the top-down and bottom-up. Future studies should compare the results of gender and diversity mainstreaming within urban design to determine a framework that could have the most positive impact.

Following the precedent set in *Urban Green Spaces and Health: A Review of Evidence* (WHO Regional Office for Europe 2016), I choose to approach the definition of the urban landscape in a holistic way. Perhaps most obviously, these terms describe the physical environments that cities occupy. This includes green spaces like parks, plazas, and nature preserves; natural features like rivers, forests, and air quality; and transportation networks like bike lanes, bus routes, and streets (WHO Regional Office for Europe 2016). Urban landscape also describes policies that direct how the physical environment takes form, for example: zoning policy, planning regulations, and design guidelines for new development (Micklow and Warner 2014). The various components of the urban landscape combine to create the overall experience of being in the city.

9.3 Women, Girls, and Disaster

The influence of gender on disaster resilience is well-documented, and it should come as no surprise that women and girls are often more marginalized and vulnerable to disaster compared to men (Ashraf and Azad 2015; Criado Perez 2019; Enarson and Chakrabarti 2009). For example, women are more likely to be affected by a disaster, and they're more likely to die or experience trauma during a disaster (O'Reilly et al. 2015). Tragically, they also experience increased rates of male violence after a disaster (Wilson et al. 1998). Women's needs are less likely to be accounted for during disaster. For example, on more than one occasion, homes rebuilt after disasters lacked kitchens (Criado Perez 2019, p. 290), and stories of shortages of menstruation and breastfeeding-related supplies in post-disaster shelters and temporary housing abound (Criado Perez 2019; Hargest-Slade and Gribble 2015). Facilities to support common income-generating opportunities, like small-scale markets and childcare, go ignored and unbuilt (Enarson and Morrow 1998b). Given these realities and the fact that women and girls represent half of everybody, a logical question to ask is: Why is this happening?

Ariyabandu (2009) says that women's vulnerability to disaster springs from pre-existing gender relations, which create differences in social and economic status, mortality rates, needs, gender-based prejudices, among other disparities. Criado Perez (2019) takes a simpler approach, distilling the reasons for women's inequal status as being related to "the female body, women's unpaid care burden, and male violence against women" (p. 49). Cannon (2002) argues that vulnerability depends on "initial conditions" like a person's health, mobility, and capacity for self-reliance. In all accounts, the circumstances that contribute to better or worse outcomes are created in our daily lives, prior to disaster. What do we know about women's existing conditions that improve or detract from their disaster resilience, and how might those conditions relate to the built environment? To answer this question, I have identified four areas of inquiry: women's transportation and economic resilience, access to safe public spaces, women's specific health needs, and inclusion and leadership through gender mainstreaming.

9.4 Transportation and Economic Resilience

Economic capacity often represents the most significant predictor of a person's resilience and ability to recover after a disaster (Ashraf and Azad 2015; Cutter et al. 2010), and women tend to start off in a worse economic position than men when a disaster strikes. In 2007, 23.8% of American women heads-of-household were in poverty (English et al. 2009). Single, working women headed one-fifth of all households with children, and they were nearly twice as likely to be unemployed as married men, possibly because of the challenge of finding work and childcare whose schedules and locations often do not complement each other (English et al. 2009). Even for those working, employment does not necessarily guarantee equitable economic opportunities. Hegewisch and Williams-Baron (2018) documented that women continue to work in female-dominated occupations, which have lower salaries on average than male-dominated occupations at similar skill levels. When women's salaries are compared to men's salaries within the same female-dominated occupations, men still out-earn women (Hegewisch and Williams-Baron 2018). Consider this: In the USA, for every one man with poverty-level wages, there are eight women also in poverty (Hegewisch and Williams-Baron 2018). With fewer and more marginalized economic opportunities, strengthening access to women's economic resources has the potential to significantly improve resilience.

Reliable transportation provides the foundation for accessing many urban resources, including economic resources, and is directly connected to women's resilience. So, struggling with limited mobility can compromise resilience due to less access to jobs and careers, professional development, economic status, and personal well-being and health (Loukaitou-Sideris et al. 2009; Madariaga 2013a). And women do struggle. Despite the fact that on average more women than men rely on public transportation in their daily lives, the design of many public transportation routes and schedules typically frustrate women (Action Aid International 2013; Loukaitou-Sideris et al. 2009; Madariaga 2013a). Their frustration stems from the fact that women and men have vastly different travel patterns, but transportation systems tend to prioritize commuter travel needs, which traditionally favor men (Micklow et al. 2015). Since women continue to shoulder the majority of unpaid care-giving and household work, commuter-centric design further puts women at a disadvantage (Khazan 2016; Madariaga 2013a). People who are responsible for care-giving and domestic work are more likely to trip-chain, a pattern of mobility characterized by connecting multiple trips into one outing (Micklow and Warner 2014). For example, a woman might drop her children off at school, go to work, leave work, pick up her kids, go grocery shopping, and check-in on an elderly parent before finally returning home at the end of her day. If the same woman relies on public transportation, a disproportionate amount of her day will be spent in transit (Micklow et al. 2015). That is in addition to the extra unpaid hours she will spend on domestic duties like child and elder care and housework, which women are more likely to be responsible for around the world (OECD 2014). While urban design is limited in its influence on the division of care-giving and household work, it has a tremendous impact on transportation options.

Even though women use public transportation more frequently than men, transportation systems have historically been designed for the commuter at the expense of the caregiver; that is, transportation design tends to focus on single-trip journeys between home and work or school (Madariaga 2013a). One reason for this is how transportation planners evaluate travel needs. Conventionally, transportation planners categorize trips into "essential" and "non-essential" groups; however, many trips that get labeled as "non-essential" are related to essential care-giving activities like grocery shopping and escorting children to school (Madariaga 2013a). When care-giving activities are grouped together, they account for about a third of all trips taken, hardly "non-essential" (Madariaga 2013a). Other mobility differences include: women tend to have less access to cars, cease driving earlier than men as they age, make more multi-modal trips, and their trips tend to be "shaped as polygons" (as opposed to the straight, two point travel associated with men) which often requires catching connecting trains or buses to get where they need to go (Madariaga 2013a; Micklow et al. 2015). Basically, women's mobility patterns are more complex than men's, but the urban landscape and transportation systems often aren't designed with women's needs in mind (Madariaga 2013a; Perera 2008).

As will be discussed more later in this chapter, safety concerns keep many women from accessing public spaces or walking alone on the street, and this includes public transportation. If the public felt more secure, public transportation ridership in the UK would increase 10.5% according to the UK Department of Transport (Loukaitou-Sideris et al. 2009). When asked how to improve their sense of safety while traveling, women often identify poor lighting as a top concern (Johnson and Miles 2014). Urban design professionals often seek to improve lighting in public places, especially on streets and around station platforms or bus stops, as an initial design intervention (Johnson and Miles 2014). However, this strategy works only when areas surrounding these transit hubs, like parking lots, are also well-lit to avoid a "fishbowl" effect (Loukaitou-Sideris et al. 2009). Additionally, women have pointed out that well-maintained, clean areas that are free of graffiti and debris feel safer than poorly maintained spaces (Loukaitou-Sideris et al. 2009).

Loukaitou-Sideris et al. (2009), who performed a comprehensive review of literature concerning safety and transit and surveyed 16 representatives of women's interest groups on the topic, found that women's and men's preferences in safety interventions for public transportation differed. Women clearly favored more visible interventions like additional security staff in public places compared to technological interventions like CCTV, which men preferred (Loukaitou-Sideris et al. 2009). Women also feel safer with clear sightlines and no corners or tight spaces where someone could hide (Loukaitou-Sideris et al. 2009). In contrast, a concurrent survey of 245 transit authorities in the USA revealed that transportation agencies are choosing the opposite of what women want, with a clear preference for technology (Loukaitou-Sideris et al. 2009). More puzzling, most of the transportation agencies that responded to the survey did not employ security officers nor have a desire to. In contrast, transportation agencies that do provide officers stationed at their facilities reported that they are "very effective and [give] you a nice, secure feeling" (Loukaitou-Sideris et al. 2009).

Improving access to transportation elevates community—especially women's—resilience by improving access to economic capital through more time available for jobs, study, and networking (Cutter et al. 2010). Urban design professionals need to create user-friendly, safe transportation systems that accommodate multiple transportation styles and patterns, paying specific attention to improving options for women.

9.5 Access to Safe Public Spaces

Transportation areas aren't the only public spaces that need to be made safer. On a global scale, violence against women remains a dire crisis, not just for women themselves, but for society as a whole, as it limits the full participation of half of the population in public life (Johnson and Miles 2014). In addition to this violence, women experience sexual harassment on a near-universal scale. A recent study from the nonprofit organization Stop Street Harassment found that 81% of women and 43% of men report experiencing some form of sexual harassment or assault at some point in their lives (Kearl 2018). Worse, official crime statistics do not represent the full scope of sexual crimes against women. A survey in New York City found that 96 percent of victims of harassment did not file a report with the police or the transportation agency (Stringer 2007). On an extreme, sexual violence has been explicitly used to prevent women from occupying public spaces. A study in Egypt found that some groups deliberately dispatched men to rape and molest women taking part in the 2013 Tahrir Square protests in Cairo (Langohr 2013). Tandogan and Ilhan (2016) speculated that because of these realities, when women discuss their fear in public spaces, what they are really discussing is women's fear of *sexual violence* in public spaces. Sexual violence, overt harassment, and crass remarks all create a hostile urban environment for many women (Tandogan and Ilhan 2016).

This information is important to contextualize broader statistics on violence in public places. While women are more likely to report violence in private spaces, it is men—not women—who are more likely to be victims of violence and crime in public

(Rollnick 2007). Despite that, women are more likely to feel unsafe in public space—two to three times as afraid by some estimates (Reid and Konrad 2004). Therefore, while men are more likely to *be* in danger, women are more likely to be *aware* of danger (Loukaitou-Sideris et al. 2009). In addition to encouraging women's use of the urban landscape, women's insights on public safety have the potential to reduce crimes against men as well since safer places will be safer for everyone (Micklow et al. 2015).

Further complicating the issue, Loukaitou-Sideris et al. (2009) found in a survey of US women that culture plays a significant role in propagating women's fear. As one interviewee observed, "[the societal dialog] has done a very good job of convincing women that [they] are unsafe in public space" (Loukaitou-Sideris et al. 2009). Parents have been found to have stricter rules for their daughters concerning curfews and mobility compared to their sons, possibly because of disproportionate media attention given to crimes against women (Loukaitou-Sideris et al. 2009). We teach our girls to believe they're vulnerable in public places, even though our boys might benefit more from the warning.

Therefore, it's not surprising that Loukaitou-Sideris et al. (2009) concluded that fear of crime is a primary barrier to women's access to the urban landscape, while Madariaga (2013b) connected the fear of public spaces to the fear of public transportation, which I've already discussed. This perceived danger—whether immediate or not—causes many women to self-limit their movements (Women 2017). Women might be reluctant to be in public space at certain hours or avoid a place altogether (Madariaga 2013a). Many women who steer clear of plazas or parks might be reducing their chances for serendipitous encounters that could build their social networks (Johnson and Miles 2014); avoiding travel at certain times of day might limit economic opportunities (Halsall 2001); spending less time walking outside could compromise health (Frank et al. 2008); and limiting use of public space might reduce women's voices in public discussions (Perera 2008). Because unsafe environments limit access to urban landscapes' wide array of resources, it is reasonable to identify a lack of safety as a key vulnerability in the urban landscape that affects women in particular. Urban design professionals must recognize that improving safety is an opportunity to support resilience, especially for women.

How can urban design professionals take these considerations into account when designing the urban landscape? The obvious answer is to ask women what they need and what they're lacking (Criado Perez 2019). For example, Toronto's Metropolitan Action Committee on Violence Against Women and Children developed the Women's Safety Audit tool for this purpose. It allows cities to undertake a critical evaluation of their built environment and has since become the most widely used tool to assess urban safety around the world (Rollnick 2007; Women in Cities International 2008). In these audits, women document specific features that either increase or decrease their sense of safety in places that they frequently visit, like bus stops and train stations (Lambrick et al. 2011). Other popular tools include focus group discussions and street surveys (Lambrick et al. 2011).

A word of caution: In documenting women's needs, urban design professionals must self-screen for benevolent sexism in their actions and words. Benevolent sexism is the well-intentioned actions or words based on inherent gendered stereotypes (Meagher Benjamin 2017); cities must avoid the "damsel in distress" rhetoric and instead encourage women to take the lead in designing their own safety measures (Rogers 2014). This is especially vital when considering controversial interventions like creating female-only spaces. In addressing safety concerns on public transportation, some trains and buses in Japan, Mexico, Germany, and Thailand have designated women-only areas, and many cities are implementing rideshares and common spaces like gyms and apartment buildings that are just for women (Hillin 2016). Train companies in Tokyo documented a 3% reduction in the number of reported cases of lewd behavior after designating cars for women, so these programs appear to offer some success in reducing harassment ("Japanese women can now travel in women-only train cars to avoid groping," 2006). Similarly, Germany, China, and Switzerland have all experimented with designated parking spots for women that place them closer to their destinations (Hillin 2016). However, despite the growing popularity of female-only spaces, some scholars and activists reject segregation as a solution to improving women's safety, advocating that the onus is on men to treat women with respect, not on women to evade men (Hillin 2016). Local women need to decide which approaches suit them, and the answers will vary across communities.

Fear of crime affects more than just access to public spaces; it takes a toll on women's health as well. Stafford et al (2007) found that people who are afraid of crime showed lower levels of mental health, exercised less, and had an overall reduced quality of life. They participated in fewer social activities and saw their friends less frequently (Stafford et al. 2007), so improving public safety will likely improve women's health as well.

9.6 Women's Specific Health Needs

A 2012 meta-analysis of men and women across 57 countries showed that women report significantly lower health than men, and this held true across different ages, socioeconomic statuses, and countries (Hosseinpoor et al. 2012). The U.S. Centers for Disease Control and Prevention estimates that illness and injury cost the economy $225.8 billion annually (CDC Foundation 2015). Poor health can also result in personal financial burdens because of missed work (Gould and Schieder 2017). Similar to how women lose time navigating lousy transportation systems, time lost to illness robs time from other activities that could benefit resilience. Chronic health concerns can negatively impact a person's overall quality of life (Ekkel and de Vries 2017). Since women manage poor health more often than men, improving health outcomes can directly contribute to women's resilience. So where do urban designers fit in?

One community-wide approach to improving health is to increase access to green space (Markevych et al. 2017; Wood et al. 2017). A recent report from the World Health Organization Regional Office for Europe (2016) summarized decades of studies that examined the relationships between urban green space and health, affirming that the available evidence supports the observation that increased exposure to green space correlates to better health and wellness. In particular, green space most significantly affects mental health compared to other aspects of health that have been studied (WHO Regional Office for Europe 2016).

While it remains unclear precisely which aspects of green space influence health, the WHO Regional Office for Europe identified nine possible pathways in which these benefits might be realized (Table 9.1). While it is possible that green spaces can contribute to some negative health impacts, it appears that the benefits of green space outweigh the negatives (WHO Regional Office for Europe 2016). Likewise, the authors note that many of these adverse effects can be mitigated with design and best practices in maintenance; for example, maintenance crews might use only the minimum amount of pesticides and herbicides, which would reduce or eliminate the risk of exposure (WHO Regional Office for Europe 2016).

From a women-exclusive perspective, studies have connected green space to improved reproductive health outcomes in pregnant women. They have found that proximity to green space predicted reduced blood pressure (Grazuleviciene et al. 2014), lower rates of depression (McEachan et al. 2016), and tendency to have children with higher birth weights (Dzhambov et al. 2014). Similarly, a Lithuanian study found that the risk of preterm birth and younger gestational age decreased as the

Table 9.1 Summary of findings from Urban green spaces and health: A review of evidence (WHO Regional Office for Europe 2016)

Pathways to improved health	Positive effects of urban green space	Negative effects of urban green space
Improved relaxation and restoration	Improved mental health and cognitive function	Risk of allergies and asthma
Improved social capital	Reduced cardiovascular morbidity	Exposure to pesticides and herbicides
Improved functioning of the immune system	Reduced prevalence of Type II diabetes	Exposure to disease vectors and zoonotic infections
Enhanced physical activity, improved fitness, and reduced obesity	Improved pregnancy outcomes	Accidental injuries
Anthropogenic noise buffering and production of natural sounds	Reduced mortality	Excessive exposure to UV radiation
Reduced exposure to air pollution		
Reduction of the urban heat island effect		
Enhanced pro-environmental behavior		
Optimized exposure to sunlight and improved sleep		

distance between residences and city parks decreased (Grazuleviciene et al. 2015). Women uniquely benefit in other ways from green space. A cross-sectional study of four European cities found that all participants who spent more time in green space reported fewer symptoms of anxiety, but only women reported significantly fewer symptoms of depression (van den Berg et al. 2016). Meanwhile, von den Bosch et al. (2015) found that serene landscapes—described as safe, calm environments—predicted significantly reduced mental health illnesses in women, supporting similar findings from Annerstedt et al. (2012). Kuo and Sullivan (2001) found in a study of inner city urban public housing residents that women whose homes were near nature were significantly less likely to report aggressive or violent behavior. A study of disadvantaged neighborhoods in Detroit, Michigan, found that people living in neighborhoods with strong local street networks and connections to the surrounding urban area reported higher levels of walking (Wineman et al. 2014); however, the authors of this study did not present gender dis-aggregated data. In contrast, a cross-sectional review of people in Atlanta showed that men tended to be thinner as connectedness increased, while women tended to be heavier (Frank et al. 2008). Frank et al. (2008) speculate that this could be related to women's perceptions of safety and use of public space. As already discussed, if women are fearful for their safety, they are less likely to walk outside. Though the authors did not address it, another possibility is that the street networks fail to address women's needs, perhaps by lacking adequate sidewalks that people with strollers, children, or mobility impairments can easily navigate (Loukaitou-Sideris et al. 2009).

Various qualities and types of green space have been shown to have different health impacts. Work by Akpinar et al. (2016) indicates that the type of green space influences its overall effect. In their analysis, they found that access to forests and urban green spaces reduced mental health complaints in Washington State more than compared to access to wetlands, rangeland, or agricultural land. Goto et al. (2017) evaluated the cognitive health impacts of experiencing Japanese-style garden design, finding that cognition and relaxation improved in patients with dementia and Alzheimer's disease when exposed to a Japanese-style garden compared to an "unstructured" garden. The team had previously found that elderly, cognitively intact individuals experienced less stress and improved overall mood when exposed to Japanese-style gardens (Goto et al. 2013).

Partly because of its correlation to mental health, green space has been found to aid in healing after disasters as well. Okvat and Zautra (2013) assert that community gardening and exposure to gardens has the potential to build resilience after a disaster. They use Common Ground Relief in New Orleans, Louisiana, as an example. Common Ground Relief is a nonprofit organization created after Hurricane Katrina to aid and assist community members in creating their own urban and community gardens, an effort that Okvat and Zautra describe as providing exercise and respite. Also in New Orleans, produce from community gardens in the local Vietnamese communities helped the neighborhood come back sooner, before grocery stores were open (Okvat and Zautra 2013). Meanwhile, Krasny et al. (2013) make the case for using nature exposure to help returning military veterans build their resilience and readjust to home life. They point to the growing popularity of horticulture therapy, employing veterans in green jobs, or encouraging them to become involved in gardening, hunting, fishing, and other outdoor activities. These examples of green space providing restorative effects to people who were traumatized in disasters are a new field of research with few studies and even fewer with gender dis-aggregated data, so more work needs to be done to learn how to tailor healing nature-based programs to different demographics in after-disaster contexts.

9.7 Inclusion and Leadership Through Gender Mainstreaming

We've examined the influence of transportation-, safety-, and health-related aspects of the built environment that impact women's resilience. At this point, you might be asking yourself *why* these inequalities continue to persist in our communities? The short answer is there has been a toxic and fundamental lack of representation of women in leadership and decision-making roles.

Why should urban design and emergency management professionals recruit more women when planning to improve their community's disaster resilience? For starters, diversity in leadership leads to more innovation, which has the potential to make communities more flexible and adaptable (Lorenzo and Reeves 2018), and if one thing is certain, it's that there is a dearth of women in the urban design world. Women are terribly underrepresented in the urban design professions, and the situation is especially bleak when examining leadership and professional roles (Rustin 2014; The American Institute of Architects 2015). Additionally, women in the community are less likely to have their voices heard in public meetings or in public design forums (Micklow et al. 2015). With decision-makers skewed heavily male, it is not surprising that an increasing body of research has documented widespread bias against women in the urban landscape (Greed 1996; Micklow et al. 2015; Urban Development Vienna 2013).

The USA is one of only seven U.N. member nations that hasn't ratified the U.N.'s Convention to Eliminate All Forms of Discrimination Against Women (Bellitto 2015), and the USA ranks 49th in the U.N.'s 2017 Global Gender Gap Report (World Economic Forum 2017). Clearly, there is plenty of room for improvement. That said, some regions—particularly Europe, Canada, and Australia—lead by example with decades of gender mainstreaming under their belts (Bacchi et al. 2010). Some of these communities have already started the transition to more-inclusive, diversity mainstreaming policies, also called diversity management (Executive Group for Organisation, Safety and Security 2011).

Male bias in the urban landscape is the tendency for cities to cater to men's needs while neglecting or sidelining women's and minority's needs (Garcia-Ramon et al. 2004). Male bias is rarely malicious; it's more a product of men in power forgetting about (and therefore missing out on) women's knowledge and capabilities (Criado

Perez 2019). Bias, at its heart, stems from a lack of female and minority representation in urban planning, city government, and community design—especially in leadership and decision-making roles (Enarson and Morrow 1998a). This lack of representation creates a blind spot for urban design professionals, and it hinders cities' ability to respond to and benefit from women's and minority's perspectives, wisdom, and insights, which are too often absent from public discussion (Greed 2007; Sham et al. 2013; Women 2017; Women in Cities International 2010). As an example of that blind spot, a survey of 624 planners in the USA found that just 2% of comprehensive plans addressed women's needs specifically, and only 7% of respondents agreed with the statement "developers are responsive to the special needs of women" (Micklow et al. 2015). The American Institute of Architects (2015) found in a 2014 survey that while women agreed that the industry had not yet achieved gender equality, only half of men surveyed held the same perspective. This implies that half of the men at the table do not recognize the gendered bias against women in their field. This does not bode well for improving representation in the USA in the near future. Simply put, women's voices aren't being heard. Without an awareness of women's needs, how can cities possibly respond to them?

Urban design professionals increasingly look to gender mainstreaming as one avenue for including women and countering male bias (Bacchi et al. 2010). Since the strategy prioritizes both women and men, gender mainstreaming actively addresses the needs of all stakeholders. The importance of direct engagement with a community's women cannot be overstated. Women have different needs globally and within the same city (Loukaitou-Sideris et al. 2009; Urban Development Vienna 2013). It's impossible to predict these needs without gender mainstreaming. Unsurprisingly, there are very few examples of gender mainstreaming in the USA (Abbey-Lambertz 2016), but that also means there's ample opportunity for improvement.

Gender mainstreaming requires a comprehensive approach to be effective (Executive Group for Organisation, Safety and Security 2011; Urban Development Vienna 2013). Praised for its inclusive process in urban planning and governance, Vienna has been actively gender mainstreaming since the early 1990s (Bellitto 2015). Their model includes guidelines for gender inclusive language on public documents, demographically diverse decision-making teams, and analysis of gendered impacts of city budgets and projects (Executive Group for Organisation, Safety and Security 2011). In their early mainstreaming efforts, women expressed concerns about their access to transportation (Executive Group for Organisation, Safety and Security 2011). By including women and asking them for insights, Vienna has been able to improve women's access to transportation through neighborhood-centric bus routes, lifts for patrons who struggle with stairs, like a person pushing a stroller, and improved lighting across the system (Criado Perez 2019). Vienna's women identified other gaps where public spaces weren't working for them—from parks to housing developments, women's voices have changed the way that Vienna is designed (Foran 2013). In recognition of its success, the program received the United Nations Human Settlements Programme Excellence in Urban Planning award in 2010 (Hassan 2010), and it still shines as a leading example of gender mainstreaming the urban landscape.

Listening to women's needs by including them into urban design processes can promote equitable access to urban landscapes. This was evident in the Nou Barris neighborhood in Barcelona where Garcia-Ramon et al. (2004) documented the transformation of Via Julia, which was accomplished with strong leadership from women. As part of the participatory process in seeking public input for the redesign, women asserted their needs for "street lights, pedestrian paths, lighting, etc." (Garcia-Ramon et al. 2004, p. 219). They demanded integrated spaces for shops, playgrounds, seating areas, and safe access to public transportation (Garcia-Ramon et al. 2004). As a result of taking women's needs into account, Garcia-Ramon, Ortiz, and Prats documented nearly balanced use between men and women of the redesigned public boulevard. They found that different demographic groups tended to filter in and out of the boulevard with time of day according to traditional schedules (Garcia-Ramon et al. 2004). The researchers conclude that the success of the space came, in part, because it was designed in negotiation with the whole community, including women, which allowed for diverse groups of people within the neighborhood to appropriate the space as their own and for their own uses.

9.8 Moving Forward

Urban landscapes contribute to and influence disaster resilience. Moreover, since everyone's personal knowledge, experiences, and skills inform their perceptions of and relationship to the urban landscape, the landscape can affect resilience in different ways among different people. Since men have been the primary shapers of the urban environment, women's needs have often been neglected in the design of their cities. Urban design professionals can use gender mainstreaming to equalize the landscape, build women's capacities across the spectrum of resilience, and increase their city's overall disaster resilience in the process. Transportation, safe access to public space, and health can all be improved through thoughtful, inclusive urban design, implying that investment in these areas could have broad effects in improving disaster resilience.

While this chapter has identified broad areas for design interventions that have the potential to improve overall resilience, cities must implement their own engagement programs to find their community's individual and unique approaches to gendering their urban landscapes and improving their disaster resilience. Administrators and leaders must commit to supporting the education and recruitment of women urban planners, landscape architects, architects, and engineers (Fleming and Tranovich 2016). They should look to global leaders in gender mainstreaming for inspiration and direction; for example, they could consult Vienna's *Gender Mainstreaming Made Easy: Practical Advice for More Gender Equality in the Vienna City Administration* (Executive Group for Organisation, Safety and Security 2011) for detailed and practical steps that cities can take to begin their journey toward gender equity. Urban design professionals should strive to collect preliminary data on their city's resilience and their city's current gender equity in order to monitor progress and to look for correlations between gendered urban landscapes and resilience that are working in their communities. Cities should emphasize that gender mainstreaming is every department's responsibility while also appointing a specific task force or gender mainstreaming officer to ensure that someone is responsible for monitoring and developing the program (Bacchi et al. 2010).

The urban landscape's role in supporting all aspects of disaster resilience is still poorly understood, and even less so when considering how it supports various demographic groups differently, but the preliminary data are promising. As urban design professionals and researchers expand the knowledge around these topics, they will find many more opportunities to maximize investments into the urban landscape so that they respond to all community members' needs and grow resilience.

References

Abbey-Lambertz, K. (2016, April 26). Cities Aren't Designed For Women. Here's Why They Should Be. *Huffington Post*. Retrieved from https://www.huffingtonpost.com/entry/cities-designed-for-women_us_571a0cdfe4b0d0042da8d264.

Action Aid International. (2013). *Women and the City II: COMBATING VIOLENCE AGAINST WOMEN AND GIRLS IN URBAN PUBLIC SPACES - THE ROLE OF PUBLIC SERVICES.*

Akpinar, A., Barbosa-Leiker, C., & Brooks, K. R. (2016). Does green space matter? Exploring relationships between green space type and health indicators. *Urban Forestry & Urban Greening, 20*, 407–418. https://doi.org/10.1016/j.ufug.2016.10.013.

Annerstedt, M., Östergren, P.-O., Björk, J., Grahn, P., Skärbäck, E., & Währborg, P. (2012). Green qualities in the neighbourhood and mental health – results from a longitudinal cohort study in Southern Sweden. *BMC Public Health, 12*, 337. https://doi.org/10.1186/1471-2458-12-337.

Annerstedt van den Bosch, M., Östergren, P.-O., Grahn, P., Skärbäck, E., & Währborg, P. (2015). Moving to Serene Nature May Prevent Poor Mental Health—Results from a Swedish Longitudinal Cohort Study. *International Journal of Environmental Research and Public Health, 12*(7), 7974–7989. https://doi.org/10.3390/ijerph120707974.

Ariyabandu, M. M. (2009). Sex, gender and gender relations in disaster. In Elaine Enarson & P. G. D. Chakrabarti (Eds.), *Women, Gender, and Disaster*. New Delhi: SAGE Publications.

Ashraf, M. A., & Azad, Md. A. K. (2015). Gender Issues in Disaster: Understanding the Relationships of Vulnerability, Preparedness and Capacity. *Environment and Ecology Research, 3*(5), 136–142. https://doi.org/10.13189/eer.2015.030504.

Bacchi, C., Eveline, J., Binns, J., Mackenzie, C., & Harwood, S. (2010). Gender analysis and social change: testing the water. In *Mainstreaming Politics: Gendering Practices and Feminist Theory* (pp. 61–85). University of Adelaide Press.

Bellitto, M. (2015). *Gender mainstreaming in the United States: a new vision of equality. 22*(2), 125–150.

Cannon, T. (2002). Gender and climate hazards in Bangladesh. *Gender and Development, 10*(2), 45–55.

CDC Foundation. (2015, January 28). Worker Illness and Injury Costs U.S. Employers $225.8 Billion Annually. Retrieved April 15, 2018, from CDC Foundation website: https://www.cdcfoundation.org/pr/2015/worker-illness-and-injury-costs-us-employers-225-billion-annually.

Criado Perez, C. (2019). *Invisible Women: Data Bias in a World Designed for Men.* New York: Abrams Press.

Cutter, S. L., Burton, C. G., & Emrich, C. T. (2010). Disaster Resilience Indicators for Benchmarking Baseline Conditions. *Journal of Homeland Security and Emergency Management*, *7*(1). https://doi.org/10.2202/1547-7355.1732.

Dzhambov, A. M., Dimitrova, D. D., & Dimitrakova, E. D. (2014). Association between residential greenness and birth weight: Systematic review and meta-analysis. *Urban Forestry & Urban Greening*, *13*(4), 621–629. https://doi.org/10.1016/j.ufug.2014.09.004.

Ekkel, E. D., & de Vries, S. (2017). Nearby green space and human health: Evaluating accessibility metrics. *Landscape and Urban Planning*, *157*, 214–220. https://doi.org/10.1016/j.landurbplan.2016.06.008.

Enarson, E., & Chakrabarti, P. G. D. (2009). *Women, Gender and Disaster: Global Issues and Initiatives* (1st ed.). Thousand Oaks, California: SAGE Publications.

Enarson, E., & Morrow, B. H. (1998a). Why Gender? Why Women? An Introduction to Women in Disaster. In *The Gendered Terrain of Disaster: Through Women's Eyes.* (pp. 1–10). Westport, Connecticut: Praeger Publishers.

Enarson, E., & Morrow, B. H. (1998b). Women Will Rebuild Miami: A Case Study of Feminist Response to Disaster. In *The Gendered Terrain of Disaster: Through Women's Eyes* (p. 275). Westport, Connecticut: Praeger Publishers.

English, A., Hartmann, H., & Hegewisch, A. (2009). *Unemployment Among Single Mother Families* (No. C369; p. 4). Institute for Women's Policy Research.

Executive Group for Organisation, Safety and Security. (2011). *Gender Mainstreaming Made Easy: Practical Advice for More Gender Equality in the Vienna City Administration* (pp. 46–46). Vienna: City of Vienna.

Fleming, Rew, & Tranovich, A. (2016, October 13). Why aren't we designing cities that work for women, not just men? Retrieved April 3, 2018, from the Guardian website: https://www.theguardian.com/global-development-professionals-network/2016/oct/13/why-arent-we-designing-cities-that-work-for-women-not-just-men.

Foran, C. (2013, September 13). How to Design a City for Women. Retrieved March 30, 2018, from CityLab website: https://www.theatlanticcities.com/commute/2013/09/how-design-city-women/6739/.

Frank, L. D., Kerr, J., Sallis, J. F., Miles, R., & Chapman, J. (2008). A hierarchy of sociodemographic and environmental correlates of walking and obesity. *Preventive Medicine*, *47*(2), 172–178. https://doi.org/10.1016/j.ypmed.2008.04.004

Garcia-Ramon, M. D., Ortiz, A., & Prats, M. (2004). Urban planning, gender and the use of public space in a peripherial neighbourhood of Barcelona. *Cities*, *21*(3), 215–223.

Goto, S., Bum-Jin Park, Tsunetsugu, Y., Herrup, K., & Miyazaki, Y. (2013). The Effect of Garden Designs on Mood and Heart Output in Older Adults Residing in an Assisted Living Facility. *Health Environments Research & Design Journal (HERD) (Vendome Group LLC)*, *6*(2), 27–42.

Goto, S., Gianfagia, T. J., Munafo, J. P., Fujii, E., Shen, X., Sun, M., … Herrup, K. (2017). The power of traditional design techniques: the effects of viewing a Japanese garden on individuals with cognitive impairment. *HERD: Health Environments Research & Design Journal*, *10*(4), 74–86. https://doi.org/10.1177/1937586716680064.

Gould, E., & Schieder, J. (2017). *Work sick or lose pay?: The high cost of being sick when you don't get paid sick days* (No. 130245). Retrieved from Economic Policy Institute website: https://www.epi.org/publication/work-sick-or-lose-pay-the-high-cost-of-being-sick-when-you-dont-get-paid-sick-days/.

Grazuleviciene, R., Danileviciute, A., Dedele, A., Vencloviene, J., Andrusaityte, S., Uždanaviciute, I., & Nieuwenhuijsen, M. J. (2015). Surrounding greenness, proximity to city parks and pregnancy outcomes in Kaunas cohort study. *International Journal of Hygiene and Environmental Health*, *218*(3), 358–365. https://doi.org/10.1016/j.ijheh.2015.02.004.

Grazuleviciene, R., Dedele, A., Danileviciute, A., Vencloviene, J., Grazulevicius, T., Andrusaityte, S., … Nieuwenhuijsen, M. J. (2014). The Influence of Proximity to City Parks on Blood Pressure in Early Pregnancy. *International Journal of Environmental Research and Public Health*, *11*(3), 2958–2972. https://doi.org/10.3390/ijerph110302958.

Greed, C. (1996). Promise or progress: women in planning. *Built Environment (1978-)*, *22*(1), 9–21.
Greed, C. (2007). *A place for everyone? Gender equality and urban planning*. Retrieved from https://www.alnap.org/resource/6972.
Halsall, D. (2001). Transport and travel: better for women ? *Geography*, *86*(1), 79–81.
Hankivsky, O. (2005). Gender vs Diversity Mainstreaming: A Preliminary Examination of the Role and Transformative Potential of Feminist Theory. *Canadian Journal of Political Science / Revue Canadienne de Science Politique*, *38*, 977–1001.
Hargest-Slade, A. C., & Gribble, K. D. (2015). Shaken but not broken: Supporting breastfeeding women after the 2011 Christchurch New Zealand earthquake. *G Review*, *23*(3).
Hassan, Z. A. (2010, October 1). UN-HABITAT Awards City of Vienna for Excellence in Urban Planning. Retrieved April 3, 2018, from https://www.unis.unvienna.org/unis/pressrels/2010/uninf391.html.
Hegewisch, A., & Williams-Baron, E. (2018). *The Gender Wage Gap by Occupation 2017* (No. C467; p. 9). Retrieved from Institute for Women's Policy Research website: https://iwpr.org/wp-content/uploads/2018/04/C467_2018-Occupational-Wage-Gap.pdf.
Hillin, T. (2016). *Do ladies-only taxis and trains empower women — or set them back?*
Hosseinpoor, A. R., Williams, J. S., Amin, A., de Carvalho, I. A., Beard, J., Boerma, T., … Chatterji, S. (2012). Social determinants of self-reported health in women and men: Understanding the role of gender in population health. *PLoS ONE*, *7*(4), e34799.
Japanese women can now travel in women-only train cars to avoid groping. (2006, August 21). Retrieved April 15, 2018, from Pravda Report website: https://www.pravdareport.com/news/society/21-08-2006/84007-groping-0/.
Johnson, A. M., & Miles, R. (2014). Toward more inclusive public spaces: Learning from the everyday experiences of Muslim Arab women in New York City. *Environment and Planning A*, *46*(8), 1892–1907. https://doi.org/10.1068/a46292.
Kearl, H. (2018). *The Facts behind the #MeToo Movement: A National Study on Sexual Harassment and Assault* (p. 41). Reston, Virginia: Stop Street Harassment.
Khazan, O. (2016). The Scourge of the Female Chore Burden.
Krasny, M. E., Pace, K. H., Tidball, K. G., & Helphand, K. (2013). Nature Engagement to Foster Resilience in Military Communities. In *Greening the Red Zone: Disaster, Resilience and Community Greening* (p. 503). New York, London: Springer Dordrecht Heidelberg.
Kuo, F. E., & Sullivan, W. C. (2001). Aggression and Violence in the Inner City: Effects of Environment via Mental Fatigue. *Environment and Behavior*, *33*(4), 543–571. https://doi.org/10.1177/00139160121973124.
Lambrick, M., Viswanath, K., Husain, S., Blanes, P., Bozhkova, E., Dalmia, S. P., … Whitzman, C. (2011). *Tools for Gathering Information about Women's Safety and Inclusion in Cities* (p. 64). Montréal, Canada: Women in Cities International.
Langohr, V. (2013). "This Is Our Square": Fighting Sexual Assault at Cairo Protests. *Middle East Report*, *268*, 18–25.
Lorenzo, R., & Reeves, M. (2018). How and Where Diversity Drives Financial Performance. *Harvard Business Review Digital Articles*, 2–5.
Loukaitou-Sideris, A., Bornstein, A., Fink, C., Samuels, L., & Gerami, S. (2009). *How to ease Women's Fear of Transportation Environments: Case Studies and Best Practices* (pp. 96–96). Retrieved from Mineta Transportation Institute website: https://transweb.sjsu.edu/MTIportal/research/publications/documents/2611-women-transportation.pdf.
Madariaga, I. S. de. (2013a). From women in transport to gender in transport: challenging conceptual frameworks for improved policymaking. *Journal of International Affairs*, *67*(1), 43–65.
Madariaga, I. S. de. (2013b). From women in transport to gender in transport: challenging conceptual frameworks for improved policymaking. *Journal of International Affairs*, *67*(1), 43–65.
Markevych, I., Schoierer, J., Hartig, T., Chudnovsky, A., Hystad, P., Dzhambov, A. M., … Fuertes, E. (2017). Exploring pathways linking greenspace to health: Theoretical and methodological guidance. *Environmental Research*, *158*, 301–317. https://doi.org/10.1016/j.envres.2017.06.028.

McEachan, R. R. C., Prady, S. L., Smith, G., Fairley, L., Cabieses, B., Gidlow, C., … Nieuwenhuijsen, M. J. (2016). The association between green space and depressive symptoms in pregnant women: moderating roles of socioeconomic status and physical activity. *J Epidemiol Community Health, 70*(3), 253–259. https://doi.org/10.1136/jech-2015-205954.

Meagher Benjamin R. (2017). Judging the gender of the inanimate: Benevolent sexism and gender stereotypes guide impressions of physical objects. *British Journal of Social Psychology, 56*(3), 537–560. https://doi.org/10.1111/bjso.12198.

Micklow, A. C., & Warner, M. E. (2014). Not Your Mother's Suburb: Remaking Communities for a More Diverse Population. *Urban Lawyer, 46*(4), 729–751. https://doi.org/10.1002/nbm.3369. Three.

Micklow, A., Kancilia, E., & Warner, M. (2015). *The need to plan for women: planning with a gender lens.*

OECD. (2014). Balancing paid work, unpaid work, and leisure. Retrieved June 3, 2019, from OECD Better Policies for Better Lives: Gender Equality website: https://www.oecd.org/gender/data/balancingpaidworkunpaidworkandleisure.htm.

Okvat, H. A., & Zautra, A. J. (2013). Sowing Seeds of Resilience: Community Gardening in a Post-Disaster Context. In *Greening the Red Zone: Disaster, Resilience and Community Greening* (p. 503). New York, London: Springer Dordrecht Heidelberg.

O'Reilly, M., Súilleabháin, A. Ó., & Paffenholz, T. (2015). *Reimagining Peacemaking: Women's Roles in Peace Processes* (p. 42). Retrieved from International Peace Institute website: https://www.ipinst.org/wp-content/uploads/2015/06/IPI-E-pub-Reimagining-Peacemaking.pdf.

Perera, G. (2008). Claiming the right to the city: A question of power. *Right to the City Alliance, 15*(1), 12–13.

Reid, L. W., & Konrad, M. (2004). The Gender Gap in Fear: Assessing the Interactive Effects of Gender and Perceived Risk on Fear of Crime. *Sociological Spectrum, 24*(4), 399–425. https://doi.org/10.1080/02732170490431331.

Rogers, R. (2014, February 17). Women are Kind and Men are Strong: How Benevolent Sexism Hurts Us All. Retrieved April 15, 2018, from https://www.themuse.com/advice/women-are-kind-and-men-are-strong-how-benevolent-sexism-hurts-us-all.

Rollnick, R. (2007). *The Global Assessment on Women's Safety* (p. 85). Nairobi, Kenya: United Nations Human Settlements Programme.

Rondeau, M. B., Brantingham, P. L., & Brantingham, P. J. (2005). The value of environmental criminology for the design professions of architecture, urban design, landscape architecture, and urban planning. *Journal of Architectural and Planning Research, 22*(4), 294–304.

Rustin, S. (2014, December). If women built cities, what would our urban landscape look like? *The Guardian*, 9–9.

Sham, R., Hamid, H. A., & Noah, R. M. (2013). Routine Activities and Crime in the City: Cases of Working Women. *Procedia - Social and Behavioral Sciences, 101*, 345–353. https://doi.org/10.1016/j.sbspro.2013.07.209.

Stafford M, Chandola T, & Marmot M. (2007). Association between fear of crime and mental health and physical functioning. *American Journal of Public Health, 97*(11), 2076–2081. https://doi.org/10.2105/AJPH.2006.097154.

Stringer, S. M. (2007). *Hidden in Plain Sight: Sexual Harassment and Assault in the New York City Subway System.* 25.

Tandogan, O., & Ilhan, B. S. (2016). Fear of Crime in Public Spaces: From the View of Women Living in Cities. *Procedia Engineering, 161*, 2011–2018. https://doi.org/10.1016/j.proeng.2016.08.795.

The American Institute of Architects. (2015). *Diversity in the Profession of Architecture* (p. 41). Retrieved from https://www.architecturalrecord.com/ext/resources/news/2016/03-Mar/AIA-Diversity-Survey/AIA-Diversity-Architecture-Survey-02.pdf.

Tidball, K. G., & Krasny, M. E. (Eds.). (2013). *Greening in the Red Zone: Disaster, Resilience and Community Greening.* New York, London: Springer Dordrecht Heidelberg.

U.N. Women. (2017). *Safe cities and safe public spaces.*

UNDRR. (2019). *Global Assessment Report on Disaster Risk Reduction* (p. 472). Geneva, Switzerland: United Nations Office for Disaster Risk Reduction (UNDRR).

Urban Development Vienna. (2013). *Gender Mainstreaming in Urban Planning and Urban Development* (Workshop Report No. 130A; p. 104). Vienna: Municipal Department 18 - Urban Development and Planning.

van den Berg, M., van Poppel, M., van Kamp, I., Andrusaityte, S., Balseviciene, B., Cirach, M., … Maas, J. (2016). Visiting green space is associated with mental health and vitality: A cross-sectional study in four european cities. *Health & Place, 38*, 8–15. https://doi.org/10.1016/j.healthplace.2016.01.003.

WHO Regional Office for Europe. (2016). *Urban Green Spaces and Health: A Review of Evidence* (p. 91). Retrieved from https://www.euro.who.int/en/health-topics/environment-and-health/urban-health/publications/2016/urban-green-spaces-and-health-a-review-of-evidence-2016.

Wilson, J., Phillips, B. D., & Neal, D. M. (1998). Domestic Violence After Disaster. In *The Gendered Terrain of Disaster: Through Women's Eyes.* (pp. 115–124). Westport, Connecticut: Praeger Publishers.

Wineman, J. D., Marans, R. W., Schulz, A. J., Westhuizen, D. L. van der, Mentz, G. B., & Max, P. (2014). Designing Healthy Neighborhoods: Contributions of the Built Environment to Physical Activity in Detroit. *Journal of Planning Education and Research, 34*(2), 180–189. https://doi.org/10.1177/0739456X14531829.

Wisner, B., Blaikie, P., Cannon, T., & Davis, I. (2003). *At Risk: Natural Hazards, People's Vulnerability, and Disasters* (2nd ed.). Retrieved from https://www.degruyter.com/view/j/jhsem.2005.2.2/jhsem.2005.2.2.1131/jhsem.2005.2.2.1131.xml.

Women in Cities International. (2008). *Women's safety audits: what works and where?* (R. Rollnick, Ed.). Retrieved from www.unhabitat.org.

Women in Cities International. (2010). *Learning from women to create gender inclusive cities: Baseline findings from the gender inclusive cities programme.* 126–126.

Wood, L., Hooper, P., Foster, S., & Bull, F. (2017). Public green spaces and positive mental health – investigating the relationship between access, quantity and types of parks and mental wellbeing. *Health & Place, 48*, 63–71. https://doi.org/10.1016/j.healthplace.2017.09.002.

World Economic Forum. (2017). *The global gender gap report: 2017*. Geneva: World Economic Forum.

Chapter 10
Compounding Impacts of Lifeline Infrastructure Failure During Natural Hazard Events

Emma A. Singh

Abstract Critical infrastructures, such as transportation systems, communication networks, power and water utilities, have become so integrated into our modern and globalised world that they are commonly taken for granted. That is, until their services are disrupted. The failure of these lifeline services during natural hazard events has the potential to impact populations by exacerbating the hazard itself and/or hindering their ability to respond to or recover from the event. The failure of lifeline infrastructure can also propagate outside the reach of the hazard footprint, causing disruption in regions not directly impacted by the event. Understanding the potential flow-on effects from lifeline failure during natural hazard events is vital for future disaster mitigation, response and recovery. The 2009 South-Eastern Australia heatwave and the 2010 Eyjafjallajökull eruption in Iceland are drawn on to highlight and discuss the vulnerability of lifelines to disruption from natural hazard shocks and the compounding impacts of lifeline failure during natural hazard events.

Keywords Critical infrastructure · Lifelines · Natural hazards · Compounding impacts · Indirect disruption · Cascading disasters

10.1 Introduction

Lifelines are the critical infrastructure and systems crucial to the distribution and continuous flow of goods and services essential for human livelihoods, the functioning of society and economic prosperity. They include, but are not limited to transportation, telecommunication and utilities such as power and water. Transportation systems are used daily to move people to and from places of work and education, and

This chapter is derived from a section of the author's Ph.D. thesis (Singh 2019); therefore, some segments of the text here are present in both bodies of work.

E. A. Singh (✉)
Department of Earth and Environmental Sciences, Macquarie University, Sydney, NSW 2109, Australia
e-mail: emma.a.singh@gmail.com

Bushfire and Natural Hazard Cooperative Research Centre, Melbourne, Australia

D. Karácsonyi et al. (eds.), *The Demography of Disasters*,
https://doi.org/10.1007/978-3-030-49920-4_10

goods between manufacturers, suppliers and customers. Telecommunication systems are used to connect people, conduct business and perform financial transactions. Electricity and gas provide power to homes and industries, and water and sewage systems contribute to the public health of communities (Murray and Grubesic 2007). These lifelines have become so integrated into our modern society that they are commonly taken for granted; that is until something goes wrong. The failure of lifelines can range from being an inconvenience to being debilitating for a population both socially and economically (Davis 1999; Hawk 1999). Being without electricity or water for less than a few hours can generally be tolerated. However, prolonged disruption of such services can lead to major economic losses, deteriorating public health that includes an increase in mortality, and eventually population migration (Rose et al. 1997; Brozovic et al. 2007; Rose and Oladosu 2008, Anderson and Bell 2012; Klinger et al. 2014; Yates 2014). Power failure in Auckland, New Zealand (Leyland 1998); the drinking water quality crisis in Sydney, Australia (Clancy 2000); and gas outages in Victoria, Australia (Dawson and Brooks 1999), that all occurred in 1998, had media pointing out how easy it was for a modern building with lights, heating and plumbing to convert into a third world slum when such services were lost (Hawk 1999). The water crisis had residents boiling drinking water and the gas outage left residents with cold showers (Hawk 1999). Moreover, the power failure in Auckland impacted business operations, with some businesses closing, some temporarily moving out of the CBD and a few continuing with generators. All suffered a significant drop in income. Those who continued to operate as usual did so at the expense of their staff that had to climb stairs and swelter in the summer heat without air-conditioning or sufficient ventilation (Leyland 1998). The impact of these utility crises could have been more severe if they had lasted for months instead of days to weeks (Hawk 1999).

Lifeline infrastructure often consists of a large number of interconnected components, which often span extensive geographic areas and, in some cases, multiple urban centres, states and international borders (Kinney et al. 2005; Guikema 2009). The concentration of mutually dependent lifeline infrastructure within urban centres results in complex and interconnected systems, creating feedback loops and complex topologies that can trigger and propagate disruptions in a variety of ways (Rinaldi et al. 2001; McEvoy et al. 2012). Consequently, a fault in one lifeline can cascade and directly or indirectly affect other lifelines or services, with impacts potentially felt not just locally but nationally or even globally (McEvoy et al. 2012). Significant power failure, for example, can result in the disruption of telecommunications and transportation systems. The most noted example of this was the Northeast blackout in August 2003 that impacted the Northeastern USA and parts of Canada (US–Canada Power System Outage Task Force 2004). This event not only cut power to over 50 million people but also disrupted Internet communications. Banks, manufacturers, business services and education institutions were severely disrupted when they were taken offline for hours to days (Cowie et al. 2003; Murray and Grubesic 2007). The event cost an estimated US$6 billion (Electricity Consumers Resource Council 2004). The blackout caused major disruptions to the lives of those impacted. Thousands of subway commuters were trapped for hours as trains stopped operating; the closure of major manufacturing plants idled thousands of workers; cars lined for blocks waiting for gas; high-rise apartments were left without water services as electric pumps used to deliver water to the upper floors shut off; mass food spoilage occurred due to the loss of refrigeration; and, hospital admissions increased due to an increase in heat exposure, exertion and stress (Electricity Consumers Resource Council 2004; Anderson et al. 2007; Lin et al. 2011). Anderson and Bell (2012) investigated the impacts of the blackout on mortality in New York and found that the event resulted in approximately 90 excess deaths (a rise of 28%), with the mortality remaining slightly higher than average throughout August. Although there was a steep rise in accidental deaths, the authors found that the majority of the excess deaths were from disease-related causes rather than non-accidental. This was due to the lack of power (and therefore the cessation of services and equipment that rely on electricity) exacerbating and complicating the management of illness. Those aged between 65 and 74 were particularly susceptible (Anderson and Bell 2012).

Lifelines can be disrupted by various shocks, including structural and technological failure, human error, targeted attack, and of interest in this chapter, natural hazard events (Murray 2013). The often-large footprint of natural hazard events has the potential to cause simultaneous failure of multiple lifeline components, across one or more networks (Erjongmanee et al. 2008). For example: storms can bring down powerlines with strong winds or fallen trees, while simultaneously cutting off transportation routes with floodwater (Chang et al. 2007; King et al. 2016); earthquakes can destroy multiple lifelines, such as pipes and roads, through ground shaking or liquefaction (Menoni 2001; Giovinazzi et al. 2011; O'Rourke et al. 2012; Lanzano et al. 2014); and only millimetres of ash fall from volcanic eruptions can disrupt most lifeline services (Blong 1984; Jenkins et al. 2014; Wilson et al. 2014).

A further complication of natural hazards is the potential for events to be prolonged. They can, for example, have an extended duration of a week or more, or they might consist of a series of events that occur in succession (Blong et al. 2017). Natural hazards can be associated with secondary hazards, which can occur simultaneously or subsequently. Earthquakes, for example, are often associated with liquefaction, landslides, fires, tsunami and aftershocks (EERI 2011; Daniell et al. 2017). Volcanic eruptions are not only associated with a range of phenomena, such as lava flows, lahars, pyroclastic density currents and ash fall (Jenkins et al. 2014; Wilson et al. 2014), but can also be sustained for weeks, months or even years (Siebert et al. 2011; Sword-Daniels et al. 2014). A prolonged natural hazard event can affect a region for an extended period, causing vast and on-going disruptions to lifeline services vital for disaster response and community recovery.

Natural hazard-prone areas, such as coastal zones and volcanic regions, have attracted large populations of people over time due to the presence of food, natural resources and fertile soils (Chester et al. 2001; Small and Naumann 2001; Klein et al. 2003). Today, some of these settlements have grown into large cities, which continue to expand with urbanisation and natural population growth. A significant proportion of these large cities are located in some of the world's lowest socially and economically developed countries (Kummu and Varis 2011) which sit within the cyclone-prone regions near the equator (10–30 degrees latitude). The increasing demand for development has seen populations and infrastructure expanding into areas more exposed to natural hazards—such as houses being built on previously avoided flood plains—despite the risk (Huppert and Sparks 2006; Cutter and Finch 2008; Lall and Deichmann 2010). The continual and growing presence of populations and infrastructure within natural hazard-prone areas, combined with climate change impacts, has the potential to intensify the exposure and vulnerability of lifelines to natural hazard shocks (Wu et al. 2002; Van Aalst 2006). Consequently, future lifeline failure, of varying degrees of seriousness, may simply have to be expected (Hawk 1999). It is therefore necessary to better understand how lifelines and their functionality might be impacted when subjected to disruption from disasters, and furthermore, the social and economic costs of lifeline failure for at-risk populations.

This chapter aims to add to this area of knowledge by looking at two case study examples: the impact of extreme heat on electricity and transportation systems during the 2009 South-Eastern Australia heatwave and the closure of the European airspace during the 2010 Eyjafjallajökull eruption in Iceland. This chapter explores the compounding effects of lifeline disruption during natural hazards and, using the learnings following these events, discusses potential changes needed to reduce future vulnerability for at-risk populations.

10.2 The 2009 South-Eastern Australia Heatwave

10.2.1 Hazard

Heatwaves are extended periods of hot days and nights resulting in little to no relief from high temperatures. According to the Australian Bureau of Meteorology (BoM), an event is considered a heatwave when there are three or more days of unusually high maximum and minimum temperatures (Bureau of Meteorology 2018). Alternatively, Scorcher (2019) defines a heatwave as '*... at least three consecutive days where the daily maximum temperature is in the top 10% of warmest temperatures for that calendar date*'.

Heatwaves are known to cause serious health, social and economic problems (McInnes and Ibrahim 2013; Wong 2016). Although not as dramatic as other natural hazards, such as wildfires and major storms, heatwaves can cause great loss of life. The 2003 European heatwave, for example, resulted in extensive loss of life with more than 40,000 heat-related deaths (García-Herrera et al. 2010). In Australia, heatwaves have been the most significant natural hazard in terms of lives lost (except for disease epidemics), causing more than 4,000 deaths over the past 200 years (see also Chap. 8) (McInnes and Ibrahim 2013; Coates et al. 2014).

An individual's ability to respond to a heatwave depends on both the degree of exposure to the heat hazard and their adaptive capacity, which is influenced by social, economic and biophysical factors and their access to resources, technology, information and infrastructure (García-Herrera et al. 2010; Reeves et al. 2010; Coates et al. 2014; Wong 2016). Elderly, very young, ill, urban, marginalised and socially isolated residents are often identified as the most vulnerable groups during a heatwave, often lacking the capacity to avoid or reduce exposure to the heat hazard (Reeves et al. 2010). Elderly populations are more likely to experience poor health, social isolation, reduced mobility and lower socio-economic circumstances (McInnes and Ibrahim 2013; Coates et al. 2014). They also tend to have a reduced thermoregulatory and physiologic heat-adaptation capability and an altered sense of thirst (McInnes and Ibrahim 2013; Coates et al. 2014). Urban populations experience higher ambient temperatures than neighbouring rural areas due to the urban heat island effect (Wong 2016; Petkova et al. 2014). The homeless are often neglected in impact assessments but are at considerable risk from heat exposure during heatwave events. This population has compromised health; limited access to resources including food, water and shelter; and, this increased risk of exposure is often coupled with the use of drugs and alcohol (Reeves et al. 2010).

Heatwaves can put a strain on medical facilities and support services, not just with an increase in morbidity but also by impacting the lifelines that support services heavily rely on such as electricity systems and transportation infrastructure (Reeves et al. 2010). Heatwaves can cause an increase in demand on electricity grids as residents turn to air-conditioning and electric fans to keep cool (Miller et al. 2008). On top of this, high temperatures can impact the electricity infrastructure directly. Most problems occur when underground transmission lines overheat and short out or when overheated above ground power lines, which have stretched and sagged, come into contact with trees and short to the ground (Palecki et al. 2001). To keep systems from overheating and to avoid complete electricity outage, the total amount of energy produced must be reduced and distributed to end-users (McEvoy et al. 2012). To mitigate overheating and potential fires produced from electrical system faults, rolling blackouts can be enforced (Broome and Smith 2012).

Extreme temperatures experienced during heatwave events can also directly impact transportation infrastructure by buckling rail lines and melting roads (Palecki et al. 2001; Zuo et al. 2015). Indirectly, transportation systems such as public transport and traffic signals, which rely on power to operate, can be impacted by the loss of electricity (Rinaldi et al. 2001). Extreme temperatures can also affect the health of public transport employees and passengers (Reeves et al. 2010; McEvoy et al. 2012).

10.2.2 Event Overview

In 2009, South-Eastern Australia experienced its most extreme heatwave on record. The states of Victoria and South Australia were exposed to severe, extensive and prolonged heat across two weeks in January and February. These conditions were caused by the stalling of a hot air mass over South-Eastern Australia due to slow-changing synoptic conditions, with sea and bay breezes providing only little relief in coastal areas (Reeves et al. 2010; McEvoy et al. 2012). During this period, new daily maximum temperatures were observed for Adelaide (45.7 °C) and Melbourne (46.4 °C) (Reeves et al. 2010), the state capital cities for South Australia and Victoria, respectively. Adelaide experienced eight consecutive days over 40 °C and Melbourne had three days over 43 °C, 12–15 °C above the seasonal average (Reeves et al. 2010; McEvoy et al. 2012). The Victorian town of Hopetoun experienced a record high of 48.88 °C (McInnes and Ibrahim 2013).

The extreme heat experienced during the 2009 heatwave resulted in a dramatic increase in mortality and morbidity. The persistent high temperatures resulted in increased cases of heat-related illness and exacerbated chronic disease. There was an increase in emergency ambulance dispatches and presentations of heat-related conditions at emergency departments (Reeves et al. 2010; Lindstrom et al. 2013). The extreme heat resulted in an estimated 374 excess deaths in Victoria and between 50 and 150 excess deaths in South Australia (Department of Human Services 2009; Reeves et al. 2010; McInnes and Ibrahim 2013). Most of the deaths that occurred in Victoria during the 2009 South-Eastern Australia heatwave were those aged 75 years or over (Table 10.1) (Department of Human Services 2009; Reeves et al. 2010; Bi et al. 2011; McInnes and Ibrahim 2013). In Adelaide, South Australia, Zhang et al. (2013) found that the populations most at risk during the heatwave were those who lived alone, had a lower socio-economic status, had renal problems or were at risk of falls and needed help from community services. It was also noted that occupational heat exposure could have been a cause of an increase in hospitalisations of working-age men (35–64 years) during the 2009 event (Reeves et al. 2010; Bi et al. 2011). However, this needs further investigation.

Table 10.1 Total mortality between 26 January and 1 February 2009: Expected deaths (derived from mean deaths in 2004–2008) versus 2009 reported deaths (based on data extracted from Department of Human Services (2009))

Age	Expected deaths	2009 deaths	Excess deaths	% Total excess	% Increase
≥75	388	636	248	66	64
65–74	99	145	46	12	46
5–64	116	180	64	17	55
0–4	4	7	3	1	–
Unknown*	–	~ 12	~ 12	~ 4	–
Total	606**	980	374***	100	62

*A small number of deaths were reported with unknown age

**Total noted in Department of Human Services (2009), sum of expected deaths here is 607. Difference may be due to rounding

***Total noted in Department of Human Services (2009), sum of expected deaths here is 373. Difference may be due to rounding

10.2.3 Lifeline Disruption and Compounding Impacts

On 29 and 30 January 2009, Melbourne experienced rolling blackouts due to a combination of failures that occurred throughout the system. Supply was impacted when the Basslink electricity supply connection between Tasmania and Victoria was shut down due to heat-related issues, and Victorian generators were unable to supply additional power. Transformer faults resulted in outages of major transmission lines, and supply loads were restricted in the Western metropolitan area. Additional faults occurred in up to 50 local voltage transformers. Load shedding was finally required to protect the security of the electricity network, further restricting supply (Reeves et al. 2010; McEvoy et al. 2012). Cumulative system faults, ageing infrastructure and a rapid increase in demand (breaking Victoria's load record by 7%), primarily from the use of air-conditioning, all contributed to the system's vulnerability to the heatwave (Reeves et al. 2010; McEvoy et al. 2012). Rolling blackouts resulted in over 500,000 residents being without power on the night of 30 January. The outages occurred for 1–2 hours, but the ripple effects lasted up to two days (McEvoy et al. 2012).

The periodic loss of electricity limited access to air-conditioning, electric fans and refrigeration, key aids that people rely on to cope with prolonged high temperatures. This was especially detrimental to those most vulnerable to heatwaves, such as the elderly, ill and isolated (Reeves et al. 2010). This event also showed how vitally important night-time relief from extreme heat is. During the 2009 heatwave, many deaths occurred overnight where residents were found in stuffy rooms with air-conditioning turned off (Reeves et al. 2010). In these cases, it was not just the blackout that hindered the use of such services, but the cost of electricity also, with many residents unable to afford to continually run air-conditioning units overnight (Reeves et al. 2010).

Transportation systems suffered minor to moderate disruptions during the heatwave, with rail being the most affected by the high temperatures. During the initial heatwave peak (27–30 January), more than a third of the train services were cancelled in Melbourne and 7% in Adelaide (Reeves et al. 2010; McEvoy et al. 2012). There were nearly 30 reported instances of track buckling in Melbourne, which slowed down or disrupted services and increased maintenance and repair expenses for the rail transport industry (McEvoy et al. 2012). Also, half of the train fleet were older stock with air-conditioning units not designed to operate above 34.5 °C. Air-conditioning failure was the leading cause for service cancellations due to industrial action taken by train drivers (Reeves et al. 2010; McEvoy et al. 2012). This is an important factor often forgotten when considering critical infrastructure vulnerability. Although there may not be any direct impacts to infrastructure components, workers themselves may not be able to continue to operate the systems under extreme heat conditions. The operation of lifelines can also rely on the operation of other lifeline infrastructures. Electricity outages, for example, resulted in the indirect disruption of transportation systems. Rolling blackouts on 30 January impaired traffic signals at 124 intersections in Melbourne and resulted in the cancellation of city loop train services, stranding a large number of inner-city commuters (Reeves et al. 2010). Overall, power outages and transport disruption resulting in an estimated financial loss of AU$800 million (Reeves et al. 2010).

Although transport was not vital in mitigating heat exposure, forcing commuters to use alternatives, such as biking or walking, could lead to increased exertion and exposure to extreme heat conditions. The loss of train services also impacted public transport users by disrupting their access to work, home or loved ones. With public transport often being an affordable form of transport for students, the elderly and poor, the loss of operation largely impacted the mobility of these demographic groups (Rodrigue et al. 2017).

10.2.4 Learnings

The 2009 heatwave was characterised by a substantial increase in health service demand and disruptions to electricity supplies and public transport systems. Governments, councils, utility providers, hospitals, emergency response organisations and the community were largely underprepared for an event of this magnitude, with the extreme conditions not anticipated in seasonal forecasts (Reeves et al. 2010). The number of excess deaths (approximately 374) demonstrates demographic impacts from severe heatwaves are most acute for older residents (at 66% of excess deaths of those 75 years or older) (see Table 10.1).

Reeves et al. (2010) concluded in their report on the 2009 event that extreme heat should be given the same prominence as high-impact natural hazards such as wildfires or flooding with regards to impacts on lifeline infrastructure. The report highlighted how vulnerable lifelines, specifically electricity and transportation systems can be to heat stress and how significant the impacts of lifeline failure, both direct and indirect, can be for people and the urban system. Overall it was noted, '*people mostly manage when they have power, when they have transport*' (Reeves et al. 2010 p. 99). The loss of these services had adverse affects on people's ability to cope and compounded the experience of the heatwave event. Broome and Smith (2012) calculated that being without power, and therefore air-conditioning, would increase the risk of those susceptible of dying from heat-related illness by 50%, especially among the elderly and those in remote rural communities. These authors estimated that if electricity was cut for an entire day in Victoria during the 2009 heatwave, 28 additional deaths could have occurred. Therefore, the urban system as a whole, including lifelines, needs to be considered in future risk assessments and mitigation measures.

Following the 2009 heatwave event, both South Australia and Victoria moved from a reactive and response-driven approach to mitigation and risk reduction. Health and emergency services developed strategies to identify and manage vulnerable groups (Reeves et al. 2010). Vulnerabilities and lack of redundancy in lifelines, such as transport and power, were highlighted. Attitudinal, socio-economic, behavioural and financial barriers to improve resilience in future heatwave events were also identified. Specifically, the increase in population vulnerability to heat-related mortality due to an ageing population and an increase in obesity in adults were highlighted as areas in need of attention (Petkova et al. 2014). It was also identified that there was an increase in the expectation and dependence on emergency services for warnings and timely advice (Reeves et al. 2010). It must be noted, however, that the unfortunate occurrence of the Black Saturday bushfires during the second phase of the heatwave overshadowed the impacts of the heatwave itself, hindering the publicity of the huge health impacts of the 2009 heatwave, lifeline vulnerabilities to extreme heat and post-event reflection.

In January 2014, South-Eastern Australia was again hit by a heatwave. Although maximum temperatures were not as high as those experienced during January 2009, mean temperatures were higher and the heatwave peak lasted for longer (four days in 2014 compared to three days in 2009) (Department of Health 2014). The 2014 heatwave resulted in an estimated 167 excess deaths in Victoria, compared to the 374 excess deaths in 2009. The decrease in estimated excess mortality was contributed to the implementation of Victoria's Heatwave Plan (Department of Health 2014). Since 2014, the Victorian State Government also focused on educating vulnerable populations on how to be prepared for electricity failure in their 'How to cope and stay safe in extreme heat' brochures (Victoria State Government 2015). Additional research is needed to determine if this information positively influences people's preparedness for, and actions during, heatwave events in the future. It would also be interesting to see if this information reached those most vulnerable.

The transportation sector also took steps to adapt to extreme heat events by: upgrading rail infrastructure to prevent track buckling; creating better heat policies for drivers and providing easier access to cold water and ice; and creating better contingency plans including providing stand-by replacement services (Chhetri et al. 2012). The 2014 heatwave resulted in minor disruption to Melbourne's public transport as trains operated at reduced speeds (Mullett and McEvoy 2014). The electricity sector also fared better in 2014 with the system avoiding load shedding during the highest temperature period. Only small local distribution outages were experienced due to distribution equipment failure; however, periods of low reserves were of concern. AEMO (2014) highlighted that any fault with an interconnector or major generator could have changed the outcome and resulted in load shedding. The system's ability to cope in 2014 was in part attributed to the contribution of embedded solar photovoltaic (PV) generation, namely in South Australia, which helped to support the peak usage period (AEMO 2014).

There is still much to be done to increase resilience and adaptation capacity to heatwaves within Australia. Although many stakeholders took steps to assure future risks were being adequately managed, they appeared to have done so in isolation (McEvoy et al. 2012). Due to the integrated nature of the urban system, there is a need for all stakeholders to take on a 'whole system' approach. McEvoy et al. (2012) concluded that sectoral segmentation and the lack of holistic urban infrastructure management are major barriers to improving urban resilience to future heatwave events.

Heat-sensitive infrastructures such as the electricity network and rail transport will continue to be vulnerable to extreme heat events in the future, compounding the impacts of events. A warming climate may make heatwaves more likely (Meehl and Tebaldi 2004; Revi et al. 2014), putting more pressure on expanding urban areas (McInnes and Ibrahim 2013). Expanding urbanisation and high-density housing will also exacerbate the situation through urban heat island effects (Coates et al. 2014; Petkova et al. 2014; Wong 2016). Furthermore, populations are increasingly living and working in climate-controlled environments, relying on air-conditioning to reduce heat stress and therefore isolating people from a changing climate and limiting their ability to acclimatise (Coates et al. 2014; Reeves et al. 2010). This further increases the dependence on air-conditioning, whose operation cannot be guaranteed during a heatwave. Air-conditioning is also counter-productive in reducing the risk of heat stress. Farbotko and Waitt (2011) note that the use of air-conditioning to decrease resident exposure to extreme heat results in a spike in energy demand during a heatwave, which contributes to system-wide blackouts, hiking energy prices as a result and exacerbating existing social-economic inequalities. On top of this, the waste heat released from air-conditioners also contributes to higher outside temperatures, worsening the urban heat island affect (Salamanca et al. 2014).

10.3 The 2010 Eyjafjallajökull Volcanic Eruption, Iceland

10.3.1 Hazard

Volcanic ash is the material produced by explosive volcanic eruptions and is made up of tiny fragments of rock and glass (Wilson et al. 2012). Volcanic ash can spread far and wide; it is hard, highly abrasive, corrosive and conductive; and only millimetres are needed to disrupt most essential lifeline services (Blong 1984; Barsotti et al. 2010; Wilson et al. 2011, 2012; Magill et al. 2013; Jenkins et al. 2014). Ash easily infiltrates openings, clogs air-filtration systems and abrades surfaces (Jenkins et al. 2014). Air transportation, in particular, can be impacted by volcanic ash both on the ground and in the air. Airline operations can be affected by ash falling on airport runways, and if volcanic ash is erupted high enough into the atmosphere, it can become a risk to flying aircraft (Casadevall 1994; Jenkins et al. 2015; Webley 2015).

The risk volcanic ash poses for modern aviation was brought to attention in 1982 when a British Airways flight flew through high concentrations of volcanic ash produced by an eruption at Mount Galunggung in West Java. The Boeing 747 lost power to all four engines and lost over 12,000 feet of altitude before restoring power and making an emergency landing in Jakarta (Lund and Benediktsson 2011; Ellertsdottir 2014; Gislason et al. 2011). The 1989 Redoubt eruption in Alaska, and 1991 Mount Pinatubo eruption in the Philippines, also caused engine failure and extensive damage to aircraft engines and windshields that came in contact with ash clouds (Przedpelski and Casadevall 1994; Casadevall et al. 1996; O'Regan 2011; Webley 2015). With the conclusion that volcanic ash can cause jet-engine failure, great financial loss and potentially loss of life, a 'no-threshold' guideline was, at the time of the 2010 Eyjafjallajökull eruption, universally adopted and no-fly zones were implemented whenever volcanic ash was detectable in airspace (O'Regan 2011; Ellertsdottir 2014).

10.3.2 Event Overview

On 20 March 2010, fire-fountain activity started along a fissure at Fimmvörðuhálsi, an ice-free area between Eyjafjallajökull and Mýrdalsjökull volcanic vents in Iceland, and ended on 13 April, leading many to believe that it was the end of the eruption (Donovan and Oppenheimer 2011; Þorkelsson et al. 2012). However, on 14 April, the eruption resumed, this time at the summit of Eyjafjallajökull. The interaction of ice and melt water with a more viscous batch of magma resulted in a more explosive eruption than the first phase (Donovan and Oppenheimer 2011; Gislason et al. 2011; Þorkelsson et al. 2012), sending ash up to 10 km in the atmosphere (Carlsen et al. 2012; Þorkelsson et al. 2012; Stevenson et al. 2012).

Ash fell to the South of the volcano impacting rural communities, causing impacts to agriculture and local tourism (Adey et al. 2011; Donovan and Oppenheimer 2011; Bird and Gísladóttir 2012; Bird et al. 2018). The eruption was accompanied by a number of glacial outburst floods, which impacted roads and farmland and resulted in the evacuation of ~800 residents (Þorkelsson et al. 2012). The majority of Iceland to the North of the volcano was relatively unaffected by the ash. Reykjavik, in particular, saw limited impacts despite being less than 150 km away from the volcano, with the airport being briefly closed on the 21 April (Lund and Benediktsson 2011; Þorkelsson et al. 2012). The most significant impact was caused by ash that remained in the atmosphere. This airborne ash was rapidly blown to the South and East towards mainland Europe and caused the largest shutdown of European airspace since World War II (Budd et al. 2011; Ellertsdottir 2014).

As the volcanic ash cloud from Iceland made its way across Europe, the European Aviation Authorities progressively closed sectors of airspace due to fears for public safety (Ellertsdottir 2014). The airspace over Scotland and Norway was first to close on the evening of 14 April. Irish, Dutch, Belgian and Swedish airspace also saw restrictions as the ash spread further South and East. By 18 April, airspace over Ireland, Ukraine, and Canary Islands was also closed (Budd et al. 2011). Flight restrictions were lifted on 21 April with close to normal air traffic resuming the next day (Ellertsdottir 2014). Over the seven-day period, more than 100,000 flights in, out and around Europe were cancelled (Budd et al. 2011, IATA 2010, Oxford and Economics 2010). The worst affected places were the United Kingdom, Ireland and Finland, who experienced a 90% decrease in air traffic (Ellertsdottir 2014). Maximum closure of airspace occurred on 18 April which grounded nearly 30 per cent of the world's scheduled flight capacity, costing the airline industry US1.7 billion in lost revenue (IATA 2010). This loss could not typically be claimed under the airlines' business interruption insurance since there was no 'material damage' (O'Regan 2011). It was reported that aviation companies risked being sent into bankruptcy if the closure had lasted much longer (Ellertsdottir 2014).

10.3.3 Flow-On Effects from Lifeline Disruption

The European airspace is one of the busiest in the world, used by 150 airlines making 9.5 million flights every year across 150,000 air routes (O'Regan 2011). The airspace and aircraft that occupy it form part of a large interconnected network that provides private, military and economic mobility. The complex network of flight paths, airways and control zones, navigated by flight crew, air traffic control, collision avoidance software and controlled by strict international regulations, is largely unseen by passengers (Budd et al. 2011). It is a system often taken for granted until flights are delayed, diverted or cancelled.

The impact of the closure of the European airspace in 2010 was far-reaching and uncovered a vast dependence on global air travel for the mobility of people and product. Air transport is relied upon by manufacturers to join spatially disaggregated operations and enables time-sensitive and perishable freight to be carried over long distances in a short time (Bowen and Leinbach 2006; Pedersen 2001; Button and Yuan 2013; Mukkala and Tervo 2013; Rodrigue et al. 2017).

Shortages of imported flowers, fruits and electronic hardware were reported in the immediate days after airspace closure. Pharmaceutical, automotive, transport and delivery companies were also impacted. The hardest hit were those supply chains that rely on airfreight for just-in-time deliveries and exporters of perishable goods. Where some products could be delivered later, perishable goods could not (Oxford Economics 2010). Flower and fruit growers in Africa were especially hard hit, with fresh products due to be air-freighted to Europe being left to rot (Oxford Economics 2010; Budd et al. 2011; Ellertsdottir 2014). The World Bank estimated that the airspace shutdown cost African countries US$65 million in exports (Oxford and Economics 2010). Kenya's flower and vegetable industry, which contributes over a fifth of the country's GDP, employs tens of thousands of workers. During the closure of the European airspace planes that were due to carry the produce to Europe remained grounded at Nairobi's international airport and the industry lost an estimated $1.3 million per day (Wadhams 2010). The unmoveable produce had to be thrown out or composted, and thousands of workers were temporarily laid off. Although working conditions on Kenyan farms has improved since the formation of the Kenya Flower Council in 1999, the majority of workers are paid below a living wage and unable to amass any savings (Nowakowska 2015; Leipold and Morgante 2013). Therefore, being without paid work for any extended period of time would vitally disadvantage people who depend on the farms for their livelihood.

The halt of air travel around Europe also led to shortages of air-freighted components necessary for electrical and car manufacturing. For example, car production at Nissan and BMW plants in the USA, Japan and Germany was temporarily suspended due to delays in the delivery of pressure sensors (Oxford and Economics 2010).

The estimated 10 million travellers that were stranded by this event faced significant delays and additional costs (Budd et al. 2011; Oxford and Economics 2010). Those unable to afford alternative transportation or accommodation were stuck at airports (Ellertsdottir 2014). Fortunately, many passengers were compensated for cancelled flights by the airlines, as required by the European Regulation 241/2004 (Ellertsdottir 2014). The low-cost airlines, which were hard hit by the event economically, contested passenger compensation requested by the European Commission arguing that the situation was outside of their control. However, the European Commission continued to hold the airlines responsible for their passengers (Ellertsdottir 2014).

The week-long cessation of air travel over the United Kingdom and Europe reduced global mobility, impeded economic activities and hindered social connections. Reducing transportation to land and sea, which is generally confined by geographic barriers, dramatically increases travel time, disrupts supply chains and results in the isolation of some populations and economies. Any extended halt to a large portion of the world's air transport capabilities would ultimately impact on development, globally. It was fortunate that the 2010 eruption was relatively short-lived, as past eruptions at Eyjafjallajökull—and other volcanoes around the world—have lasted for months to years (Gertisser 2010).

10.3.4 Learnings

With around 60 volcanic eruptions occurring world-wide every year, volcanic ash, or the potential for it, causes airspace restrictions on an almost daily basis (Donovan and Oppenheimer 2011). Iceland's volcanoes, in particular, produce around 20 eruptions per century, many involving significant explosive activity (Langmann et al. 2012). Tephra from numerous eruptions from Icelandic volcanoes over the past 7,000 years were documented to have reached Europe (Swindles et al. 2011; Stevenson et al. 2012). The 2010 eruption at Eyjafjallajökull itself was not an extraordinary event. However, the disruption it caused, to an industry that should have been prepared for its occurrence, was.

Although the risk of volcanic ash was recognised by scientists, operational meteorological institutes and aviation authorities, it was considered a relatively low risk and therefore had not penetrated into policy (Adey et al. 2011; Donovan and Oppenheimer 2011). Because of this, the relatively small eruption of 2010 required a hasty and extreme response, which resulted in the reactive formation of advisory committees and meetings throughout Europe (Bonadonna and Folch 2011a, b; Donovan and Oppenheimer 2011; Bonadonna et al. 2012).

The heavy reliance on the airline industry for world trade and transportation resulted in the 'no-threshold' guideline, which was in place at the time of the 2010 eruption, being questioned. Decision-makers were caught between public safety concerns and the demanding need for global mobility (Lund and Benediktsson 2011). The complete closure of European airspace was seen as an overreaction, and a review of the normal procedures was demanded. A study by Gislason et al. (2011) on the physical and chemical properties of the Eyjafjallajökull ash showed that very sharp, hard and fine-grained particles could have put aircraft at risk from abrasion and melting in jet engines. However, at the time, what constituted a safe concentration of volcanic ash was highly contested among international safety regulators, airlines, aircraft engineers and manufactures (Budd et al. 2011; Ellertsdottir 2014). As a result, major airlines, including Lufthansa, KLM and British Airways, performed a series of test flights and determined that 2,000 microg of ash per cubic metre was an accepted threshold through which aircraft could fly (Budd et al. 2011; O'Regan 2011).

The response to the 2010 eruption and airspace closure was hindered by the lack of data detailing ash tolerance of aircraft and engines and the inability of international safety regulators, airlines, aircraft engineers and manufactures to agree on a 'safe' concentration of atmospheric ash (Budd et al. 2011). Budd et al. (2011) also highlighted that national policies were not aligned with other countries, such as the USA. Had the creation of advisory groups occurred before the eruption, rather than a reactive scramble, preparations could have been made both politically and financially, and safe ash thresholds predetermined (Donovan and Oppenheimer 2011).

This event highlighted the dependency on global mobility and the need to recognise that natural hazards and their impacts are not always confined to geographical or political boarders. Globalisation turned an ordinary geological event into near worldwide chaos. Economies around the world have been made more vulnerable to disruptions to the normal flow of goods by increased globalisation in trade, outsourcing and the advent of leaner global supply chains, focused on cutting costs (Besedeš and Murshid 2017). In the aftermath of the 2010 Eyjafjallajökull eruption, there was a call for increased supply chain flexibility through adaptive logistic capacities and sourcing products from various locations. In their paper on the trade-effects of airspace closures in the aftermath of Eyjafjallajökull Besedeš and Murshid (2017) found that some supply chains were able to combat some of the disruptions. The authors found that the US market, in particular, was able to form trading relationships with countries outside of Europe to supplement some of the lost supplies.

The African flower and vegetable growers, on the other hand, did not have fallback options. The flower market in Kenya depends almost wholly on Europe for its exports. With a low domestic demand, the industry (Kenya's top exchange earner) is at the peril of fluctuating foreign demand and the availability of airfreight into Europe (Leipold and Morgante 2013; Kargbo et al. 2010). At the time, Kenyan horticulture was still recovering from previous impacts from domestic post-election violence in 2008/2009 (Justus 2015). These disruptive events resulted in the call for Kenya's flower and vegetable industry to reduce over-dependency on foreign global markets and diversify into local and regional alternative market outlets to provide a short-term market shock cushion (Justus 2015; Leipold and Morgante 2013). Once some airports reopened tons of produce was flown to Spain and put on trucks to Northern Europe (Gettlemen 2010).

The 2010 Eyjafjallajökull eruption was an expensive lesson to learn but, undoubtedly, it better prepared the airline industry and global businesses for future volcanic eruptions. This was highlighted in 2011 when another Icelandic volcano, Grímsvötn, began to erupt. At its peak, the eruption column reached a height of 20 km, compared to approximately 10 km of Eyjafjallajökull's and produced nearly twice as much volcanic ash (Stevenson 2012). Although the 2011 eruption of Grímsvötn was nearly 100 times larger in magnitude than that of Eyjafjallajökull in 2010, it did not have the same impact. The situation was different, with different eruption parameters and weather conditions, and prevailing winds prevented large amounts of ash from entering international airspace. Ash that entered the airspace was better dealt with. The new rules around safe ash concentrations meant the disruption in Europe was relatively minor, with just 900 out of 90,000 scheduled flights cancelled during the first three days of the eruption compared to 42,600 flights cancelled in the first three days in 2010 (European Commission 2011; Parker 2015).

10.4 Discussion

The failure of lifeline networks during natural hazard events has the potential to impact populations by exacerbating the hazard itself and/or hindering the ability to respond to or recover from the event. The primary demographic impact seen in the case of lifeline failure in 2009 in South-Eastern Australia was excess mortality, particularly in older age groups. Lifeline failure can also propagate outside the reach of the hazard footprint, causing disruption in regions not directly impacted by the event. In the case of the 2010 Eyjafjallajökull eruption, one of the industries hardest hit was the flower and fruit growers in Africa, ~8,500 kms away.

The outcomes of both these case study events were influenced by unpreparedness. For South-Eastern Australia it was the magnitude of the event itself. South-Eastern Australia has suffered from heatwave events previously (e.g. 1908 and 1939) but the heatwave of 2009 was extraordinarily extensive, long-lasting and severe compared to previous events (Chhetri et al. 2012). Meanwhile, potential impacts from Icelandic volcanos for the United Kingdom/Europe were a scenario deemed unlikely and hence previously ignored. Both events became the much needed wakeup call to spark major reviews of management plans and policies. Implementation of various recommendations has since proven to be beneficial, improving the management of subsequent events.

The improvements made in each case study are commendable. However, particularly in the case of South-Eastern Australia, the approach to hazard mitigation was disjointed with minimal sector cross-over. In their report on Australia's response and adaptation to major weather events, Chhetri et al. (2012) found that individual post-event actions usually only result in small improvements in overall resilience. To improve urban-wide resilience to natural hazard events, there is a need to improve communication, information sharing, collaboration and coordination between all sectors. The fragmented nature of critical infrastructure sectors, like transportation and electricity, and the spanning of lifeline infrastructure across local or state government borders can make the development of coordinated policies and planning frameworks a challenge, hindering a structured response to natural hazard shocks.

The world has never relied on lifelines as much as it does now, and our reliance on lifeline services and technology will only continue to grow alongside increasing urbanisation and ageing populations. The increasing concentration of populations and built environments in hazard-prone regions combined with the impacts of climate change are likely to increase the exposure of critical infrastructure and essential services to natural hazard shocks. Future natural hazard events can almost always be expected to result in lifeline disruption or failure. Not only is it important to make lifeline infrastructure more resilient to disruption from future shocks, there is also a need to increase resilience by preparing communities to better cope with service outages.

Furthermore, with an increasingly connected world there is potential for the cost of indirect disruption—both economic and social—to match, if not surpass, the cost of direct damage from hazard events and the case studies in this chapter show how costly it is to be caught unprepared. To be resilient to disruption in this globalised world, we need to look 'beyond our backyards' and acknowledge that lifelines can often span large geographic areas, with the ripple effects of disruption felt beyond regional and national borders. The vulnerability of lifelines to disruption from disasters is also dependent on the quality and management of infrastructure. The impacts of lifeline failure on a population depend on the time it takes for the lifeline to return to operation and who is responsible for the cost of repairs and resulting disruption.

Societal factors such as increasing disparities in wealth and health can contribute to greater vulnerability of particular sections of the population (Cutter et al. 2000). It has been demonstrated time and again how demographics—age, ethnicity, income, gender and housing—can influence (amplify or reduce) overall vulnerability to natural hazards (Donner and Rodríguez 2008; Wisner et al. 2004; Cutter et al. 2003, 2000). This chapter has shown that the ability to cope with lifeline failure is also dependent on these factors. Therefore, in preparation for the true impacts of future natural hazard events on modern society, we need to better understand the interconnectedness and behaviour of lifeline networks and to identify vulnerable populations that rely on their operation. Future disaster mitigation plans not only need to incorporate lifeline infrastructure systems but also take into account the ability of the population to adapt or cope with lifeline disruption. To do this, more work needs to be done in measuring the social impacts of lifeline failure. In their review on power outages during extreme natural events, Klinger et al. (2014) found that power outages resulted in health impacts by inhibiting people accessing healthcare but noted there were few attempts to quantify health impacts in terms of morbidity, mortality or quality of life in the literature. The authors note that this area of research can be difficult but stress that more work needs to be done so we can learn from past events and improve resilience for the future. It is not only the failings that we should look at either but also successes, where communities manage to avoid lifeline failure during extreme events or were able to come up with solutions for outages. More work is needed in collating effective strategies so that these experiences and knowledge can be shared and utilised (Klinger et al. 2014).

Acknowledgments The author thanks Dr. Christina Magill and Dr. Deanne Bird for editing this chapter. The author also acknowledges financial support from the Bush-fire and Natural Hazards Cooperative Research Centre, Macquarie University Research Excellence Scholarship and a scholarship top top-up from Risk Frontiers.

References

Adey, P., B. Anderson & L. L. Guerrero (2011) An ash cloud, airspace and environmental threat. *Transactions of the Institute of British Geographers,* 36, 338–343.

AEMO. (2014) Heatwave 13–17 January 2014. Australian Energy Market Operator.

Anderson, C. W., J. R. Santos & Y. Y. Haimes (2007) A risk-based input-output methodology for measuring the effects of the August 2003 Northeast Blackout. *Economic Systems Research,* 19:2, 183–204.

Anderson, G. B. & M. L. Bell (2012) Lights out: Impact of the August 2003 power outage on mortality in New York, NY. *Epidemiology (Cambridge, Mass.),* 23, 189–193.

Barsotti, S., D. Andronico, A. Neri, P. Del Carlo, P. Baxter, W. Aspinall & T. Hincks (2010) Quantitative assessment of volcanic ash hazards for health and infrastructure at Mt. Etna (Italy) by numerical simulation. *Journal of Volcanology and Geothermal Research,* 192, 85–96.

Besedeš, T. & A. P. Murshid (2017) *Experimenting with Ash: The Trade-Effects of Airspace Closures in the Aftermath of Eyjafjallajökull.* Working paper. Georgia Institute of Technology. Available at: https://besedes.econ.gatech.edu/wp-content/uploads/sites/322/2017/08/besedes-volcano.pdf Accessed 22/05/2019.

Bi, P., S. Williams, M. Loughnan, G. Lloyd, A. Hansen, T. Kjellstrom, K. Dear & A. Saniotis (2011) The effects of extreme heat on human mortality and morbidity in Australia: implications for public health. *Asia Pacific Journal of Public Health,* 23, 27S–36S.

Bird, D. K. & G. Gísladóttir (2012) Residents'attitudes and behaviour before and after the 2010 Eyjafjallajökull eruptions—a case study from southern Iceland. *Bulletin of volcanology,* 74, 1263–1279.

Bird D.K., Jóhannesdóttir G., Reynisson V., Karlsdóttir S., Gudmundsson M.T., Gísladóttir G. (2018) Crisis Coordination and Communication During the 2010 Eyjafjallajökull Eruption. In: Fearnley C.J., Bird D.K., Haynes K., McGuire W.J., Jolly G. (eds) Observing the Volcano World. Advances in Volcanology (An Official Book Series of the International Association of Volcanology and Chemistry of the Earth's Interior – IAVCEI, Barcelona, Spain). Springer.

Blong, R., C. Tillyard & G. Attard. (2017) Insurance and a Volcanic Crisis—A Tale of One (Big) Eruption, Two Insurers, and Innumerable Insureds. In *Advances in Volcanology.* Berlin, Heidelberg: Springer.

Blong, R. J. (1984) Volcanic hazards: a source book on the effects of eruptions. Sydney: Academic Press.

Bureau of Meteorology (2018). About the Heatwave Service. https://www.bom.gov.au/australia/heatwave/about.shtml Accessed 5/6/2018.

Bonadonna, C. & A. Folch. (2011a) Ash Dispersal Forecast and Civil Aviation Workshop - Consensual Document.

Bonadonna, C. & A. Folch. (2011b) Ash Dispersal Forecast and Civil Aviation Workshop - Model Benchmark Document.

Bonadonna, C., A. Folch, S. Loughlin & H. Puempel (2012) Future developments in modelling and monitoring of volcanic ash clouds: outcomes from the first IAVCEI-WMO workshop on Ash Dispersal Forecast and Civil Aviation. *Bulletin of Volcanology,* 74, 1–10.

Bowen, J. T. & T. R. Leinbach (2006) Competitive advantage in global production networks: air freight services and the electronics industry in Southeast Asia. *Economic Geography,* 82, 147-166.

Broome, R. A. & W. T. Smith (2012) The definite health risks from cutting power outweigh possible bushfire prevention benefits. *Medical Journal of Australia,* 197, 440.

Brozović, N., D. L. Sunding & D. Zilberman (2007) Estimating business and residential water supply interruption losses from catastrophic events. *Water Resources Research,* 43, W08423.

Budd, L., S. Griggs, D. Howarth & S. Ison (2011) A fiasco of volcanic proportions? Eyjafjallajökull and the closure of European airspace. *Mobilities,* 6, 31–40.

Button, K. & J. Yuan (2013) Airfreight Transport and Economic Development: An Examination of Causality. *Urban Studies,* 50, 329–340.

Carlsen, H. K., T. Gislason, B. Benediktsdottir, T. B. Kolbeinsson, A. Hauksdottir, T. Thorsteinsson & H. Briem (2012) A survey of early health effects of the Eyjafjallajökull 2010 eruption in Iceland: a population-based study. *BMJ open,* 2, e000343.

Casadevall, T. J. (1994) *Volcanic ash and aviation safety: proceedings of the first international symposium on volcanic ash and aviation safety.* US Government Printing Office.

Casadevall, T. J., P. Delos Reyes & D. J. Schneider (1996) The 1991 Pinatubo eruptions and their effects on aircraft operations. *Fire and Mud: eruptions and lahars of Mount Pinatubo, Philippines,* 625–636.

Chhetri, P., A. Hashemi, F. Basic, A. Manzoni & G. Jayatilleke. (2012) Bushfire, heat wave and flooding - case studies from Australia. *Report from the International Panel of the Weather project funded by the European Commission's 7th framework programme.* Melbourne.

Clancy, J. L. (2000) Sydney's 1998 water quality crisis. *American Water Works Association. Journal,* 92, 55.

Chang, S. E., T. L. McDaniels, J. Mikawoz & K. Peterson (2007) Infrastructure failure interdependencies in extreme events: power outage consequences in the 1998 Ice Storm. *Natural Hazards,* 41, 337–358.

Chester, D. K., M. Degg, A. M. Duncan & J. E. Guest (2001) The increasing exposure of cities to the effects of volcanic eruptions: a global survey. *Global Environmental Change Part B: Environmental Hazards,* 2, 89–103.

Coates, L., K. Haynes, J. O'Brien, J. McAneney & F. D. De Oliveira (2014) Exploring 167 years of vulnerability: An examination of extreme heat events in Australia 1844–2010. *Environmental Science & Policy,* 42, 33–44.

Cowie, J. H., A. T. Ogielski, B. Premore, E. A. Smith & T. Underwood (2003) Impact of the 2003 blackouts on Internet communications. *Preliminary Report, Renesys Corporation (updated March 1, 2004).*

Cutter, S. L., J. T. Mitchell & M. S. Scott (2000) Revealing the vulnerability of people and places: a case study of Georgetown County, South Carolina. *Annals of the association of American Geographers,* 90, 713–737.

Cutter, S. L., B. J. Boruff & W. L. Shirley (2003) Social vulnerability to environmental hazards. *Social science quarterly,* 84, 242–261.

Cutter, S. L. & C. Finch (2008) Temporal and spatial changes in social vulnerability to natural hazards. *Proceedings of the National Academy of Sciences,* 105, 2301–2306.

Daniell, J. E., A. M. Schaefer & F. Wenzel (2017) Losses Associated with Secondary Effects in Earthquakes. *Frontiers in Built Environment,* 3.

Davis, G. (1999) The Auckland electricity supply disruption 1998: emergency management aspects. *Australian Journal of Emergency Management,* 13, 44–46.

Dawson, D. M. & B. J. Brooks. (1999) *The Esso Longford gas plant accident: report of the Longford Royal Commission.* Government Printer, South Africa.

Department of Health (2014) *The health impacts of the January 2014 heatwave in Victoria.* State Government of Victoria Melbourne.

Department of Human Services (2009) January 2009 heatwave in Victoria: An assessment of health impacts. *Victorian Government Department of Human Services.*

Donner, W. & H. Rodríguez (2008) Population composition, migration and inequality: The influence of demographic changes on disaster risk and vulnerability. *Social forces,* 87, 1089–1114.

Donovan, A. R. & C. Oppenheimer (2011) The 2010 Eyjafjallajökull eruption and the reconstruction of geography. *The Geographical Journal,* 177, 4–11.

EERI. (2011) Learnings from earthquakes - the March 11, 2011, Great East Japan (Tohoku) Earthquake and Tsunami: societal dimensions. EERI Special Earthquake Report - August 2011.

Electricity Consumers Resource Council (ELCON) (2004) *The Economic impacts of the August 2003 Blackout.* February 2004.

Ellertsdottir, E. T. (2014) Eyjafjallajökull and the 2010 closure of European airspace: crisis management, economic impact, and tackling future risks. *The Student Economic Review* XXVIII, 129–137.

Erjongmanee, S., C. ji, J. Stokely & N. Hightower. (2008) *Large-Scale Inference of Network-Service Disruption upon Natural Disasters.*
European Commission. (2011) Volcano Grimsvötn: how is the European response different to the Eyjafjallajökull eruption last year? Frequently Asked Questions. Brussels.
Farbotko, C. & G. Waitt (2011) Residential air-conditioning and climate change: voices of the vulnerable. *Health Promotion Journal of Australia,* 22, 13–16.
García-Herrera, R., J. Díaz, R. M. Trigo, J. Luterbacher & E. M. Fischer (2010) A review of the European summer heat wave of 2003. *Critical Reviews in Environmental Science and Technology,* 40, 267–306.
Gertisser, R. (2010) Eyjafjallajökull volcano causes widespread disruption to European air traffic. *Geology Today,* 26, 94–95.
Gettleman, J. (2010) With flights grounded, Kenya's produce wilts. New York Times. https://www.nytimes.com/2010/04/20/world/africa/20kenya.html. Accessed 22/05/2019.
Giovinazzi, S., T. Wilson, C. Davis, D. Bristow, M. Gallagher, A. Schofield, M. Villemure, J. Eidinger & A. Tang (2011) Lifelines performance and management following the 22 February 2011 Christchurch earthquake, New Zealand: highlights of resilience. *Bullentin of the New Zealand Society for Earthquake Engineering,* 44, 402–417.
Gislason, S. R., Hassenkam, T., Nedel, S., Bovet, N., Eiriksdottir, E. S., Alfredsson, H. A., Hem, C.P., Balogh, Z.I., Dideriksen, K., Oskarsson, N. and Sigfusson, B. (2011). Characterization of Eyjafjallajökull volcanic ash particles and a protocol for rapid risk assessment. *Proceedings of the National Academy of Sciences,* 108(18), 7307–7312.
Guikema, S. D. (2009) Natural disaster risk analysis for critical infrastructure systems: An approach based on statistical learning theory. *Reliability Engineering & System Safety,* 94, 855–860.
Hawk, F. (1999) You Would Cry Too If It Happened to You: The Legal and Insurance Implications of Major Utility Failure. *Australian Mining & Petroleum LJ,* 18, 42.
Huppert, H. E. & R. S. J. Sparks (2006) Extreme natural hazards: population growth, globalization and environmental change. *Philosophical Transactions of the Royal Society of London A: Mathematical, Physical and Engineering Sciences,* 364, 1875–1888.
IATA. (2010) The impact of Eyjafjallajokull's volcanic ash plume. In *IATA Economic Briefing.*
Jenkins, S., R. Spence, J. Fonseca, R. Solidum & T. Wilson (2014) Volcanic risk assessment: Quantifying physical vulnerability in the built environment. *Journal of Volcanology and Geothermal Research,* 276, 105–120.
Jenkins, S., T. Wilson, C. Magill, V. Miller, C. Stewart, R. Blong, W. Marzocchi, M. Boulton, C. Bonadonna & A. Costa (2015) Volcanic ash fall hazard and risk. *Global Volcanic Hazards and Risk, Cambridge University Press, Cambridge,* 173–222.
Justus, F.K. (2015) Coupled effects on Kenyan horticulture following the 2008/2009 post-election violence and the 2010 volcanic eruption of Eyjafjallajökull. *Natural Hazards,* 76(2), 1205–1218.
Kargbo, A., J. Mao and C. Y. Wang (2010) The progress and issues in the Dutch, Chinese and Kenyan floriculture industries. *African Journal of Biotechnology,* 9(44), 7401–7408.
King, A., D. McConnell, H. Saddler, N. Ison & R. Dargaville (2016) What Caused South Australia's State-Wide Blackout? *The Conversation.*
Kinney, R., P. Crucitti, R. Albert & V. Latora (2005) Modeling cascading failures in the North American power grid. *The European Physical Journal B - Condensed Matter and Complex Systems,* 46, 101–107.
Klein, R. J., R. J. Nicholls & F. Thomalla (2003) Resilience to natural hazards: How useful is this concept? *Global Environmental Change Part B: Environmental Hazards,* 5, 35–45.
Klinger, C., O. Landeg & V. Murray (2014) Power Outages, Extreme Events and Health: a Systematic Review of the Literature from 2011–2012. *PLOS Currents Disasters.* 2014 Jan 2. Edition 1.
Kummu, M. and O. Varis (2011) The world by latitudes: A global analysis of human population, development level and environment across the north–south axis over the past half century. *Applied geography,* 31(2), 495–507.
Lall, S. V. & U. Deichmann (2010) Density and disasters: economics of urban hazard risk. *The World Bank Research Observer,* 27, 74–105.

Langmann, B., A. Folch, M. Hensch & V. Matthias (2012) Volcanic ash over Europe during the eruption of Eyjafjallajökull on Iceland, April–May 2010. *Atmospheric Environment,* 48, 1–8.

Lanzano, G., E. Salzano, F. Santucci de Magistris & G. Fabbrocino (2014) Seismic vulnerability of gas and liquid buried pipelines. *Journal of Loss Prevention in the Process Industries,* 28, 72–78.

Leipold, B. and F. Morgante (2013) The impact of the flower industry on Kenya's sustainable development. *International Public Policy Review*, 7(2), 1–31.

Leyland, B. (1998) Auckland central business district power failure. *Power Engineering Journal,* 12, 109–114.

Lin, A., B. A. Fletcher, M. Luo, R. Chinery & S. Hwang (2011) Health impact in New York city during the Northeastern Blackout of 2003. *Public Health Reports,* 126, 384–393.

Lindstrom, S., V. Nagalingam & H. Newnham (2013) Impact of the 2009 Melbourne heatwave on a major public hospital. *Internal medicine journal,* 43, 1246–1250.

Lund, K. A. & K. Benediktsson (2011) Inhabiting a risky earth: The Eyjafjallajökull eruption in 2010 and its impacts. *Anthropology Today,* 27, 6–9.

Magill, C., T. Wilson & T. Okada (2013) Observations of tephra fall impacts from the 2011 Shinmoedake eruption, Japan. *Earth, Planets and Space,* 65, 18.

McEvoy, D., I. Ahmed & J. Mullett (2012) The impact of the 2009 heat wave on Melbourne's critical infrastructure. *Local Environment,* 17, 783–796.

McInnes, J. A. & J. E. Ibrahim (2013) Preparation of residential aged care services for extreme hot weather in Victoria, Australia. *Australian Health Review,* 37, 442–448.

Meehl, G. A. & C. Tebaldi (2004) More intense, more frequent, and longer lasting heat waves in the 21st century. *Science,* 305, 994–997.

Menoni, S. (2001) Chains of damages and failures in a metropolitan environment: some observations on the Kobe earthquake in 1995. *Journal of Hazardous Materials,* 86, 101–119.

Miller, N. L., K. Hayhoe, J. Jin & M. Auffhammer (2008) Climate, Extreme Heat, and Electricity Demand in California. *Journal of Applied Meteorology and Climatology,* 47, 1834–1844.

Mukkala, K. & H. Tervo (2013) Air Transportation and Regional Growth: Which Way Does the Causality Run? *Environment and Planning A: Economy and Space,* 45, 1508–1520.

Mullett, J. & D. McEvoy (2014) With more heatwaves to come, how will our cities hold up? . *The Conversation,* 17 January 2014.

Murray, A. T. (2013) An overview of network vulnerability modelling approaches. *GeoJournal,* 78, 209–221.

Murray, A. T. & T. Grubesic. (2007) *Critical infrastructure: Reliability and vulnerability*. Springer Science & Business Media.

Nowakowska, M. (2015) Safety of Supply Chain of Kenya's Cut Flower Industry. Case of Kongoni River Farm. *Logistics and Transport*, 26(2), 43–48.

O'Regan, M. (2011) On the Edge of Chaos: European Aviation and Disrupted Mobilities. *Mobilities,* 6, 21–30.

O'Rourke, T., S. Jeon, S. Toprak, M. Cubrinovski & J. Jung. (2012) Underground lifeline system performance during the Canterbury earthquake sequence. In *Proceedings of the 15th world conference on earthquake engineering, Lisbon, Portugal*, 24.

Oxford Economics (2010) The economic impacts of air travel restrictions due to volcanic ash. A report prepared for airbus.

Palecki, M. A., S. A. Changnon & K. E. Kunkel (2001) The nature and impacts of the July 1999 heat wave in the midwestern United States: learning from the lessons of 1995. *Bulletin of the American Meteorological Society,* 82, 1353–1367.

Parker, C. F. (2015) Complex negative events and the diffusion of crisis: lessons from the 2010 and 2011 Icelandic volcanic ash cloud events. *Geografiska Annaler: Series A, Physical Geography,* 97, 97–108.

Pedersen, P. O. (2001) Freight transport under globalisation and its impact on Africa. *Journal of Transport Geography,* 9, 85–99.

Chapter 11
Communities in Fukushima and Chernobyl—Enabling and Inhibiting Factors for Recovery in Nuclear Disaster Areas

Tetsuya Okada, Serhii Cholii, Dávid Karácsonyi, and Michimasa Matsumoto

Abstract This chapter provides case studies on disaster recovery in the context of community participation. It presents two cases that explore, compare and contrast the nuclear disasters in Chernobyl and Fukushima. Despite differences in the socio-economic circumstances between the Soviet Union (Soviet–Ukraine) in 1986 and Japan in 2011, the Chernobyl and Fukushima disasters provide an opportunity to discuss power relations in disaster management and the role of local communities. These large-scale nuclear disasters are amongst the most traumatic experiences for the disaster-impacted communities worldwide. This chapter discusses the implementation of relocation and resettlement measures with socio-political power relations within and between the stakeholders. The combination of these is shown to significantly affect the everyday lives of those within the communities throughout the recovery process. Along with government documentation, the interviews with evacuees, community leaders and decision-makers conducted between 2012 and 2016 form the basis of the case studies discussed in this chapter.

Keywords Community · Power relation · Chernobyl · Fukushima · Evacuation · Resettlement

T. Okada
School of International Studies, Faculty of Arts and Social Sciences, University of Technology Sydney, 235 Jones St., Ultimo, NSW 2007, Australia
e-mail: tetsuya.okada@uts.edu.au

S. Cholii
Igor Sikorsky Kyiv Polytechnic Institute, Kyiv, Ukraine
e-mail: scholij@ukr.net

D. Karácsonyi (✉)
Geographical Institute, CSFK Hungarian Academy of Sciences, Budaörsi út 45, Budapest 1112, Hungary
e-mail: karacsonyi.david@csfk.mta.hu

M. Matsumoto
International Research Institute of Disaster Science, Tohoku University, Sendai, Japan
e-mail: matsumoto@irides.tohoku.ac.jp

D. Karácsonyi et al. (eds.), *The Demography of Disasters*,
https://doi.org/10.1007/978-3-030-49920-4_11

11.1 Introduction

This chapter provides case studies on disaster recovery in the context of community participation. It presents two cases that explore, compare and contrast the nuclear disasters in Chernobyl and Fukushima. Despite differences in the socio-economic circumstances between the Soviet Union (Soviet–Ukraine) in 1986 and Japan in 2011, the Chernobyl and Fukushima disasters provide an opportunity to discuss power relations in disaster management and the role of local communities. These large-scale nuclear disasters are amongst the most traumatic experiences for the disaster-impacted communities worldwide (Josephson 2010).

This chapter discusses the implementation of relocation and resettlement measures with socio-political power relations within and between the stakeholders. The combination of these is shown to significantly affect the everyday lives of those within the communities throughout the recovery process. Along with government documentation, the interviews with evacuees, community leaders and decision-makers conducted between 2012 and 2016 form the basis of the case studies discussed in this chapter.

11.2 Community Participation in Disaster Recovery[1]

The organisational cultures of authorities in disaster management often follow a top-down approach and downplay or ignore local people's needs when managing disaster recovery and risk reduction (Cannon 2015; Gaillard 2008; Manyena 2006; McEntire et al. 2002). As a consequence, affected communities were excluded from decision-making during disaster recovery.

Maintaining social network by maintaining relocated communities is necessary for revitalisation for individuals along with the physical recovery and reconstruction. Even so, the importance of communities came to the fore just recently during disaster mitigation and reconstruction (Oliver-Smith 2013). A community's capacity in disaster recovery assists its members to work together and solve problems leading to collective actions and decision-making (Norris et al. 2008). Community capacity builds up through the development of local knowledge, and it balances individual and collective interests (Patterson et al. 2010). However, authorities holding socio-political power often overlook or disregard the significance of local knowledge and experiences imposing mainstream values with expertise developed outside the disaster-affected communities (Bird et al. 2009; Haalboom and Natcher 2012; Howitt et al. 2012). Organisations responsible for disaster management are often found to implement policies using a top-down approach while downplaying or even ignoring the local community's needs and preferences in managing disaster recovery and risk

[1]Literature review included in this chapter was also included in the Ph.D. thesis (Okada 2017). Material drawing on my research on the Fukushima (Namie) case study was included in the thesis (Okada 2017) submitted to and accepted by Macquarie University in April 2017.

reduction (Cannon 2015; Gaillard 2008; Manyena 2006; McEntire et al. 2002). As a result, implemented measures and policies miss local needs, bring harmful consequences and/or create excessive dependency in local communities, while parties with power remain unaware or indifferent to these situations (Haalboom and Natcher 2012).

Centring communities within a risk reduction processes before a disaster minimises risks, ensuring effective partnerships in the process while incorporating local knowledge (Hayashi 2007; Ingram et al. 2006; Pandey and Okazaki 2005; Patterson et al. 2010). By assigning clear roles and responsibilities, decision-makers promote locally self-sustained recovery and risk reduction (Ahrens and Rudolph 2006). Community participation in disaster management develops a greater understanding among community members concerning their responsibility in recovery and risk reduction processes (Aguirre 1994; Pearce 2003). Community participation also facilitates a suitable balance in the interrelationship between all actors involved (Berke et al. 1993; Davidson et al. 2007). All in all, community members should play an active role in disaster recovery and risk reduction as key drivers, instead of remaining as passive receivers of services and information provided by authorities (Pearce 2003; Usamah and Haynes 2012; Bird et al. 2011; Haynes et al. 2008).

11.3 Data and Methods

In the Chernobyl case, we investigated the process of community relocation and resettlement in the post-disaster recovery, based on a series of integrated analyses from two main information sources. The first source is the official historical documents, including government policies collected from local governments and archives. These documents allow us to focus on the regulations (resolutions and decisions) legislated and implemented by the central Soviet and Soviet–Ukrainian governments. This is because the actions to deal with the widespread consequences of the disaster were mainly coordinated in Kiev (or Kyiv, the capital city of Ukraine) due to the location of Chernobyl inside Soviet–Ukraine. However, some neighbouring districts of Belarus and Russia, which were at that time also part of the USSR, also received high amounts of nuclear fallout.[2]

The second information source was based on 30 in-depth interviews with eyewitnesses of the relocation process. These interview sessions were conducted in Zhytomyr and Kiev regions of Ukraine with families displaced from the contaminated areas, exclusively from villages. Most sessions were structured interviews conducted with individuals and focus-groups between 2012 and 2015. The majority were chosen based on availability and random criteria because of high mortality rate of relocated persons during the previous decades and psychological distress among them. The 30 interviewees represent a vast geographical diversity of settlements of

[2]The actual size of radioactively contaminated area is larger in Belarus than in Ukraine; the power plant just located 8 km from the Belarusian-Ukrainian border.

origins (Fig. 11.1). Most of the interviewees were in their productive age, with 80% of respondents were in the 20–40 age cohort during the course of the disaster (1986).

The comparison of Soviet government documents with in-depth interviews of evacuees enabled us to understand the power balances and community responses to the recovery processes.

Fieldworks in Japan were conducted between 2011 and 2017 consisting of 49 interviews with evacuated individuals, community groups, support organisations and government officials. Research participants were recruited through direct communication. This was largely developed with snowball sampling as residents in the communities knew each other well and were interconnected, although their residences after the evacuation were spread across Fukushima Prefecture and beyond.

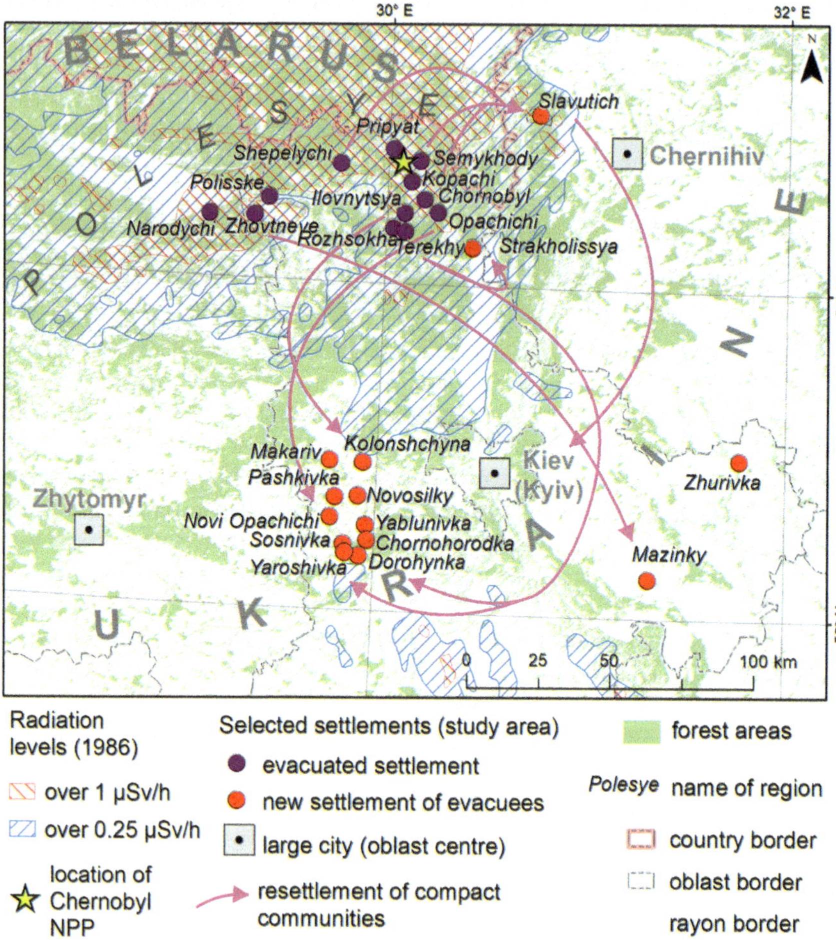

Fig. 11.1 Study area in Ukraine (Author Cholii, Karácsonyi, cartography by Karácsonyi)

A wide variety of participants were involved who were impacted by the disaster and the recovery process in different ways and extents.

The semi-structured interviews were conducted as qualitative in-depth research. Hence, they were not intended to quantify responses or generalise across the community. Instead, the interviews aimed to (and this book chapter aims to) understand and explore some of the issues being experienced during the recovery process within the Namie community in a qualitative manner.

11.4 Recovery Process After Chernobyl Disaster in Frame of Gradual Democratisation and Increasing Public Participation

The Chernobyl nuclear power plant is located in Ukraine, part of the Soviet Union (hereafter USSR) at the time of the disaster (1986). From the mid-1980s, during the Gorbachev era the USSR was undergoing gradual democratisation and marketisation processes (the so-called *Perestroika*, restructuring). The centralised and planned command economy was in deep crisis already in mid-1980s. The economic reforms, however, resulted in chaos and accelerated the decline. This affected the availability of funds to deal with the aftermath of the Chernobyl disaster. The estimated costs were enormous, which surpassed the economic benefits of the entire civil nuclear power industry of the USSR since its establishment in the 1950s (Bekar 2014, p. 19; Prister et al. 2013). The rise of *Glasnost* (openness, transparency)[3] revealed a more realistic and complete picture of the disaster, as one of the largest catastrophes in world history. On the other hand, the changing societal, political and economic circumstances also gradually influenced the community participation in the recovery processes over time with more and more involvement of affected people.

11.4.1 An Overview of Chernobyl Evacuation and Resettlement Measures

The central administration of USSR in Moscow envisioned the complete clean-up of the aftermath of the Chernobyl catastrophe as its main goal of the recovery efforts (Leukhina 2010; Vendland 2011). However, at the same time, responsibility for the recovery efforts was often imposed on local authorities in Kiev. Coordination of mitigation measures between these local authorities was a political challenge. According to general statistics (based on a document collection issued on the 18th

[3]Glasnost is a state-led policy of USSR during 1985–1991, aimed at gradual democratization of society. Its main principle was gradually lifting the limitation on basic democratic rights of its citizens such as freedom of speech, assembly, consciousness with simultaneous abolition of censorship and organization of first (partly) democratic elections in the USSR (in 1989).

anniversary of the disaster in 2004, referred hereafter as '18th anniversary 2004'), 164 thousand people were relocated from the contaminated areas to 204 resettlement sites across the country, where more than 29 thousand dwelling units were constructed.

An important feature of the relocation caused by the Chernobyl disaster was that most of the affected population had lived in rural areas (except those evacuated from Pripyat town, see Fig. 11.1). Many of these people were settled in rural communities such as Zhurivka district in Kiev region, which were set up exclusively to host evacuees from the same villages. Many of them suffered from the fragmentation of their community, lack of job opportunities and difficulties to adapt to the environment in the new place. The new settlements were created, sometimes hundreds of kilometres away from the disaster-prone area mostly, using available lands adjacent to already existing towns and villages. The new environment in these locations made it difficult to integrate the newcomers into the already existing local communities. Most of these resettlement measures were planned to complete in haste following tight schedules based on the command economy practised in USSR to facilitate the progress under the challenging situation. However, the economic crisis and political collapse of the USSR coupled with the negative consequences from the disaster (radiation risk, tremendous costs of reconstruction, relocation of a large number of people) caused major delays. These delays stretched the schedules of the clean-up and recovery measures, which eventually took several decades.

11.4.2 Evacuation and Resettlement Stages

The resettlement process consisted of three stages (The Human Consequences 2002). The first was an immediate evacuation of people during April–May 1986, who lived in the areas contaminated by the highest level of radiation mostly in the alienation/exclusion zone (about zoning see Chap. 2). Alternative residential areas for these people were arranged in Kiev and its surrounding areas (Belyakov 2003). Evacuees were often provided with already occupied apartment units, sharing a unit with other families for some time.[4] Some industrial properties were also turned into short-term residential facilities.

Although the evacuation order was made within 24 h after the explosion, only Pripyat town (the nuclear power station worker's city) responded to it quickly (27 April 1986). Villages in the alienation zone were evacuated later (e.g. Semykhody, Shepelychi and Kopachi on 2–3 May), exposing their residents to high levels of radiation due to failures of decision-making and because of political reasons (to avoid panic and hide the real extent of the disaster). Evacuees had a very limited

[4]The apartments were owned by the state or city council in the USSR and they were rented out to families or individuals. It was not because of the emergency situation to have more families in one apartment, or in one detached house. The communal apartment, the so called *kommunalka* (one family per one room and sharing kitchen and bathroom of the apartment) introduced in 1917 by the Bolsheviks to overcome housing shortages, was still usual in large Soviet metropolises in the late 1980s.

time to organise and carry necessary things and were forcibly evacuated. Along with the alienation zone (10-km radius), the compulsory evacuation order was also issued for other areas located further away but with radiation levels assessed as dangerous. These areas included Chernobyl town, which was evacuated on May 6 (Belyakov 2003; Prister et al. 2013, pp. 66–78).

Prompt evacuation of the entire population within the 30-km zone was challenging because of the very large area and population involved (see details in Chap. 2). This resulted in delays and may have facilitated the political decision, later on, to send those evacuees back home who were relocated from areas where the radiation levels were less-severe.

Children and their mothers who had lived in the 30-km zone were temporary evacuated from their place of residence. After a medical check-up, they were sent to hospitals or sanatoriums distant from the contaminated zone and stayed at those facilities for three months (until the end of summer 1986). By contrast, most males in the contaminated areas stayed at home to continue agricultural work or look after stock in the local collective farm (*kolkhoz*). This may have contributed to high mortality rates of male population in the following decade of the disaster. This male-victimising situation was created by the authorities' order designed to avoid panic and abandonment of local agricultural production. As it was still under the Soviet command economy, such emergency situation mixed with panic was concerned to shake the Soviet system (Ioffe 2007; Romashko 2016), which the authorities apparently wanted to avoid disrupting at any cost.

The second wave of evacuations was the immediate resettlement (1986–1987) after authorities recognised that evacuees could perhaps never return to their homes. Preparation for relocation from the compulsory resettlement zone (30-km zone) started in May 1986. This included investigation of possible construction sites for housing the evacuees, planning different types of buildings, estimation of the number of houses needed and planning construction works. The construction had to be done in a very short period. For example, only three months were allocated for the construction in Kolonshchyna and Novi Opachichi villages in the Kiev region housing resettled people. This led to some people resettling in unfinished buildings and squeezing two or three families into one house. Some waited in nearby villages until the construction works finished (but not necessarily completed). It took approximately a year for the state to resolve these issues in early 1987.

In most cases in the immediate resettlement stage, whole communities were relocated without breaking existing local social connections. Newly constructed settlement sites of this stage were typically large-scale with a hundred or more houses in each estate (*Chornobyls'kyi posyolok*). The first and second stages of resettlement occurred within a short period, between April 1986 and by the end of 1987. Around 90 thousand residents were involved in the relocation and the subsequent resettlement just inside the territory of Ukraine.

The third stage was organised resettlement (1988–2002), which aimed to relocate people who lived in the guaranteed voluntary resettlement zone (outside the 30-km zone). In contrast to the quick implementation of the compulsory resettlements between 1986 and 1987, there was a debate on whether the government should extend the resettlement efforts beyond the 30-km zone or not (Tykhyi 1998; Malko 1998). This resettlement stage took a long time and exposed the local population in the contaminated areas to radiation for a decade after the catastrophe. The main reason for such delay was the deep crisis of the disintegrating USSR and the economic problems during the post-Soviet transformation in newly emerged independent countries, such as Ukraine.

The organised resettlement stage consisted of two sub-stages, which were defined by government regulations, the so-called Chernobyl construction programs. The first sub-stage was based on government regulation No. 333 (1989), commencing on 30 December 1989. It included plans to construct 2318 houses, 18 blocks of flats with a combined 1052 apartments, 17 kindergartens, 11 schools and 210 km of gas pipelines. These works were in anticipation of 3370 families relocating from Narodychi district (*rayon*) of Zhytomyr region (*oblast*) and Polissky district of Kiev region. The second sub-stage was based on regulation No. 228 (1990), commencing on 23 August 1990. This included the construction of housing estates for another 14,700 families from the guaranteed voluntary resettlement zone. The organised resettlement stage in Ukraine continued until the early 2000s, when the government stopped constructing resettlement estates and offered additional compensation money for those impacted by disaster and who had not received government assistance (Baranovska 2011, p. 32; regulations Nr. 333, 1989; Nr. 228 1990).

These resettlement processes allowed elderly people to return to their home (Davies and Polese 2015). They were often referred to as *samosely* (self-settlers), who lived and still live in their home town even inside the alienation zone. Other residents repeatedly experienced relocation and resettlement.[5]

Many of those who resettled in apartment blocks in big cities such as Kiev were also separated from their original communities (regulations Nr. 333, 1989; Nr. 228 1990). Many interviewees stated that their communities were often split and/or destroyed through the resettlement processes in the organised resettlement stage. Although the government tried to resettle communities as whole, each community was often split into two or more (The Human Consequences 2002, p. 14) because of their large population size. After relocating more than 90 thousand residents in the evacuation and immediate resettlement stages, it was impossible for the government to keep tailoring resettlement land for each community to fit its large population. In other cases, people from different communities were transferred to and mixed in a new settlement. These new settlers experienced challenges in various aspects of their everyday lives. For example, in the case of Novosilky, a family relocated and lived in

[5]For example, a family, originated in Terehy village, firstly resettled in the neighbouring village of Straholissya located outside the 30-km zone. After a while, the family returned to their original home in Terehy. The next resettlement was conducted end of summer 1986, when all residents of Terehy and Straholissya were transferred to the new housing estates built in Fastiv district of Kiev region. But, this family among several others returned to Terehy by end of 1986. From 1986 to 1991, many of these families lobbied governments in Kiev and Moscow to stay in Terehy, but they, including this family, were all forcibly transferred to different locations of Makariv district of Kiev region.

a new settlement with other evacuees from very different regions, experienced additional socialising problems on top of their psychological problems that they were already suffering from settling in a totally new environment after a tragic event.

11.4.3 Resettlement and Community in the Context of Changing Political Structures

The relationships between society and government in the late Soviet era also influenced the resettlement processes. So-called *Perestroika*, a series of democratisation processes in the Gorbachev-era of the Soviet Union, rapidly progressed in parallel with the post-disaster relocation processes. This rise of democratisation may have diminished the political influence of Soviet authorities (Lane 1992; Pickett 2016; Plokhiy 2014). During the third organised resettlement stage, private and community initiatives were increasingly established with government supervision, and private industries were involved in construction work associated with the government's resettlement efforts. This situation is quite a contrast to the first two stages of relocation: the evacuation and immediate resettlement stages.

However, this democratic stage was still influenced by some dichotomies and contradictions. Although USSR gradually incorporated democratic ideas in its governance, its citizens' democratic rights such as freedom of action were still limited (Marples 1988). For example, although many resettled residents participated in local governance activities like sending their representatives to their councils, the centralised political system strictly limited the decision-making power of these councils. While these local councils were only allowed to deal with local issues, democratisation continued to be absent at higher levels of politics in the late 1980s. Despite this, almost all interviewees recognised a high level of support from their new local governments (village or collective farm administrations) such as integrating the resettled people transferred from different areas, the new settlers were not entitled to directly influence the state that retained a dominating power.

Communities were also divided by the rise of individual and collective initiatives around resettlement. Residents of the guaranteed voluntarily resettlement zone continued discussions on their relocation and resettlement options. As the repeated relocation and resettlement of mass population became increasingly challenging, the state actively invested to (re-)develop local infrastructure and services inside the guaranteed voluntary resettlement zone hoping that many local people would choose to stay in their original areas (regulations Nr. 315, 1989; Nr. 1006–286 1986). This would often divide local communities into two groups—the conservative group (older generations), who were willing to stay, and the opposition group who wanted to leave. In particular, the opposition groups often formed by young people and/or families with small children, as they feared radiation effects on their health (The Human

Consequences 2002, p. 14).[6] Communities were often informed about the levels of radiation in their areas of the guaranteed voluntarily resettlement zone and this also triggered arguments for or against relocation and resettlement. The state provided options for the most active groups, especially families with pregnant women and small children, to relocate and resettle quickly, but this support was provided inconsistently (regulations Nr. 115, 1990; Nr. 886, 1989). These debates ended around 1989 and 1990, when the government eventually decided to evacuate everyone from the compulsory resettlement and guaranteed voluntarily resettlement zones.

11.5 Fukushima Recovery—Disintegrating Communities Under Uncertainties

The second section of this chapter addresses the formation and transition of communities of evacuees who were displaced by the nuclear accidents at the Fukushima Daiichi Nuclear Power Plant following the magnitude 9.0 earthquake off the Pacific coast of Tohoku occurred on 11 March 2011 (see also Chap. 2). According to the website of Fukushima Prefecture (2013), approximately 154 thousand residents of Fukushima Prefecture were evacuated, of which 109 thousand were mandatorily evacuated by 19 February 2013. The number of evacuees of the Prefecture as of January 2017 was approximately 81 thousand (Fukushima Prefecture 2017).

In this section, we detail results of interviews and document analyses related to the impacts brought by the relocation or resettlement processes on the disaster-impacted people's everyday lives. Semi-structured interviews were conducted with community members and stakeholders from Namie, Naraha and Tomioka towns and Iwaki city in Fukushima Prefecture (Fig. 11.2). Each of the semi-structured interviews focused on the disaster and the recovery processes. Namie and Tomioka towns are different in location and land size, but both were under evacuation order (which was partially lifted in March and April 2017) and contain areas where residents will not be able to return *at least* for five years[7] (Cabinet Office 2013). Naraha town lifted its evacuation order in September 2015. Iwaki city did not issue an evacuation order, but experienced

[6]Here we have symptomatic example of Zhovtneve area outside the 30-km zone that was divided by resettlement into fractions of resettled in 1990–1991 to different locations, mostly in Makariv district of Kiev region and the main cluster of population, resettled later to Mazinky village in Pereyaslav district of Kiev region several years later. It demonstrates the situation when initiative groups were exempted from the community and dispersed while the main part of community was later resettled compact that partially ruined the community unity after resettlement.

[7]Areas under evacuation order were divided into three zones. Difficult-to-return zone contains the annual integrated doses of more than 50 mSv, where residents cannot return to live at least for five years. Restricted residence zone has the radiation level between 20 and 50 mSv, where daytime business activities without overnight staying are generally permitted. Evacuation order cancellation preparation zone's radiation level is less than 20 mSv. These zones have been re-shaped along with decontamination outcomes (see https://www.pref.fukushima.lg.jp/site/portal-english/en03-08.html).

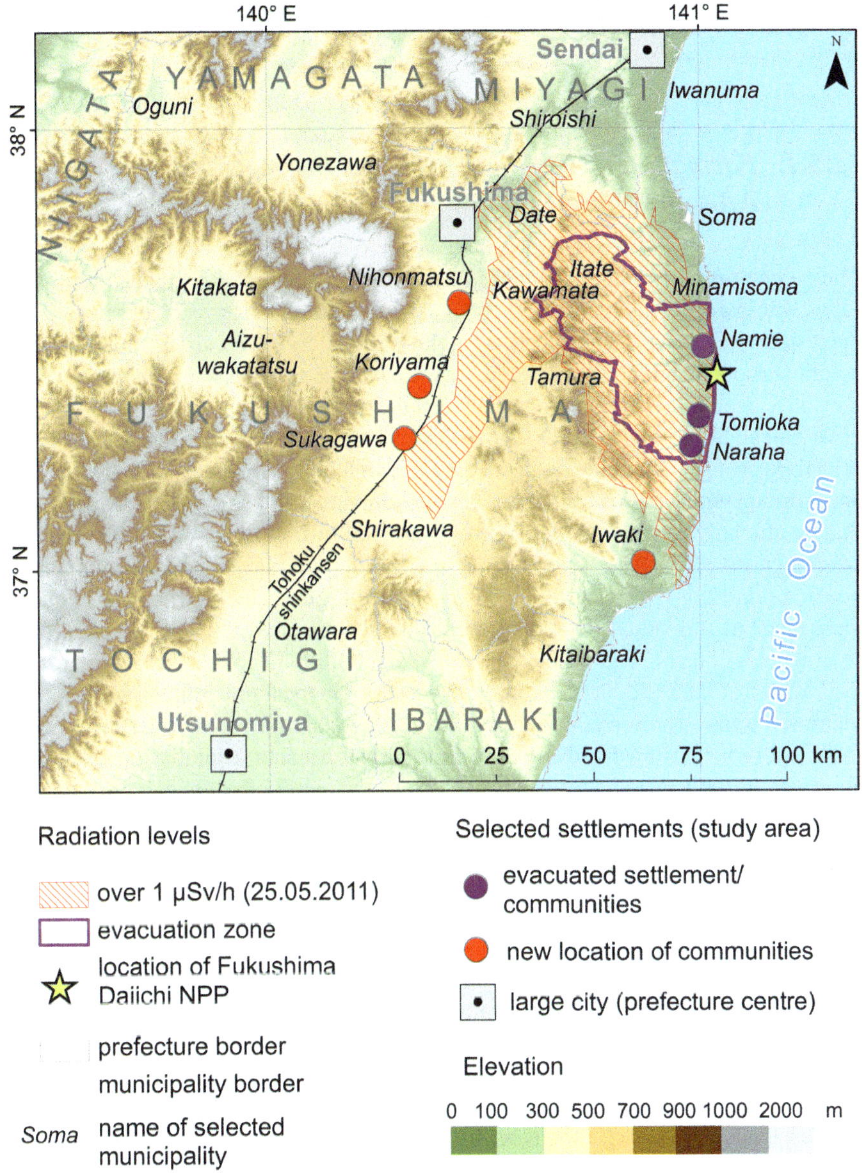

Fig. 11.2 Study area in Japan (Author/cartography by Karácsonyi)

severe tsunami damage on the coastal areas and received influx of evacuees from other towns.

11.5.1 Formation of Residents' Groups During First Year After the Disaster

The neighbourhood councils called *jichikai* are a form of self-organised groups in Japan who are officially recognised by local governments and have a lobbying role for the residents living in their respective neighbourhood. Evacuees also established their neighbourhood councils at each temporary housing units (*kasetsu* in Japanese) where they have been evacuated, to liaise between the residents of their unit and their local government. Some interviewees highlighted that the effectiveness of these bottom-up organised groups was more effective compared to individual voices when negotiating and working with authorities. For example, a temporary housing council leader stated that he officially established the evacuee's council in July 2011, which was the first official group of evacuees from Namie. He recalled that his requests on temporary housing conditions were all rejected by the local government when the requests were made as individuals, but then, were mostly addressed after the establishment of evacuee's council. The official *jichikai* (neighbourhood council) status increased the validity of the residents' collective voices for the local government to recognise the raised issues as more community-based rather than individually generated.

Temporary housing for evacuees from Naraha became available in July 2011. However, initially, Naraha evacuees in some of the temporary housing units did not establish their neighbourhood councils, because decisions on the establishment were left to the residents. Eventually, each estate formed its evacuee's councils once both the town office and the residents recognised the necessity of an official liaison. In Tomioka, the town office maximised its pre-existing ties with residents to establish temporary housing evacuee's councils (*kasetsu jichikai*-s), consulting with local leaders in the early stage of evacuation. Therefore, the establishment of such councils in Tomioka (May 2011—only two months after the disaster) was quicker than that in Naraha. This led to the outcome that the scattered Tomioka residents played a leading role in establishing these evacuee's councils. In any case, evacuee's councils were expected to process requests and announcements from both the government and residents, although the focus was mostly on dealing with residents collectively rather than individually. With the help of town office meeting venues for evacuee's councils, which became available from October 2012, extensive networks among these councils were quickly and successfully enabled.

The evacuee's council (*jichikai*) structure was also helpful for those who lived outside temporary housing units, including evacuees who were living in rental housing subsidised by local governments (*kariage* in Japanese).[8] Namie set up

[8]Kariage housing means a rental apartment, where local government reimburse the rent for the tenants regardless that it is a commercial property or public housing.

evacuee's councils for themselves in various locations following the series of establishments of temporary housing councils. Many interviewees highlighted the importance of looking after the evacuees outside the temporary housing units as much as those inside. This is because, in general, people outside temporary housing units attracted less attention than those inside due to their less visible presence. Temporary housing units were physically visible and identifiable living as clusters (see Chap. 2, Fig. 2.4), but subsidised rental accommodations (*kariage*) were geographically woven into host communities. The situation of public housing evacuees from Naraha and Tomioka was similar to that of Namie evacuees. Evacuees living in subsidised rental housing received services such as town magazines, rental tablet devices and regular visits of social welfare council staff to maintain basic communication.

Staff members of the social welfare council of Namie pointed out the challenges of sharing information evenly and timely with evacuees scattered in subsidised rental housing. Because of unfamiliar surroundings that these evacuees suddenly faced, they often became reluctant to go out, and they eventually became isolated. According to Shinmachi-Namie, a citizen-led support group of Namie, these *kariage* evacuees had little opportunities to gather or interact with other Namie residents and share their feelings and experiences. It was one of major challenges for Namie community in the early stage of temporary settlements.

Retaining broader connections between evacuees was an important issue too, because several residents evacuated to not only within the Fukushima region but also across Japan and beyond. Shinmachi-Namie was established by some members of the local merchants' association of Namie. According to a leading staff member of Shinmachi-Namie, the organisation adopted a form of NPO (Non-Profit Organisation) to maximise the social and financial opportunities in recovery such as grant application. It held evacuees' networking events and supported the establishment of evacuee's councils for those living in subsidised rental housing in the cities of Nihonmatsu, Koriyama and Sukagawa and set-up of a community-based recovery committee of Namie.

11.5.2 Emerging Communities and Issues in Recovery—Second and Third Years Post Disaster

Many interviewees living in temporary housing units strongly valued the sense of belonging to their new communities at each estate. These emerging communities were the place where they could live together after being transferred from one place to another many times. While this sense of community developed, various changes kept widening gaps between individuals and communities despite that evacuees' everyday lives seemed less chaotic than the initial phase of evacuation.

Staff members of the Namie social welfare council were concerned that some residents at temporary housing units, particularly the most vulnerable, were feeling left behind. These residents often would have nowhere to go when, one day, their units were closed, while others gradually started moving out from these estates. Adding to this view, a staff member of Shinmachi-Namie warned that this does not mean that those who moved out from the estates achieved their recovery because moving out was simply one of several stages that they would still have to go through in their recovery. Evacuees' views on receiving external supports also varied. During the interview sessions in 2012 in Naraha and Tomioka, evacuee's council leaders often focused on self-reliance in their management, anticipating their future lives after moving out from temporary housing. However, some other temporary housing residents demanded more external supports. This shows that the views on their future lives already varied between residents at this stage.

The Namie social welfare council staff achieved trustful relationships with many evacuees through their long-term regular visits but also identified that some requests from the evacuees were becoming overly reliant on external supports. For example, some family members refused to take care of their elderly parents for nursing. This was because they feared that they might get too impatient and aggressive with parents as a result of excessive stress from their recovery issues. Such problems previously were dealt with from within local communities, but the former neighbourhoods disappeared. As a result, their mutual support at a micro-local scale did not function as effectively as it once had.

Of course, some residents retained their sense of community both before and after the disaster. They often kept community-ties having regular meetings and events during the post-disaster. This seemed to ease residents' uncertain feelings over prolonged evacuation to some extent. Some of the evacuee's councils developed friendships with their host communities, co-organising socialising events and festivals. Based on this development, new neighbourhood ties were established between evacuees and host communities. Some other evacuee's councils, such as Namie So-So Jichikai participated in and facilitated cleaning activities in the local areas. Aizu-Miyasato Jichikai joined local sporting events. Evacuees' Namie Sasaya-Tobu Jichikai and host communities' Fukushima Kita-Nakajo Jichikai jointly held regular socialising events.

However, in general, some of the evacuee's councils maintained regular meetings and gatherings with their host communities. Tanaka et al. (2013) pointed out that local communities existed but did not exist in Fukushima. This meant that although the communities existed in local people's post-disaster lives, pre-disaster personalisation of people's lifestyle continued to diminish their community lives.

Regarding the development of broader works on evacuees' connections, the citizen-based NPO Shinmachi-Namie further expanded its projects with other support groups such as academics and businesses. For example, the NPO and a research group at Waseda University, Tokyo, jointly introduced a semi-public transport system for Namie evacuees in Fukushima Prefecture and beyond using large passenger vehicles funded by external businesses.

At the same time, the NPO started experiencing a new difficulty in maintaining the organisation as a result of the prolonged uncertainty in recovery. Between 2013 and 2015, the number of staff at the NPO dropped from ten to three. One main reason pointed by some of the members were that many evacuees could not settle in one area to work for the NPO because of their unstable life circumstance. Prolonged evacuation without certainty discouraged them working, when budget for personnel costs was often unavailable due to the organisation's status as NPO. Therefore, even if the organisation won valuable project grants, they often could not work on those projects due to staff shortages.

11.5.3 Return or No Return with Widening Gaps—Four Years After the Disaster and Beyond

The unstable situation in recovery developed further, although reconstruction progressed and lifting of evacuation orders came into affect (at least politically). In addition, effects of a serious lack of sharing information, projects and goals in recovery started unfolding between authorities, support groups and residents. Political uncertainty further complicated to the situation.

A big gap in recognition of and expectation for the future between governments and the residents of their jurisdiction emerged in case of Namie. Generally, two major ideas of semi-permanent residence had been publicly discussed and shared in parallel: (1) to collectively settle outside the disaster-impacted town and redevelop evacuees' everyday lives there; (2) to build public apartments inside and outside the disaster-impacted areas and return (Namie Town 2012). However, in 2015, the central government (and the local government) was almost solely focusing on the latter idea of 'return' and the local government followed it, although this shift of focus was not clearly announced. Namie and Tomioka lifted their evacuation orders (except the difficult-to-return zones) on 31 March and 1 April 2017, respectively (Namie Town 2017; Tomioka Town 2017).

This shift of focus also showed in Namie town office's responses to residential developments. According to a Namie town officer interviewed, the town office approved Namie residents to move into public disaster housing provided by Koriyama and Fukushima. There were also extension plans to build more public disaster housing on the same site. However, Namie town office did not approve another extensive residential development plan, which was a non-governmental project but endorsed by Fukushima city office. Possible factors that contributed to the difference in Namie town office's decisions were the driving factors of the projects (either government-led or others), scale of the projects (about 20 houses in Koriyama, 400 blocks in Fukushima) and items that they were dealing with (either public housing or land blocks). This seemingly relates to the government's current intention to lead Namie residents to return in a coerced manner.

Regarding the residents' view on returning or not returning, a member of the NPO Shinmachi Namie pointed out that, in 2015, 80% of the recent government survey respondents said that they would or could not return anytime soon for various reasons such as safety and availability of social services (Reconstruction Authority et al. 2016). However, another town officer, during the interview, interpreted this survey result differently; that 80% of non-returning did not necessarily mean giving up on the town. The officer also explained that building Namie communities in other cities and towns would have several difficulties such as taxation, public services and facilities to use.

In Tomioka, the evacuation order (except for the difficult-to-return areas) was lifted in March–April 2017. Public housings for returning evacuees were being built; infrastructures such as medical and commercial facilities were being redeveloped. However, similar to Namie, many of the residents already secured their houses outside the town. Therefore, it would be difficult to expect a high return rate.

In Naraha, the whole town evacuation order was lifted on 5 September 2015. However, the return rate as of September 2016 was less than 10% of the pre-disaster population. One of the potential reasons for this low return rate is that there was no any difficult-to-return area designated in Naraha (unlike Namie and Tomioka). Because of the absence of the difficult-to-return areas, Naraha residents are not entitled to the disaster public housings located outside the town, while utilities and essential services have not been organised sufficiently. At the same time, approximately 80% of the town's population evacuated to the nearby Iwaki city, and they often live their lives in Iwaki without too much inconvenience. These factors may have prompted Naraha residents to individually secure their houses outside the town, which caused low return rates (see Fukushima Minpo 2016).

Evacuees at subsidised rental housing also started accepting the reality and trying to find ways to cope with challenging situations to settle in their current places of living, although they often faced difficulties living in and with the host communities. For example, host community members reportedly criticised Namie evacuees for daily issues (such as managing household waste dump sites). A leader of their evacuee's council explored, at their regular socialising occasions, with other evacuees how they would better deal with such situations so both the evacuees and host communities could maintain and improve their lives rather than conflicting to each other. Media have reported increasing cases of bullying and discrimination against Fukushima evacuees across Japan (see Japan Times 2017; Asahi Shimbun 2017).

Conflicting relationships between evacuees and host communities in Iwaki city were being moderated due to better mutual understanding brought by progress of time and interaction, according to the interviewees. This moderation was also facilitated by the city's recognition of an important challenge to restore its decreased population after the disaster. Iwaki lost its large coastal population following the severe impacts caused by the 2011 tsunami, similar to the coastal areas of Miyagi and Iwate Prefectures in the Northern Tohoku region.

In exploring the evacuees' connections, a number of interviewees (mainly residents) expressed their dilemma that they wanted to return, but knew practically that they cannot. This was mainly because there would not be the everyday community lives that they had become use to. Furthermore, some reported that those who were determined to return often changed their minds at some points, because the evacuation was going on for too long. They could not decide, because they didn't know when or if they could return. They often could not help thinking about it and got depressed, because there was no clear answer available.

A Namie interviewee stated that she was feeling guilty because realistically she could not return. Another interviewee identified uncertainty as the most difficult issue of Namie's recovery, because local people's recovery and re-establishment of lives were heavily and repeatedly affected by the factors that were beyond their control. These included environmental conditions that brought radiation to Namie and the decisions and intentions of authorities at the time, whose information was often not clearly announced to or shared with public, most importantly with the local residents.

In Naraha, where the evacuation order was lifted for the entire town in September 2015, interviewees' attitudes varied roughly in three categories: (1) return, (2) do not return and (3) have no means to return. The 'do not return' group often had already bought houses outside the town such as Iwaki and secured foundations of their everyday life including schools and jobs. The 'have no means to return' group included those who ran out of compensation money and only hoped public disaster housings in Naraha to be available (see Sect. 11.4.2). This group would likely stay in the emergency housings as long as these are available, which is likely to become a social issue later on. Considering these stories in this sub-section, lifting evacuation order may also be a trigger point that divides residents into a greater number of different groups and widens the gaps between them.

11.6 Differences, Commonalities and Lessons

Our two case studies demonstrated the importance of socio-political systems, financial capacity, mass migration and uncertainty, which affected local communities in different and similar ways through the post-disaster times.

The political framework during the Chernobyl and Fukushima disaster recovery cases were very different. The Chernobyl disaster occurred under the communist system of former USSR, which rapidly collapsed and shifted towards democratic directions in the early post-disaster period. The communist political system, despite the rapid democratic changes, avoided or considerably limited the opportunities for the disaster-impacted citizens to contribute to their recovery and relocation processes. By contrast, the Fukushima disaster and its recovery and relocation processes have been responded to by continuous government systems of Japan, which are generally democratic and relatively stable compared to that of the Chernobyl disaster. This at least provided the impacted citizens of the Fukushima disaster with rights to be actively involved in the recovery process.

Financial capacities of the governments also varied between the two cases. Due to the declining economy of the USSR and the down-scaled capacities of the new states, the large-scale relocation and recovery efforts increasingly experienced financial limitations over time. The Japanese economy was also not in a position to respond to the scale of damages. The state's economy had been stagnant for 20 years. However, the stable political system allowed Japan to sustain its capacity to provide funding support and reconstruction.

Despite these differences in political–societal system and financial capacity, the overwhelming uncertainty that was largely generated and amplified by governmental and political responses without centring local communities further impacted the lives of disaster-impacted citizens. The uneven policy-making, budget-availability and implementation of the relocation in the former USSR were affected by the progress of democratisation destroyed pre-disaster local communities and continued restricting emerging communities. The lack of foresight also exacerbated hardships in everyday life and life-planning of the disaster-impacted local people of Fukushima, pushing the most vulnerable groups into an even more vulnerable positions. The uncertainty inhibited various stakeholders from sharing visions and efforts; the poor levels of understanding between them increased the uncertainty as a vicious circle.

This uncertainty was further complicated by other factors. For example, the difficulty of sudden, mass relocation and resettlement. Despite the initial speedy relocation and resettlement of 90 thousand citizens without damaging community structures, inhibited authorities of the Chernobyl case from continuing the same level of service due to the reduced availability of land and resources, possibly affected by the drastic political changes as well. The Fukushima case, Namie in particular, also struggled with the limited availability of land and housing and had to utilise scattered temporary shelters such as school gymnasiums and motels to accommodate hundreds of thousands of evacuees all at once. This meant evacuees kept moving until temporary housing and public housing arrangements become available and brought some host communities surged property prices, reduced land availability and strained public resources and services. Another example was the lack of knowledge about the real scale of the radiation threat, which allowed the Chernobyl authorities to expose citizens to high-level radiation. In Fukushima, although the stakeholders were more aware of measurement and potential risks of radiation, the fundamental lack of knowledge about radiation, such as long-term presence and health effects, continues distressing the evacuees in their decision-making.

Uncertainty was commonly identified as a major challenge at a local-scale recovery following the two nuclear disasters. In particular, the large scale of affected areas and populations and several unknown and unresolved aspects of radiation effects considerably increased the uncertainty such as inability to stay and work on-site or nearby, varied effects of decontamination efforts, enormous amounts of human, financial and land resources needed. In order to avoid exacerbating the uncertainty and associated social damages among disaster-impacted citizens, it is important to centre the impacted citizens in disaster management processes. These case studies illuminated the importance of sharing ideas, issues, options and limitations between all stakeholders centring disaster-impacted citizens so that the recovery processes and outcomes gain community members' awareness of their community lives, reduce vulnerability and increase resilience at individual and community scales.

References

18th anniversary of Chernobyl disaster 2004. 18 richnytsya Chornobyl′s′koyi katastrofy, Pohlyad u maybutnye (18th anniversary of the Chornobyl disaster. Looking to the Future). Verkhovna rada Ukrayiny, Parliamentary Publishing House, Kiev

Aguirre, B. E. 1994. Planning, warning, evacuation, and search and rescue: A review of the social science research literature, Hazard Reduction Recovery Center

Ahrens, J. & Rudolph, P. M. 2006. The Importance of Governance in Risk Reduction and Disaster Management. Journal of Contingencies and Crisis Management, 14, 207–220

Asahi Shinbun, 2017. Bullying against evacuees of nuclear accident. *Asahi Shimbun*, 8 March. https://www.asahi.com/articles/ASK385TZVK38ULOB01M.html (Japanese)

Baranovska, N. 2011. Chornobyl′s′ka trahediya, Narysy z istoriyi (Chornobyl tragedy, Essays on history). Institute of Ukrainian History, Kiev

Bekar, A. 2014. State Failure: Environmental Crisis of the Chernobyl Accident, its political repercussions and its impacts on the Soviet Collapse. https://academia.eu. Accessed 10 January 2016

Belyakov, S. 2003. Lykvydator. Accessed on 3 December 2016 https://lib.ru/MEMUARY/CHERNOBYL/belyakow.txt

Berke, P. R., Kartez, J. & Wenger, D. 1993. Recovery after Disaster: Achieving Sustainable Development, Mitigation and Equity. *Disasters,* 17, 93–109

Bird, D. K., Gisladottir, G. & Dominey-Howes, D. 2009. Resident perception of volcanic hazards and evacuation procedures. *Nat. Hazards Earth Syst. Sci.*, 9, 251–266

Bird, D. K., Gisladottir, G. & Dominey-Howes, D. 2011. Different communities, different perspectives: issues affecting residents' response to a volcanic eruption in southern Iceland. *Bulletin of Volcanology,* 73, 1209–1227

Cabinet Office (of Japan), 2013. Regarding the difficult-to-return zone. Accessed on 3 December 2016. https://www.mext.go.jp/b_menu/shingi/chousa/kaihatu/016/shiryo/__icsFiles/afieldfile/2013/10/02/1340046_4_2.pdf (Japanese)

Cannon, T. 2015. Disasters, climate change and the significance of 'culture'. *In:* Krüger, F., Bankoff, G., Cannon, T., Orlowski, B. & Schipper, E. L. F. (eds.) *Cultures and disasters.* Routledge

Davidson, C. H., Johnson, C., Lizarralde, G., Dikmen, N. & Sliwinski, A. 2007. Truths and myths about community participation in post disaster housing projects. *Habitat International,* 31, 100–115

Davies, Th., & Polese, A. 2015. Informality and survival in Ukraine's nuclear landscape: Living with the risks of Chernobyl. *Journal of Eurasian Studies*, 6, 34–45

Fukushima Minpo, 2016. Affected by the rules. *Fukushima Minpo,* 15 March. Accessed on 3 December 2016 https://www.minpo.jp/pub/topics/jishin2011/2016/03/post_13473.html (Japanese)

Fukushima Prefecture, 2013. Records and recovery progress of the Great East Japan Earthquake. Accessed on 3 December 2016 https://www.pref.fukushima.lg.jp/sec_file/koho/e-book/HTML5/pc.html#/page/130 (Japanese)

Fukushima Prefecture, 2017. Situation of the areas under evacuation order, and support for evacuees. Accessed on 2 December 2019 https://www.pref.fukushima.lg.jp/site/portal/list271.html (Japanese)

Gaillard, J.-C. 2008. Differentiated adjustment to the 1991 Mt Pinatubo resettlement program among lowland ethnic groups of the Philippines. *Australian Journal of Emergency Management,* 23, 31–39

Haalboom, B. & Natcher, D. C. 2012. The Power and Peril of "Vulnerability": Approaching Community Labels with Caution in Climate Change Research. *Arctic,* 65, 319–327

Hayashi, H. 2007. Long-term Recovery from Recent Disasters in Japan and the United States. *Journal of Disaster Research* 2, 413–418

Haynes, K., Barclay, J. & Pidgeon, N. 2008. The issue of trust and its influence on risk communication during a volcanic crisis. *Bulletin of Volcanology,* 70, 605–621

Howitt, R., Havnen, O. & Veland, S. 2012. Natural and Unnatural Disasters: Responding with Respect for Indigenous Rights and Knowledges. *Geographical Research*, 50, 47–59

Ingram, J. C., Franco, G., Rumbaitis-Del Rio, C. & Khazai, B. 2006. Post-disaster recovery dilemmas: challenges in balancing short-term and long-term needs for vulnerability reduction. *Environmental Science and Policy*, 9, 607–613

Ioffe, G. 2007. Belarus and Chernobyl: Separating seeds from chaff. *Post-Soviet Affairs*, 23(2), 1–14

Japan Times, 2017. Six years on, Fukushima child evacuees face menace of school bullies. *Japan Times*, 10 March. Accessed on 2 December 2019 https://www.japantimes.co.jp/news/2017/03/10/national/social-issues/six-years-fukushima-child-evacuees-face-menace-school-bullies/#.WOW2EEaGNhE (Japanese)

Josephson, P. 2010. War on Nature as Part of the Cold War: The Strategic and ideological Roots of Environmental Degradation in the Soviet Union. In Mcneil, J., & Unger C. (Eds.). Environmental Histories of the Cold War. Washington: Cambridge University Press. pp. 21–50

Lane, D. 1992. Soviet Society under Perestroika. Routledge: Rev Sub edition

Leukhina, A. 2010. Ukrainian Environmental NGOs after Chernobyl Catastrophe: Trends and Issues. *International Journal of Politics and Good Governance*, 1, 1–13

Manyena, S. B. 2006. The concept of resilience revisited. *Disasters*, 30, 434–450

Malko, M. V. 1998. Social Aspects of the Chernobyl Activity in Belarus – Research Activities about the Radiological Consequences of the Chernobyl NPS Accident and Social Activities to Assist the Sufferers by the Accident, Kyoto Accessed on 2 December 2019 https://www.rri.kyoto-u.ac.jp/NSRG/reports/kr21/kr21pdf/Malko3.pdf pp. 24–37

Marples, D. R. 1988. The social impact of the Chernobyl disaster. Houndmills, Basingstoke, Hampshire and London: Macmillan press

McEntire, D. A., Fuller, C., Johnston, C. W. & Weber, R. 2002. A Comparison of Disaster Paradigms: The Search for a Holistic Policy Guide. *Public Administration Review*, 62, 267-281

Namie Town, 2012. Namie Town Recovery Plan, October 2012

Namie Town, 2017. Namie Town at a glance. Accessed on 2 December 2019 https://www.town.namie.fukushima.jp/soshiki/2/namie-factsheet.html (Japanese)

Norris, F. H., Stevens, S. P., Pfefferbaum, B., Wyche, K. F. & Pfefferbaum, R. L. 2008. Community Resilience as a Metaphor, Theory, Set of Capacities, and Strategy for Disaster Readiness. *American Journal of Community Psychology*, 41, 127–150

Nr. 115, 1990. Decision of 21 May 1990. Pro zabezpechennya zhytlom hromadyan, yaki pidlyahayut′ dodatkovomu pereselennyu z terytoriy, shcho zaznaly radioaktyvnoho zabrudnennya v rezul′tati avariyi na Chornobyl′s′kiy AES (On the provision of housing to citizens subject to additional resettlement from the territories affected by the Chernobyl disaster). Accessed on 10 January 2016 https://zakon.rada.gov.ua

Nr. 228, 1990. Decision of 23 August 1990 Pro orhanizatsiyu vykonannya Postanovy Verkhovnoyi Rady Ukrayins′koyi RSR "Pro nevidkladni zakhody shchodo zakhystu hromadyan Ukrayiny vid naslidkiv Chornobyl′s′koyi katastrofy" (On the organization of implementation of the Verkhovna Rada of the Ukrainian SSR "On urgent measures to protect the citizens of Ukraine from the consequences of the Chornobyl disaster"). Accessed on 10 January 2016 https://zakon.rada.gov.ua

Nr. 315, 1989. Decision of 14 December 1989. Pro dodatkovi zakhody shchodo posylennya okhorony zdorov'ya ta polipshennya material′noho stanovyshcha naselennya, yake prozhyvaye na terytoriyi, shcho zaznala radioaktyvnoho zabrudnennya v rezul′tati avariyi na Chornobyl′s′kiy AES (On additional measures to strengthen health care and improve the financial situation of the population living in the territory, which was the subject of radioactive contamination as a result of the Chernobyl accident.). Accessed on 10 January 2016 https://zakon.rada.gov.ua

Nr. 333, 1989. Decision of 30 December 1989. Pro pereselennya zhyteliv z naselenykh punktiv Narodyts′koho rayonu Zhytomyrs′koyi oblasti i Polis′koho rayonu Kyyivs′koyi oblasti, a takozh budivnytstvo dlya nykh ob'yektiv sotsial′no-pobutovoho i vyrobnychoho pryznachennya (On the resettlement of residents from the settlements of the Narodychi district of Zhytomyr region and

Polissya district of Kiev region, as well as the construction of objects of social, domestic and industrial purpose for them). Accessed on 10 January 2016 https://zakon.rada.gov.ua

Nr. 886, 1989. Resolution of 20 October 1989. O dopolnitel'nykh merakh po usileniyu okhrany zdorov'ya i uluchsheniyu material'nogo polozheniya naseleniya, prozhivayushchego na territorii, podvergsheysya radioaktivnomu zagryazneniyu v rezul'tate avarii na Chernobyl'skoy AES (On additional measures to strengthen health protection and improve the financial situation of the population living in the territory subjected to radioactive contamination as a result of the Chernobyl accident). Accessed on 10 January 2016 https://www.libussr.ru

Nr. 1006–286, 1986. Resolution of 22 August 1896. Ob uluchshenii material'nogo polozheniya naseleniya, prozhivayushchego v naselennykh punktakh s ogranicheniyem potrebleniya sel'skokhozyaystvennoy produktsii mestnogo proizvodstva v svyazi s avariyey na Chernobyl'skoy AES (On improving the financial situation of the population living in settlements with limited consumption of locally produced agricultural products in connection with the Chernobyl accident). Accessed on 10 January 2016 https://www.libussr.ru

Okada, T. 2017. Acknowledging local sociality in disaster recovery: a longitudinal, qualitative study. PhD thesis. Macquarie University, Syndey

Oliver-Smith, A. 2013. Catastrophes, Mass Displacement and Population Resettlement. in: Bissel, R, 2013. Preparadness and Response for Catastrophic Disasters. CRC Press, Taylor&Francis Group. Boca Raton, London, New York. 185–224

Pandey, B. & Okazaki, K. 2005. Community-based disaster management: Empowering communities to cope with disaster risks. *Regional Development Dialogue,* 26

Patterson, O., Weil, F. & Patel, K. 2010. The Role of Community in Disaster Response: Conceptual Models. *Population Research and Policy Review,* 29, 127–141

Pearce, L. 2003. Disaster Management and Community Planning, and Public Participation: How to Achieve Sustainable Hazard Mitigation. *Natural Hazards,* 28, 211–228

Pickett, N. R. 2016. The Chornobyl' disaster and the end of the Soviet Union. In krakh radyans'koyi imperiyi: Anatomiya katastrofy. Sotsialistychnyy tabir, SRSR ta postradyans'kyy prostir u druhiy polovyni XX – na pochatku XXI st. (In the collapse of the Soviet Empire: The anatomy of disaster. The socialist camp, the USSR, and the post-Soviet space in the second half of the twentieth-early twentieth centuries.) Nizhin State University named after Gogol, Nizhin, Ukraine, 85–91

Plokhyi, S. 2014. The Last Empire: The Final Days of the Soviet Union. New York: Basic Books

Prister, B., Klyuchnikov, A., Shestopalov, V., & Kukhar V. 2013. Problemy bezopasnosti yadernoy energetiki, Uroki Chernobylya (Nuclear safety issues. Chernobyl lessons). Chernobyl

Reconstruction Authority, Fukushima Prefecture & Namie Town 2016. Report of a survey on residents' intention in Namie Town

Romashko, E. 2016. Religion and 'Radiation Culture': Spirituality in a Post-Chernobyl World. Material Religions. Accessed on 1 June 2016https://www.materialreligions.blogspot.com.au/2016/05/religionandradiationculture.html

Tanaka, S., Hunabashi, H. & Masamura, T., 2013. True and false in community, The Great Eastern Japan Earthquake and Japanese Sociology, 47–70, 2013

The Human Consequences of the Chernobyl Nuclear Accident. A Strategy for Recovery 2002. A Report Commissioned by UNDP and UNICEF with the support of UN-OCHA and WHO

Tomioka Town, 2017. Lifting evacuation orders. Accessed on 2 December 2019 https://www.tomioka-town.jp/living/cat25/2017/03/003374.html (Japanese)

Tykhyi, V. 1998. Chernobyl Sufferers in Ukraine and Their Social Problems: Short Outline. Research Activities about the Radiological Consequences of the Chernobyl NPS Accident and Social Activities to Assist the Sufferers by the Accident, Kyoto Accessed on 2 December 2019 https://www.rri.kyoto-u.ac.jp/NSRG/reports/kr21/kr21pdf/Tykhyi.pdf pp. 235–245

Usamah, M. & Haynes, K. 2012. An examination of the resettlement program at Mayon Volcano: what can we learn for sustainable volcanic risk reduction? *Bulletin of Volcanology,* 74**,** 839–859
Vendland, A. V. 2011. Povernennya do Chornobylya. Vid natsional′noyi trahediyi do predmeta innovatsiynykh dystsyplin v istoriohrafiyi ne til′ky Ukrayiny (Return to Chernobyl. From national tragedy to the subject of innovative disciplines in historiography not only of Ukraine). *Ukrayina moderna*, 18, 151–185

Chapter 12
Exchanging Disaster Science Expertise Between Countries—A Japanese Personal Perspective

Osamu Murao

Abstract Having experienced firsthand the catastrophic Great East Japan Earthquake and Tsunami of 2011, Tohoku University founded the International Research Institute of Disaster Science (IRIDeS) in 2012. IRIDeS staff, with a broad array of relevant specializations, conducts world-class research on disaster science and disaster mitigation in collaboration with organizations from many countries. As a member of IRIDeS, Prof. Osamu Murao, the founder and manager of the International Strategy for Disaster Mitigation Laboratory (ISDM), has conducted several international collaborative research projects. This chapter briefly reports on the activities of the IRIDeS and ISDM and highlights key factors for successful international collaborative research and exchange experiences with other countries. The author recounts his initial collaborative research experience in a long-term project examining Taiwan's recovery from the impact of the 1999 Chi-Chi Earthquake which was the foundation of the international research collaboration at ISDM. The chapter concludes with a summary of the valuable lessons learned from the author's participation in this research.

Keywords IRIDeS · Chi-Chi earthquake · Taiwan · Japan · Partnership · Local communities

12.1 Introduction

Stories of natural calamities and misfortunes appear almost daily throughout the world. In many cases, these are relatively minor events such as heavy rains or power failure. However, occasionally a major disaster occurs so seriously that it affects millions of people directly or indirectly, as happened with the 2004 Indian Ocean Tsunami or the 2011 Great East Japan Earthquake or in case of 1999 Chi-Chi Earthquake. Damage conditions and post-disaster recovery situations affected by these major events inevitably differ according to regional and social characteristics as well

O. Murao (✉)
International Research Institute of Disaster Science, Tohoku University, Sendai, Japan
e-mail: murao@irides.tohoku.ac.jp

D. Karácsonyi et al. (eds.), *The Demography of Disasters*,
https://doi.org/10.1007/978-3-030-49920-4_12

as the nature and severity of the hazard. However, similarities can be usefully examined. In fact, it is crucial that we learn from our experiences with urban and regional disasters in order to develop and refine disaster risk reduction measures for the future. This is particularly true for artificial environments such as cities.

When a manufacturer develops a product that might pose a risk to the consumer, the maker carries out repeated tests until the safety and reliability of the product can be assured. The development of a city is far different. The typical city is an artificial environment that has formed and evolved rather than a product that has been thoroughly planned, tested, and delivered with safety guarantees. In most cases, serious safety weaknesses go unrecognized until the city faces a disaster (see also Chap. 9 by Barnes on designing cities considering disaster and gender aspects). It is for this reason that past disasters and recoveries need to be examined and analyzed carefully and the lessons learned from the experience disseminated in order to reduce urban disaster risks.

Japanese society has accumulated valuable knowledge and experiences from past disasters. In recognition of this, the United Nations held influential international events for disaster risk reduction in Japan at Yokohama in 1994, Kobe in 2005, and Sendai in 2015. Emanating from the 3rd World Conference on Disaster Risk Reduction was the Sendai Framework for Disaster Risk Reduction 2015–2030 (UNISDR 2015) (hereafter, the Sendai Framework), which describes the importance of international cooperation to reduce disaster risk around the world:

> It is necessary to continue strengthening good governance in disaster risk reduction strategies at the national, regional and global levels and improving preparedness and national coordination for disaster response, rehabilitation and reconstruction, and to use post-disaster recovery and reconstruction to "Build Back Better", supported by strengthened modalities of international cooperation.

The Sendai Framework, which succeeds the Hyogo Framework for Action 2005–2015, continues to provide global guidelines for disaster risk reduction from 2005 to 2015 (UNISDR 2005). The Sendai Framework suggests gaps exist "in addressing the underlying disaster risk factors, in the formulation of goals and priorities for action, in the need to foster disaster resilience at all levels and in ensuring adequate means of implementation," as well as identifying "a need to develop an action-oriented framework" (see UNISDR 2015; p. 11). International cooperation and global partnerships are seen as significant factors in the implementation of effective disaster risk reduction measures.

The International Research Institute of Disaster Science (IRIDeS) is designed to address some of the gaps outlined in the Sendai Framework. The IRIDeS was founded in April 2012 by Tohoku University following the costly and catastrophic disaster of the Great East Japan Earthquake and Tsunami in 2011. The mission of the IRIDeS has been to create a new academic area of disaster mitigation that subsumes the lessons of the Great East Japan Earthquake and Tsunami and the findings of world-class researchers. It aims to establish social systems capable of responding promptly, sensibly, and effectively to disasters, withstanding adversities with resiliency, and passing on and extending the lessons learned to future disaster management.

The IRIDeS is committed to assisting on-going recovery and reconstruction efforts in affected areas, conducting action-oriented research, and pursuing effective disaster management to build sustainable and resilient societies. Together with collaborating organizations from many countries and highly specialized experts, the IRIDeS conducts world-class research on natural hazard science and disaster mitigation. The following section describes the IRIDeS, its activities, and outcomes.

12.2 Activities at International Research Institute of Disaster Science (IRIDeS)

The IRIDeS aims to become a world center for the study of disasters and disaster mitigation, learning from and building on past lessons in disaster management from Japan and around the world. To achieve this, the IRIDeS undertakes a range of action-orientated research and collaborative activities.

In the frame of action-oriented research, the IRIDeS has been committed to enhancing cooperation with local municipalities and governments in areas affected by the 2011 Great East Japan Earthquake and Tsunami as part of its contribution to recovery and reconstruction efforts. The IRIDeS established a number of agreements with local governments affected by the 2011 earthquake and seriously hit by the following tsunami in Iwate (Rikuzentakata City) and Miyagi Prefectures (Kesennuma City, Ishinomaki City, Higashi-Matsushima City, Tagajo City, Sendai City, Natori City, Iwanuma City, Watari Town, and Yamamoto Town). The long-term objective of these cooperations is to create disaster-resilient societies that can manage the complex and diverse challenges associated with disasters. Not only doing so by identifying and implementing prevention measures but also by preparing and responding to the challenges they present and achieving recovery and renovation. Broadly speaking, the goal is to engender a culture of disaster-resiliency and incorporate it into our social systems.

The action-oriented research from the IRIDeS (2012) focuses on:

1. Investigating the physics of global scale disasters such as mega-earthquakes, tsunamis, and extreme weather.
2. Reconstructing disaster response and mitigation technologies based on the lessons of the 2011 Great East Japan Earthquake and Tsunami disaster.
3. Inventing "Affected Area Supportology" in the aftermath of disasters.
4. Enhancing the disaster-resiliency and performance of multiple fail-safe systems in regional and urban areas.
5. Establishing disaster medicine and medical service systems for catastrophic disasters.
6. Designing disaster-resilient societies and developing a digital archive system to pass along the lessons of past disasters.

The partnerships with local governments have produced co-organized seminars open to the public, tsunami evacuation drills, efforts to devise a post-tsunami recovery plan, educational activities related to disaster risk reduction, and an array of additional activities. Based on its close relationships with local residents, the IRIDeS opened a branch office in Kesennuma to organize periodic disaster risk reduction seminars.

The IRIDeS has prioritized partnerships with other academic organizations. Such partnerships are crucial to the conduct of cutting-edge research and the collection of pertinent information for disaster risk reduction. The IRIDeS has established international academic exchange agreements with the Center for Weather, Climate and Disaster Research at the National Taiwan University, the German Aerospace Center, the Faculty of Mathematics and Natural Sciences, the Edwin O. Reischauer Institute of Japanese Studies at Harvard University (USA), Project NOAH at the Department of Science and Technology, the Cabinet of the Philippines, the University of the Philippines, Manila, the Angeles University Foundation (Philippines), the Institute of Geological and Nuclear Sciences Limited (New Zealand), the United Nations Development Program, the Global Risk Forum GRF Davos (Switzerland), the Tsunami and Disaster Mitigation Research Center (TDMRC) and Syiah Kuala University (Indonesia), the Aceh Tsunami Museum (Indonesia), the Institute of Medicine, Tribhuvan University (Nepal), the United States Geological Survey (USGS), and the University of Moratuwa (Sri Lanka).

In collaboration with international and domestic organizations, the IRIDeS investigated damage conditions and recovery processes in areas affected such as the 2013 Typhoon Yolanda (Haiyan), the 2015 Nepal Earthquake (Fig. 12.1), the 2014 Heavy Rain in Japan, and the 2016 Kumamoto Earthquake. The collaborative research on

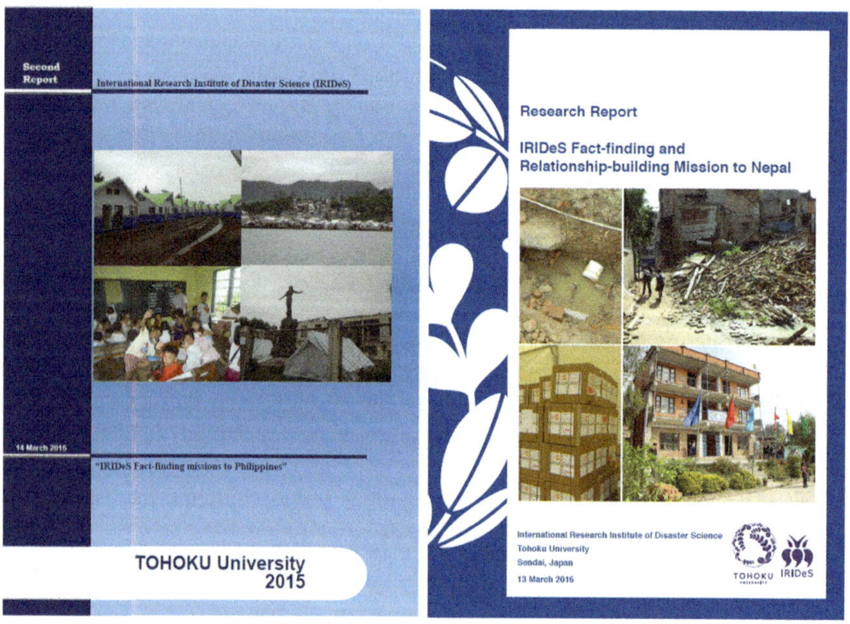

Fig. 12.1 Report on 2013 Typhoon Yolanda (L) and 2015 Nepal Earthquake (R)

the 2013 Typhoon Yolanda was the first big international research project for the IRIDeS.

Using established networks, the IRIDeS embraces opportunities to share and exchange knowledge, experience, and learnings from past disasters among practitioners, government officers, researchers, and ordinary people from Japan and other countries. The 3rd World Conference of Disaster Risk Reduction held in Sendai in March 2015, mentioned earlier, was one of the most impactful events in the relatively brief history of the IRIDeS. Tohoku University, as the most affected Japanese national university by the 2011 Great East Japan Earthquake, joined with Sendai City and the Japanese Government to support and co-organize the main conference. This resulted in two significant outputs. Firstly, the conference symposium entitled *Resilient Communities: Our Home, Our Communities, Our Recovery* was organized by the IRIDeS in conjunction with UN-HABITAT members. The symposium invited community leaders and mayors from several areas affected by the 2004 Indian Ocean Tsunami, the 2013 Typhoon Yolanda, and the 2011 Great East Japan Earthquake to discuss vulnerability reduction in local communities. Secondly, the IRIDeS announced the establishment of its Global Center for Disaster Risk Reduction. The Center was created to play an important coordinating role, collaborating with several United Nations agencies, including United Nations Development Program and the International Strategy for Disaster Reduction (ISDR), to help monitor progress on the new global disaster risk reduction framework.

Triggered by this conference in 2015, the IRIDeS held the inaugural World Bosai Forum in November 2017, the first of a series of biyearly International Disaster and Risk Conferences developed in cooperation with the Global Risk Forum held in Davos, the city of Sendai, and others. The word *Bosai* is a traditional Japanese term implying a holistic approach to reducing human and economic losses from disasters. It represents activities in all disaster phases, including prevention, recovery, response, and mitigation. With 947 participants from 44 countries, the inaugural Bosai Forum marked the successful beginning of what is likely to be a highly productive series of conferences (Fig. 12.2).

12.2.1 International Collaborative Research Projects at the International Strategy for Disaster Mitigation Laboratory (ISDM)

The International Strategy for Disaster Mitigation Laboratory (ISDM), founded and managed by the author, belongs to the Regional and Urban Reconstruction Research Division at the IRIDeS since 2013. Before that it was part of Tsukuba University. The ISDM seeks to provide practical international strategies for disaster mitigation or post-disaster recovery as well as to develop international frameworks that enable the realization of those strategies based on field surveys and data analysis. In order

Fig. 12.2 Poster of World Bosai Forum 2017 (L) and the opening ceremony (R) (Sendai, 2017, taken by Osamu Murao)

to clarify existing problems and make recommendations for future disaster reduction, the ISDM researches the relationship between disaster management and urban-regional space through case studies of vulnerable areas, including disaster-affected cities.

One of the ISDM's most significant research areas is urban vulnerability evaluation. After the 1995 Great Kobe Earthquake, a research associate at the Institute of Industrial Science within the University of Tokyo clarified the relationship between seismic ground motion and building damage using actual damage data provided by Kobe City. As a result, building vulnerability functions were formulated (Yamazaki and Murao 2000) and a method for conducting building collapse risk evaluation for Tokyo was proposed (Murao et al. 2000). Those research activities on urban vulnerability evaluation continued for the research associate.

While the ISDML monitors urban recovery in areas affected by disasters across the world, it quantitatively evaluates urban recovery processes for planning future disaster risk reduction strategies. To date, international field research has taken place in Taiwan, Turkey, Sri Lanka, Thailand, Indonesia, Peru, China, Hawaii, New York, Bangladesh, and Myanmar (Fig. 12.3). Significant domestic field research sites include Kobe, Tokyo, Kanagawa, and the Sanriku coastal areas affected by the 2011 Great East Japan Earthquake and Tsunami. These field experiences, which have included extensive communication with local residents and negotiations to acquire critical data, have produced significant lessons regarding the conduct of international collaborative research.

The following section is a first-person summary of the author's introduction to international collaborative research in a project focused on urban recovery following the 1999 Chi-Chi Earthquake in Taiwan while a member of the University of Tsukuba. This experience with the project served as the foundation for the authors current work in international research collaboration at ISDML.

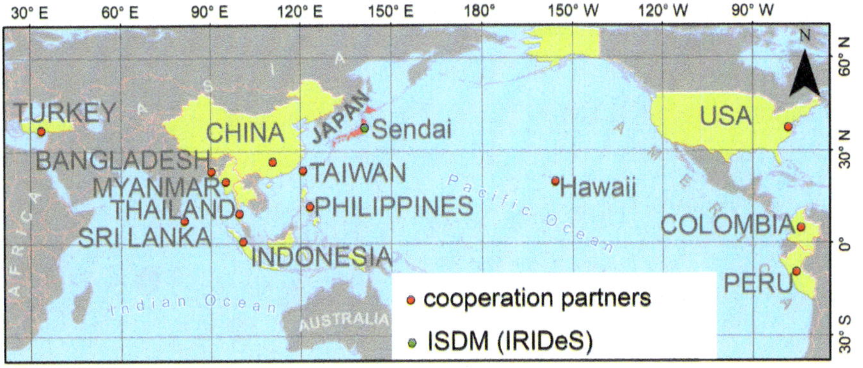

Fig. 12.3 ISDM (IRIDeS) research fields in the world (Author Murao, cartography by Karácsonyi)

12.3 Collaborative Research on Recovery Processes from the 1999 Chi-Chi Earthquake in Taiwan

12.3.1 Chi-Chi Township and the 1999 Chi-Chi Earthquake in Taiwan

Chi-Chi Township (shown in Fig. 12.4) is located in Nantou County in the central part of Taiwan. Early in the twentieth century during Japanese colonial rule, it prospered as a center of traffic, commerce, and politics. This was largely due to the building of the Chi-Chi railroad, the construction of the city hall, and a successful banana industry. The township currently has a population of approximately 12,000 across 11 villages.

The Chi-Chi Earthquake, with its epicenter near Chi-Chi, occurred on September 21, 1999 (see Chap. 6, the case of Christchurch earthquake and its population impact for comparison). It caused damage to more than 106,000 buildings and with an estimated 2500 casualties throughout Taiwan. In Chi-Chi itself, 1736 buildings were seriously damaged, 792 buildings were moderately damaged, and 42 people died. It was among one of the most serious disasters in the history of Chi-Chi (and Taiwan). It destroyed tourist attractions such as the traditional Japanese-style Chi-Chi station, historical temples, traditional pottery, and various important public facilities. These cultural resources, representing the area's historically unique background, were significant elements in Chi-Chi's economic and industrial recovery.

An event of personal significance occurred on October 1, 1999, ten days after the earthquake. My colleagues and I visited Chi-Chi to conduct a damage survey. It was my first field survey outside Japan and the town was littered with collapsed buildings (Fig. 12.5). Everywhere we looked, people were responding to emergencies. Because of its close proximity to the epicenter of the earthquake, the town had been devastated.

This time in Chi-Chi was relatively brief, but deeply affecting. It was a wonder how this town could ever recover. Just prior to this visit, I had finalized my doctoral

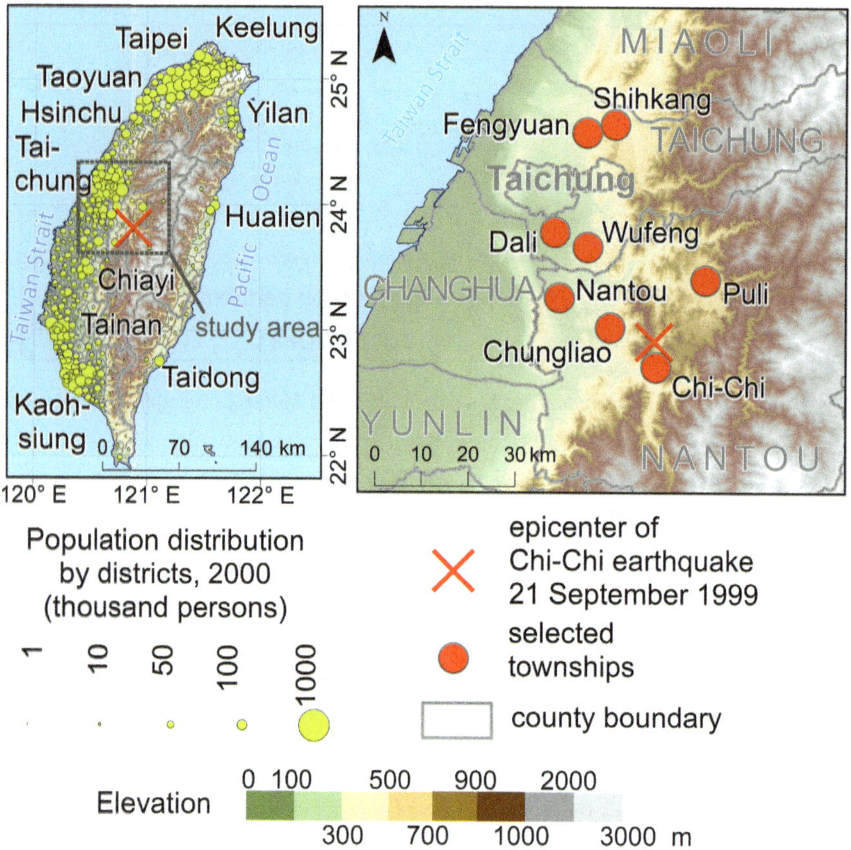

Fig. 12.4 Location of Chi-Chi and the epicenter of 1999 Taiwan earthquake (Authors Murao, Karacsonyi, cartography by Karácsonyi, population distribution based on census data)

Fig. 12.5 Collapsed buildings (L) and Wuchang temple (R) in Chi-Chi, Taiwan, due to the 1999 Chi-Chi Earthquake (Murao 1999)

thesis, "Study on Building Damage Estimation based on the Actual Damage Data due to the 1995 Hyogoken Nanbu Earthquake," and was looking for the next research topic. As a researcher with a background in architecture and city planning, I became immediately interested in the post-earthquake recovery of Chi-Chi. This experience turned out to be the trigger for later post-disaster recovery research.

12.3.2 *Continuous Surveys in Chi-Chi and Building Ties*

At the beginning of 2000, the Research Committee on Urban Planning and Community Development for Disaster Reduction (the Committee) was established within the City Planning Institute of Japan (CPIJ). As member of the Committee, I met many young like-minded Japanese researchers and a Taiwanese student, Hsueh-Wen Wang, who were motivated to understand the process of urban recovery in Taiwan. The importance of on-site field surveys was advocated by this group who successfully gained CPIJ's support.

As a delegate member of the Committee, I visited Chi-Chi again in April 2000, approximately six months after the earthquake and set about investigating recovery conditions in the township, as well as other affected areas in Nantou County and Taichung County. On this visit, we gathered materials and information on damage (walk through surveys, photo recording, interviews, and collecting data on housing stock) that would help in better understanding the recovery situation and the urban recovery strategies that were being developed and implemented.

Numerous stakeholders and specialists affected by the tragedy or involved in the recovery were engaged in the course of conducting the survey. The stakeholders included faculty members of the National Taiwan University and Feng-Chia University, government officials, NPO members in charge of rebuilding communities in the affected districts, architects, planners, and local residents. Among our many interactions, meetings with Prof. Liang-Chun Chen of the National Taiwan University were perhaps the most significant. They enabled us to maintain continuous contact with the Taiwanese recovery situation over a number of years. These important meetings were arranged by Ms. Wang, who played a critical role helping our study survey.

In April 2001, the Committee received a three-year research grant entitled *A Comparative Study on Disaster Management and Reconstruction Strategy among Earthquake Disasters of Hanshin (Japan), Kocaeri (Turkey), and Chi-Chi (Taiwan),* from the Japan Society for the Promotion of Science (JSPS). The 1999 Kocaeri Earthquake, also known as the 1999 Izmit Earthquake, occurred in Turkey on August 17 1999, approximately one month before the 1999 Chi-Chi Earthquake. Our intention was to compare the Kocaeri and Chi-Chi urban recovery processes to the 1995 Great Kobe (Hanshin) Earthquake recovery.

Supported by the grant, frequent visits were made to affected areas in Turkey, Taiwan, and Japan for surveys with particular research concerns and continued to collect related materials and information through to March 2004. The research

Fig. 12.6 Permanent housings (L) and childcare activities (R) in Kocaeli, Turkey (Murao 2001)

covered a wide range of topics that included urban recovery planning, temporary housing, permanent housing (Fig. 12.6, left), construction methods, economic recovery, debris management, community building, childcare activities (Fig. 12.6, right), and more. Given that the field of post-disaster urban recovery research in at this time was still in its infancy, the interviews and discussions with local government officials and recovery specialists in Turkey and Taiwan were found to be extremely helpful in shaping our thoughts regarding what is needed to conduct an insightful comparative study of post-disaster urban recovery.

During this time, I visited Chi-Chi several more times meeting two key persons who would be instrumental in advancing my research. An Internet search for an interpreter who could translate from Chinese to English led us to Ms. Yayoi Mitsuda (Yoyo), a multilingual Japanese student who was studying cultural anthropology at National Tsing Hua University. It was my great fortune, for Yoyo became an indispensable partner in conducting surveys.

Random good fortune also led us to a restaurant owner in Chi-Chi referred here to as David. While out conducting field surveys, we stopped at David's restaurant for dinner. It was a small, unassuming restaurant of the type we would often see along the streets of Taiwan. We quickly discovered that the restaurant operated as a community hub. It turned out that David was a community leader in Chi-Chi, with close relationships with the current and former mayors and other local people of influence. Since that first night, David's restaurant became my base for conducting surveys in Chi-Chi. My time there yielded an exceptional amount of information on the recovery process, local history, key persons, politics, culture, human relationships, and much more.

Soon after the project ended in April 2004, I managed another JSPS grant to pursue research in Chi-Chi entitled *Architecture of Reconstruction Process of Chi-Chi Area (Taiwan) and Archives Related to Urban Reconstruction in the World*. This meant that the research ideas, which had been forming since my first days in Chi-Chi, would continue to develop and mature.

12.3.3 Research on Post-earthquake Recovery in Chi-Chi

My research activity in Chi-Chi went on until 2008. It allowed for continuous surveys and research activities over an extended ten year period, mostly in a trial-and-error fashion. This work is published in several academic journals (e.g., Murao 2006a, b, Murao et al. 2007) and international conference proceedings. Some of the research findings are described below.

Through the field surveys, I was able to monitor and record the urban recovery conditions of Chi-Chi, which were in a continuous state of change. In order to clearly understand Chi-Chi's transition, a proper town map was necessary. In 2000–2001 however, such a map was difficult if not impossible to obtain. Consequently, I decided to create my own digital base map. Walking around and surveying the town with lab students using IKONOS (satellite) imagery, I continued to digitize the research area until a suitable GIS base map was completed. The map functioned to chronologically record the building demolition and reconstruction conditions that I had been monitoring since the earthquake struck. The recovery processes of Chi-Chi could be representable visually, as illustrated by the maps in Fig. 12.7.

Through the study of the Hanshin (Japan), Kocaeri (Turkey), and Chi-Chi (Taiwan) earthquake disasters and recoveries, I compared post-disaster urban recovery processes in cities with such different social backgrounds with my collaborators in Taiwan, Turkey, and Japan. Based on this cooperation experience, I recognized the need to develop a quantitative evaluation method but was yet to determine how to assess the recovery process quantitatively. This became an important research question during my time in Chi-Chi.

The question continued to taunt me, until one day, in a research meeting at the National Taiwan University, I was struck with the idea to represent the progress of the recovery by creating "recovery curves" based on building construction data. This process is essentially counting the number of building completions of various types over time. Later, with the support of Yoyo our translator and several officials, I was able to obtain statistical data on post-earthquake construction from the Nanto Government. From this data, I was able to create recovery curves for the various building types (temporary houses, rebuilt buildings, and new buildings) that are shown in Fig. 12.8.

The recovery curves show the reconstruction of buildings in Chi-Chi began approximately six months after the earthquake and continued for a further three years. The construction of new buildings began only after a delay of 1.5 years. This component of my research was challenging, but the method for constructing recovery curves would later be applied to good effect to my other research cases in Sri Lanka, Thailand, Indonesia, and the Japanese coastal areas affected by the 2011 Great East Japan Earthquake. The approach helped to conveniently compare recovery processes in areas affected by disasters in the four countries under study.

Having the opportunity to monitor the reconstruction process over an extended period and applying the idea of recovery curves contributed greatly to the research outcomes in Chi-Chi. It should be noted that while the research was carried out to

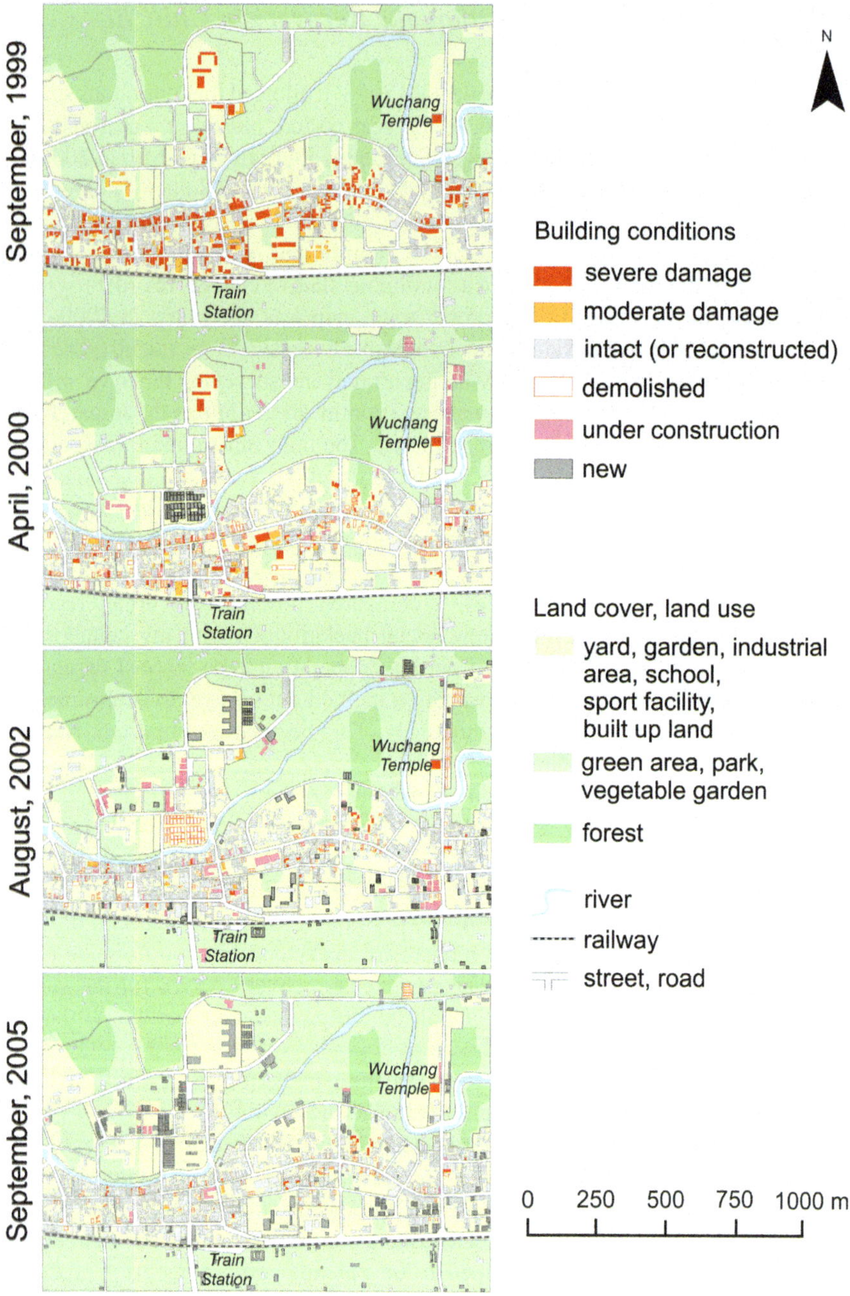

Fig. 12.7 Change of post-earthquake recovery conditions of Chi-Chi (Murao 2006a, cartography by Murao and Karácsonyi)

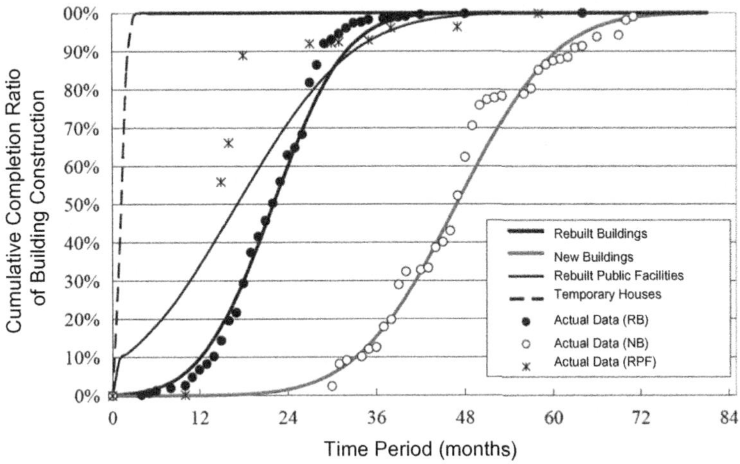

Fig. 12.8 Recovery curves in terms of building types (Murao et al. 2007)

understand Chi-Chi's recovery process, in terms of changes in the area's physical environment such as building reconstruction and the restoration of housing, we also recognized that the observable recovery was the product of human activity. Accordingly, I often sought out and interviewed key persons in the recovery effort. They included mayors, local government officials, shop owners, and victims. They were interviewed about Chi-Chi's history and cultural background, the behavior of individuals from the initial emergency response stage to the reconstruction stage, and the various recovery strategies that were being employed. Based on the opinions and concerns expressed by residents in these interviews, I modeled the post-earthquake recovery process by using the simplified model shown in Fig. 12.9. This systematic model indicates the sequential process of recovery from the viewpoint of the victims, from the early catastrophic moments until the time of resettlement in permanent housing.

By 2006, I had monitored the ever-changing status of Chi-Chi Township and spoken extensively with local residents for several years. I began to consider the importance of producing a permanent record of Chi-Chi's urban disaster recovery processes. By this time, I had already gathered a great number of pictures, movies, and considerable other data. From this, I conceived the idea of creating a digital model of Chi-Chi Township to preserve at least a segment of the recovery process. To this end, together with my laboratory students, I took pictures of every building elevation in the area and ultimately completed the model called *Digital Chi-Chi City* on the Google Earth platform. With information on the recovery conditions of important facilities in the township, Fig. 12.10 illustrates the product of our efforts.

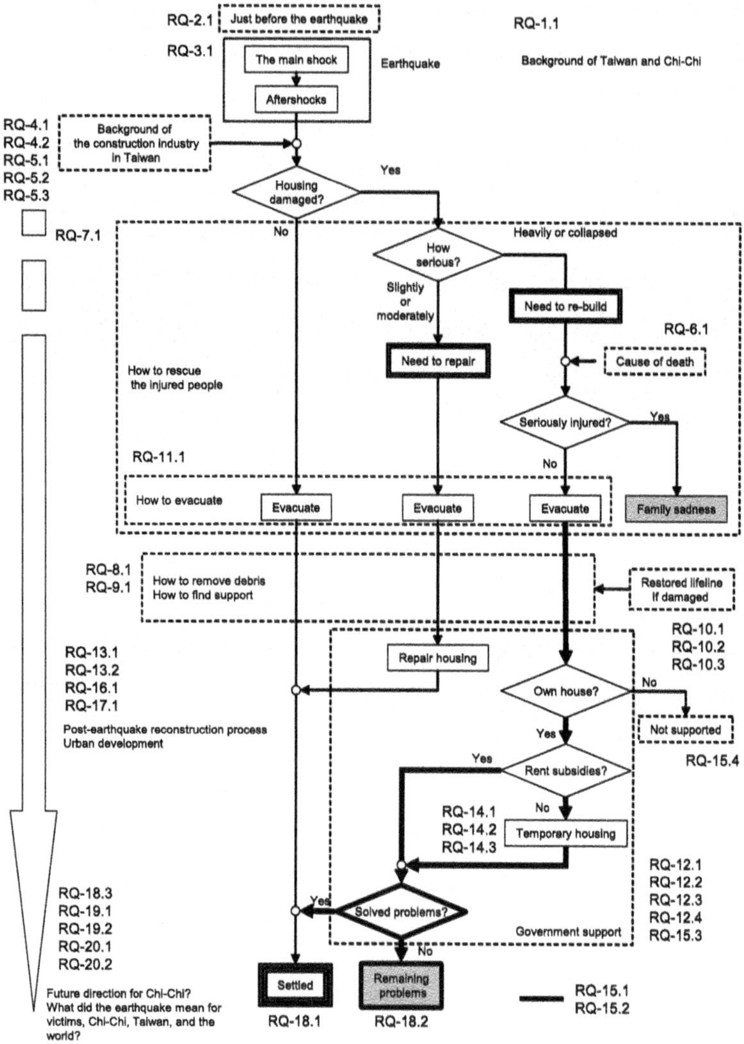

Fig. 12.9 Structure of post-earthquake recovery process (Murao 2006b)

12.3.4 Lessons from Post-earthquake Recovery in Chi-Chi

The Chi-Chi experience provided a number of important lessons regarding long-term post-disaster research in foreign countries. From the first visit to Chi-Chi in 1999, it was clear that the support of others was needed to conduct effective surveys and reach useful research outcomes. The support of Prof. Liang-Chun Chen, David, and Yoyo was crucial. The following describes their contributions as a way of underlining

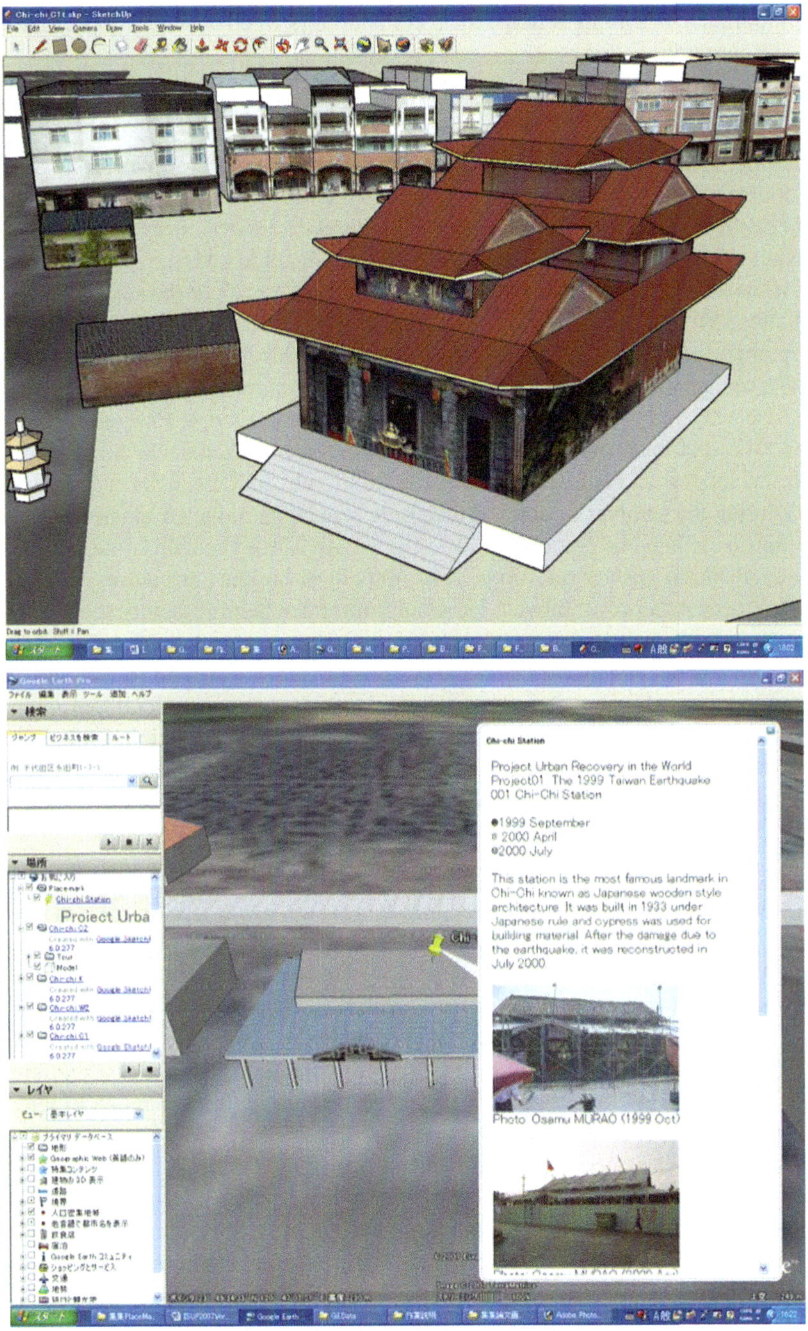

Fig. 12.10 Digital Chi-Chi City on Google Earth as recovery digital archives (Murao 2007)

the importance of finding reliable local collaborators when conducting international disaster research.

I. *Seek out collaborative relationships with local specialists/researchers*

I first met Prof. Lian-Chun Chen (2001) at the National Taiwan University during the course of developing my survey for the *Comparative Study on Disaster Management and Reconstruction Strategy among Earthquake Disasters of Hanshin (Japan), Kocaeri (Turkey), and Chi-Chi (Taiwan).* Prof. Chen graduated from Waseda University in Japan, which made it easy for us to communicate about the survey's purpose and subsequent information requests. Following our first meeting, we had many more opportunities to exchange information and ideas on the recovery in Taiwan. Prof. Chen generously provided all requested information.

Under the 2004 JSPS grant *Architecture of Reconstruction Process of Chi-Chi Area (Taiwan) and Archives Related to Urban Reconstruction in the World* and with the support of Prof. Chen, I spent three months in Taiwan (primarily in Chi-Chi) during the summer of 2005 as a visiting researcher at the Graduate Institute of Building and Planning, National Taiwan University. Throughout this time, Prof. Chen was available to answer questions about the cities, building structures, or disaster management systems of Taiwan. He would introduce helpful faculty members and suggest suitable paths for this research.

His invaluable assistance made me realize that when traveling to affected areas after a disaster, and especially in the field of urban recovery, it is difficult for foreigners to successfully conduct field surveys using suitable research questions without solid knowledge of the social context. Thus, it is extremely important to have a specialist or local researcher in the same research field who can provide critical support.

II. *Be a friend and build good relationships with the local community*

Whenever I returned to Chi-Chi, David's restaurant was always the first place I visited. There I would have dinner and get the latest information on Chi-Chi Township, the recovery situation, politics, the new mayor, new restaurants, business conditions, visitors to Chi-Chi, and the current circumstances of individual residents. Because of his numerous connections, I was able to conduct interviews with many of the local people, which would eventually lead to my structural model of the post-earthquake recovery process (Murao 2006b). His contacts opened the door to much of the data and materials needed in this research.

As noted earlier, meeting David was by sheer accident, yet he became precious to my research. It was important to continue to visit David's restaurant, particularly in the early stages of my long-term work in Chi-Chi. With each visit, exchanging stories about our families and ourselves, drinking together, or singing loudly brought our relationship closer. The rapport that we built led David to introduce me to many of Chi-Chi's stakeholders and played a key role in my ability to finalize my research there.

When a disaster occurs, the temptation for disaster recovery researchers is to immediately survey those affected by the tragedy. However, essential information for comprehensive recovery research, including knowledge of the social context, cannot be acquired without taking the time to establish a strong rapport with the local people.

III. *Find the best partner possible*

Yoyo was a doctoral student at the Institute of Anthropology, Tsing Hua University, who was conducting research on the survival strategies of the Thao people living near Sun Moon Lake in Nantou County, Taiwan. She became critical to my research in Chi-Chi. I have an engineering background with a focus on architecture and city planning. However, post-disaster urban recovery activities are, in a sense, comprehensive phenomena of society as a whole. Given this conception, I intended to take an interdisciplinary approach to my research in Chi-Chi—an approach that included sociology and anthropology. Yoyo fully understood my intention and advised me on the most appropriate ways to conduct my local surveys.

In her capacity as interpreter, noting that I cannot speak or understand Chinese, Yoyo would often pose more questions to the interviewee than I had requested in order to gather the kind of information that she knew I needed. This ability arose from her sense of anthropology and intimate knowledge of society in central Taiwan, as well as her deep understanding of my long-term research goals. As a young anthropologist whose doctoral thesis focused on the indigenous people of Taiwan, Yoyo's complementary knowledge and instincts were indispensable to my work. I met her in 2002 quite by accident, but the successful outcomes of my research could not have been produced without her advice and support.

I can easily say, no one could replace her. Good research in overseas fields can sometimes come down to one happy chance encounter. Such continuous surveys are basic activities to obtain proper datasets for the demography of disasters.

12.4 Summary and Conclusions

This chapter describes the Sendai Framework for Disaster Risk Reduction, introduced the ISDML within IRIDeS at the Tohoku University, and highlighted their international collaborative activities. It discussed some of the research activities overseen by the author and took a personal lense through this experience. The importance of international collaboration and harnessing opportunities for effective partnerships has been stressed throughout. The author's extensive recovery research in Chi-Chi, Taiwan, was presented in some detail as an example of international collaborative activities.

References

International Research Institute of Disaster Science (IRIDeS) (2012), "About IRIDeS," https://irides.tohoku.ac.jp/eng/outline/index.html. [accessed on December 8, 2017]

Murao, O., Tanaka, H., and Yamazaki, F. (2000), "Risk Evaluation Method of Building Collapse from the Experience of the Kobe Earthquake," 12th World Conference on Earthquake Engineering (WCEE) (CD-ROM), No.2312, 8p, Auckland, New Zealand

Murao, O. (2006a), "Reconstruction Process based on the Spatial Reconstruction Model in Chi-Chi Area after the 1999 Chi-Chi Earthquake," Journal of Architecture and Planning, No.607, 95–102. (in Japanese)

Murao, O. (2006b), "Structure of Post-earthquake Recovery Process after the 1999 Chi-Chi Earthquake-A Case Study of Chi-Chi," Proceedings of the International Symposium on City Planning, 164–175, Taipei, Taiwan

Murao, O., Mitsuda, Y., Miyamoto, A., Sasaki, T., Nakazato, H., and Hayashi, T. (2007), "Recovery Curves and Digital City of Chi-Chi as Urban Recovery Digital Archives," Proceedings of the 2nd International Conference on Urban Disaster Reduction (ICUDR) (CD-ROM), Taipei, Taiwan

UNISDR 2005 United Nations International Strategy for Disaster Reduction (UNISDR), "Hyogo Framework for Action 2005–2015: Building the Resilience of Nations and Communities to Disasters," https://www.unisdr.org/we/coordinate/hfa, 2005. [accessed on April 6, 2011]

UNISDR 2015 United Nations International Strategy for Disaster Reduction (UNISDR), "Sendai Framework for Disaster Risk Reduction 2015–2030," https://www.unisdr.org/we/coordinate/sendai-framework, 2015. [accessed on July 2, 2015]

Yamazaki, F., and Murao, O. (2000), "Vulnerability Functions for Japanese Buildings based on Damage Data due to the 1995 Kobe Earthquake," Implications of Recent Earthquakes on Seismic Risk, Series of Innovation in Structures and Construction, Vol.2, pp.91–102, Imperial College Press

Chapter 13
The Ontological Praxis Between Disaster Studies and Demography—Extension of the Scope

Dávid Karácsonyi and Andrew Taylor

Abstract This chapter serves as a summary of the learnings from the present volume and an extension of the scope on disaster-demography nexus. We outline the benefits of exploring the disaster-demography nexus and develop a categorisation summarising seven different approaches to the interlink of disasters with demography from examinations of existing literature. These are: disaster impacts on population, measuring vulnerability, mass displacement, spatial-regional approach, climate change, urbanisation and an applied approach. These seven approaches are our attempt to highlight the complex and multifarious nexus between demography and disasters which may not simply be linked to vulnerability. It is recognised that others may separate or merge some of these approaches in different ways.

Keywords Disaster-demography nexus · Climate change · Urban vulnerability · Geographic possibilism · Mass displacement

13.1 Introduction

This chapter serves as a summary of the learnings from the cases presented in this volume and an extension of the scope on disaster-demography nexus. We overviewed in Chap. 1 the two perspectives on a disaster; the vulnerability school (social embeddedness) and the holistic school (non-routineness). While we stressed in Chap. 1 that the disaster-demography nexus should immanently be part of the 'social embeddedness' perspective, the majority of existing demography studies has a' non-routineness' outlook on disaster, and feature the disaster-demography nexus through population change as a consequence to disaster. Even the term 'demography of disasters'

D. Karácsonyi (✉)
Geographical Institute, CSFK Hungarian Academy of Sciences, Budaörsi út 45, Budapest 1112, Hungary
e-mail: karacsonyi.david@csfk.mta.hu

A. Taylor
Northern Institute, Charles Darwin University, Darwin, Ellengowan Dr, Casuarina NT 0810, Australia
e-mail: andrew.taylor@cdu.edu.au

D. Karácsonyi et al. (eds.), *The Demography of Disasters*,
https://doi.org/10.1007/978-3-030-49920-4_13

reflected demographic outcomes when it (according to our knowledge) first appeared as the title of the work by Smith (1996); *Demography of Disasters: Population estimates after Hurricane Andrew.*

The nexus was also observed by Schultz and Elliott (2012) in a literature summary of studies of local demographic consequences of disasters in the USA dating back to the early 1980s. In addition, the edited volume by Kurosu et al. (2010) provided a wide range of case studies on demographic responses to environmental crises in the past, such as famines and weather fluctuations in rural societies and epidemic diseases (including smallpox and Spanish flu). However, this volume failed to summarise findings from the cases and provide their theoretical implications for the disaster-demography nexus. There are a plethora of case studies on the disaster-demography nexus in developing countries, especially those concerned with climate change-induced mass relocations. The summarising works, however, are mostly policy-oriented documents with a few exceptions (see for instance Martine and Schensul 2013). This gap in the literature on summarising the multifarious approaches on the demography-disaster nexus was partially filled by two theoretical papers by Donner and Rodríguez (2008) and Hugo (2011) which focused on vulnerability, migration and climate change.

On that basis, the present book and this chapter are not the first attempt to provide a framework on the disaster-demography nexus. Still, our ambition is to traverse the disaster-demography nexus from both 'non-routineness' (holistic) and 'social embeddedness' (vulnerability) perspectives. These two perspectives cover various forms of the disaster-demography nexus, and, based on the existing literature, we could discern seven different subthemes which are summarised in the following paragraphs and in Table 13.1.

13.2 Seven Approaches of Disaster-Demography Nexus

The first and the most commonly used approach to the disaster-demography nexus focuses exclusively on the *consequences from disasters. The demography of disasters* by Frankenberg et al. (2014) defined how demographers should study disasters and the essay gives a summary on the various ways disasters affect demographic processes. For instance, a simple demographic technique of measuring disaster impacts is the enumeration of the casualties. Lindell (2013, p. 4) provided the demographic balance equation for this enumeration, basically using the population number before and after the disaster event, subtracting natural population increase and migration to obtain the difference to approximate the impact from the disaster. More complex methods are elaborated by Nobles et al. (2015) to investigate the fertility and natural reproduction response, in particular for baby booms and replacement fertility after disasters. In Bourque's et al. (2007), understanding a disaster is an unanticipated mortality shock; hence, the subject of demographic analysis is the response to mortality within the community where births represent renewal and the return to 'normal'. It is important to add that, according to Naik et al. (2007), studies

Table 13.1 Demographical approaches of disasters

Approaches	Keywords	References	Chapters
Disaster impacts on population incl. disaster epidemiology	Population enumeration, death toll, health impacts fertility/migration response community, regional, country impacts delayed or indirect impacts of disaster: disaster epidemiology increased suicide rate caused by post-traumatic stress	Frankenberg et al. (2014) Nobles et al. (2015) Oliver-Smith (2013) Lindell (2013) Bourque et al. (2007) Veenema et al. (2017) Lechat (1979)) Noji (1995) Briere and Elliott (2000) Krug et al. (1999)	2, 3, 5, 7
Measuring vulnerability (demography is a root cause)	Age, gender, ethnic and social composition of the population, disaster affected special groups—such as disabled people, females, children, elderly people or people on move, refugees, tourists people living under hazard risk adaptive capacity	Malone (2009) Flangan et al. (2011) Fothergill et al. (1999), Fothergill and Peek (2004) Wisner et al. (2004) Bolin (2007) Enarson et al. (2007) Orum et al. (2014) Donner and Rodríguez (2008) Friedsam (1960) Jia et al. (2010) Stough and Mayhorn (2013) Peacock et al. (1997) Zhou et al. (2014))	4, 8, 6
Mass displacement	Forced migration, displaced communities, social cohesion and (local) identity	Oliver-Smith (2013) Naik et al. (2007) Cernea and Guggenheim (1993) Cernea (2004) Gray and Mueller (2012) Levine et al. (2007)	2, 11
Spatial/ regional approach	Development inequalities, scales between community and global	Schultz and Elliott (2012) World Bank (2005) Naik et al. (2007)	2, 4, 11
Climate change	Climate change-induced migration, vulnerability	Oliver-Smith (1996, 2012, 2013) Lavell et al. (2012) Lavell and Ginnetti (2013) de Sherbinin et al. (2011) Bouwer (2011) Martine and Schensul (2013)	8

(continued)

Table 13.1 (continued)

Approaches	Keywords	References	Chapters
Urbanisation	Urban concentration and their risks and vulnerabilities to hazard events	Gencer (2013) Armenakis and Nirupama (2013)	6, 9, 10, 12
Practical-applied	Demographic techniques in emergency assessment substitutional practices	Kapuchu and Özerdem (2013) Robinson et al. (2003) Wilson et al. (2016) Brown et al. (2001)	2, 3

such as these mostly focus on developing countries because usually there are higher death tolls during disasters compared to developed ones.

Disaster epidemiology can be also understood as part of 'consequences from disasters', we discussed in the previous paragraph. It should be added that disaster epidemiology and its demographic consequences have been well discussed in the literature (see Lechat 1979; Noji 1995). Recently, Veenema et al. (2017) provided a systematic review on studies related to the climate change-induced hydrological and meteorological hazard events which caused epidemics through lack of access to drinking water. According to Bissel (1983), epidemic diseases appear several months after hazard events because of crowded and inappropriate temporary housing or water transmitted pathogens. These 'delayed' deaths are often excluded from the disaster death toll. In addition to post-disaster epidemics, increased suicide rates are also apparent 3–4 years after disaster events and are related to the post-traumatic stress (Krug et al. 1999), distorted life courses or failed, delayed post-disaster recovery.

By extension, demographic consequences from epidemics can be considered as part of a wider disaster-demography nexus, for example, HIV altered demographic trajectories in southern African countries (Nicoll et al. 1994; Gould 2005). The presence of infectious diseases such as HIV, plague, cholera and Ebola (see the 2014 Ebola emergency in Western Africa, Briand et al. 2014) can be explained by poor governance, education, low living standards and lack of modern medicine in the developing world and hence by global social and spatial inequalities. Still antibiotic resistance (WHO 2018), and changing social attitudes (for example, anti-vaccinationism) challenge and may impact demography of developed nations in the future (Kata 2010; Casey 2015).

Text Box 13.1 the Relationship of Violent Conflicts to the Disaster-Demography Nexus

Some may think about a third angle of the 'consequences from disaster' approach as well, which, instead of being part of the holistic school, is related to the social vulnerability paradigm. This third angle may be related to the demographic consequences of malfunctions in society, particularly to economic crises, violence and 'bad governance' (Moore 2001). In Wisner's and his colleague's (2004) radical embeddedness perspective, all disasters have a root cause in societal failures related to inequalities and spatial exclusion, and hence, violent conflicts are also considered as disasters. But we argue these crises (ethnic cleansings, genocides, sabotages, terrorist actions and violent crimes) are not immanently part of the disaster-demography nexus. As an illustration for the reason for this position, we take the example of outmigration from urban crime hotspots (Foote 2015) which in the *Wisnerian* logic would be considered as part of disaster-demography nexus since violent crime is a kind of malfunction in society. But studies on crime are indeed far from the disaster study field and likely to be related to social sciences other than disaster studies, particularly to criminology, sociology and urban studies. The notion '*famine demography*' by Dyson and Ó Gráda (2002) does not reflect a direct link to disaster. Study of famines has a much stronger link to other fields such as history and economics.

It should be added however that there are often interactions between a violent societal downturn and a coincident natural hazard event which increase population impact which reflects the complexness of a disaster (Robinson 2003; Barton 2005; Cutter 2005). For instance, the *GortaMór* (The Great Irish Famine) killed around one million people and caused mass emigration of the Irish population from Ireland to North America during 1845–1849 (Dyson and Ó Gráda 2002) where the British laissez-faire capitalism, the dependence on one food source and the land rent system interacted with the potato blight. Disasters other than famine are also likely to occur because of violence and war, such as the 1918–1920 Spanish flu pandemic following World War One which caused more death than the war itself (Johnson and Mueller 2002). The 1953 North Sea flood in the Netherlands caused more than 2 000 deaths and was a result of dilapidated state of physical flood defence and warning systems due to the consequences of World War Two and coinciding with an extreme high spring tide and storm surge (Hall 2013).

In some cases, natural hazard events interacting with violent societal downturns are (and were) used to cover up responsibilities for these events (Smith 2014). This is illustrated by the scientific dispute on the causes of high death tolls among indigenous peoples during the colonisation era in several countries, when ethnic compositions of entire continents have been changed. While Crosby (1976) and McNeil (1976) considered a decisive role of the effects of 'virgin soil epidemics' for these high death tolls due to lack of immunity of indigenous people to the infectious diseases introduced by Europeans, their view is now strongly disputed (Jones 2003). Recently, most of authors explaining death tolls through genocide (see Lemkin 2012; Curthoys 2005; Jones 2017) and indigenous population decline through loss of livelihoods during colonisation (Smith 1989).

The *applied demography approach* is also strongly affiliated with the 'consequences from disaster' paradigm, but, in contrast to disaster epidemics and societal downturns, we conceptualise it as a separate theme (the second one among the seven approaches of disaster-demography nexus summarised in Table 13.1). The applied approach was emphasised by Robinson et al. (2003) who summarised the demographic means in relation to disaster mitigation in their work *Principles and uses of demography in emergency assessment*. The applied demographic techniques in disaster situations can help to measure a disaster's impact on the affected population. Moreover, according to Lindell (2013), applied demographic methods can be used in every stage of the disaster cycle. Despite the widespread use of demographic techniques to study disasters, Frankenberg et al. (2014) have emphasised that, due to a lack of adequate, spatially and timely detailed data, there are not many studies which would interrelate demography and disasters through mortality and fertility changes to provide an integrated analysis. Hence, Robinson et al. (2003) provided an overview of substitutional procedures to be used for disaster assessment where there is not an adequate dataset. To fill the immediate knowledge gap about the size of affected populations, the area sampling method is usually applied to estimate the numbers impacted in developing countries (Brown et al. 2001). In developed nations, mobile phone location data can also substitute for adequate datasets for disaster impact assessment for population, as it is discussed in Chap. 2 and by Wilson et al. (2016). Among administrative data, school enrolments can provide an estimate on population displacement (Plyer et al. 2010) as it also presented in Chap. 3. Longitudinal practices and the application of supplementary data on disaster impacts, for example, housing damages, are also introduced in Chap. 12.

While the assessment of the demographical consequences of disasters in the previous paragraph has strong links to the non-routineness (holistic) school, the third approach which addresses *demography as root cause* of the disaster is clearly part of the 'social embeddedness' perspective. As an illustration for social, or more precisely 'demographic embeddedness', Malone (2009) called attention to the importance of using demographic analysis to measure vulnerability. In fact, socially created vulnerabilities are difficult to quantify (James 2012) because they are a combination of different factors (Wisner et al. 2004). Furthermore, Malone (2009, p. 13) provided a method and a group of indicators, including demographic data to measure vulnerability and resilience in the form of a vulnerability-resilience indicators model. Malone suggested to use detailed socio-demographic analysis, for example, population distribution and density, births and mortality in different areas characterised by different livelihoods. These demographic analyses throw light on the social context that allows analysts to see how households are constituted, the elements that affect their functioning and disrupts them. Other similar indexes have been developed recently to measure social vulnerability (see Flanaganet al. 2011; James 2012). As an example, by using a combination of multivariable and spatial analysis in vulnerability assessment, Zhou et al. (2014) used factor analysis to create a complex disaster vulnerability index at a county level for China (2361 units in total) based on population census data. Zhou and his co-authors analysed also the spatial variation of index values obtained from factor analysis by using local and global autocorrelations which helped them to identify vulnerability hotspots to hazard events.

Vulnerability assessment is often part of planning for large industrial investments. As an illustration, Orum et al. (2014) provided an estimate of vulnerability using demographic data of people living in the zones around more than three thousand chemical plants in the USA. They focused primary on social status and ethnicity and found that residents of chemical facility vulnerability zones belong mostly to minority groups (African Americans or Latinos). These populations have higher rates of poverty and cheaper housing, lower incomes, and education levels than the national average. The case of Hurricane Katrina (2005, New Orleans, USA) also drew attention to race and ethnic inequalities in the USA (Bolin 2007 p. 113, or Chap. 6), because the lower laying flood-prone areas were mostly inhabited by poorer African Americans. According to Peacock et al. (1997), ethnic segregation also occurred during the post-Hurricane Andrew relocation (1992, Florida) which caused social change and put African American communities into a more vulnerable position.

As Peacock's study suggests, demographic vulnerabilities are not only root causes, they can deteriorate disaster consequences as well. For instance, along with ethnicity (Fothergill et al. 1999; Fothergill and Peek 2004), gender (see Enarson 2000; Enarson et al. 2007, or Chap. 9) and age composition can make communities more vulnerable. In particular, population ageing in developed countries is establishing age-related vulnerable population enclaves (Fernandez et al. 2002). As an illustration, Isoda (2011) pointed out that, during the 2011 Great East Japan Earthquake and Tsunami, the death toll was higher among the elderly who typically stayed at home when the tsunami hit and were not able to escape. This is because the tsunami hit Japan during working hours, and in contrast to schools and job places, private homes were neither designed to resist the tsunami nor provided with shelters. Consequently, the death toll was in general higher in those rural communities on the coastal areas where the proportion of elderly people in the population were higher. A high death toll among the elderly was also observed in the case of 2008 Sichuan earthquake (Jia et al. 2010). Friedsam (1960) in his early literature review on disasters in the USA and on impacts of World War Two bombings in Europe distinguished direct (older people are more likely to be hit since they have limited mobility) and indirect or secondary effects (lack or poor level of medical treatment for people in need during the emergency). Furthermore, indirect effects include a shortening of life expectancy for people living with diabetes (Fonseca et al. 2009) or for those requiring any kind of regular medical treatments such as haemodialysis. Ironically, despite the higher death tolls among elderly, they are the most likely to be 'post-disaster returners' to an area because of their stronger attachment to place through their longer life experience (see Chaps. 2 and 10).

Along with static demographic aspects such as age, race and sex composition, we summarised in the previous two paragraphs, dynamic demographic aspects, in particular migration, should be also considered when discussing vulnerabilities. For instance, people on move, such as refugees, internally displaced or tourists are particularly vulnerable to hazards (Robinson 2003; Donner and Rodríguez 2008). Large numbers of tourists were impacted during the 2004 Indian Ocean tsunami, since they had a high concentration in coastal areas during the Christmas high season (Becken et al. 2014). Of course, refugees are generally more vulnerable when compared to tourists because of their social status. According to Naik et al. (2007) excess populations, such as visitors can easily become the 'forgotten group' in the course of a disaster because of a lack of response planning (ibid. p. 57).

The vulnerability of people on the move links us to the fourth meta-approach to the disaster-demography nexus which focuses on *migration and mass displacement*. According to Hugo (2008), migration has always been one of the most important survival strategies adopted by people facing disasters. As an illustration to Hugo's point, King and Gunter in Chap. 6 highlighted population loss due to post-disaster outmigration in case of New Orleans (Hurricane Katrina), Christchurch (2011 earthquake) and Innisfail (2006 and 2011 cyclones). Lavell and Ginnetti (2013) suggest, the demographic profile of entire regions can be altered over a long period of time as consequence of disaster-induced mass displacements. Furthermore, these mass displacements can be short or long distance, temporary or permanent (Cernea and Guggenheim 1993; Cernea 2004). Regarding research on long-term and long-distance displacements after disasters, Levine et al. (2007) pointed out a gap in the literature due to lack of data and the difficulty to follow up such migration.

Furthermore, the non-spontaneous character of mass displacement is stressed by Oliver-Smith (1996); the relocation or resettlement of disaster-stricken populations is a common strategy pursued by planners in post-disaster reconstruction efforts. Displacement is also selective based on vulnerabilities and the connection between migration and hazard impact is not always clear. As an illustration, Gray and Mueller (2012) brought to the fore that those families impacted directly by a disaster are less likely to move out compared to those in the disaster-prone area not impacted directly. This is because the latter group has the means to fund their move, while the former, who may have lost everything, are 'stuck' in the disaster-prone area. It is important to add that displacement is not only related to disaster, it is a broader and more common phenomenon. That is, displacement could take place because of occurrence of a disaster, climate change inducted environmental change (see Chap. 8), violent conflict or a development project (Oliver-Smith 2013). Furthermore, Oliver-Smith (ibid.) as well as Scudder and Colson (1982) highlighted that development projects cause much more displacements than all disasters combined.

Disaster-induced spatial movement brings the fifth approach to the fore which is the disaster-demography nexus in the *spatial-geographical context*. The spatial-geographical approach has roots in disaster studies in the first half of twentieth century (see White 1945) through the theory of environmental adaptation (Alexander 2001). Environmental adaptation reflects the conviction of *geographic possibilism*, a dominant paradigm of human geography in the first half of twentieth century and itself rooted in the French regional geography (see Vidal de la Blache 1911). The concept of geographic possibilism means that the diversity of the natural environment provides different opportunities and constraints where people react to their environment and make their own choices. These choices, their ways of life (*genres de vie*), are the manifestation of their culture in the Vidalian regional geography.

Geography and space also play an important role in the social embeddedness perspective on disasters as well. But instead of the now outdated geographic possibilism, it is connected to the neo-Marxist critical geography paradigm (see Harvey 1996). Within the embeddness viewpoint, space is understood to be an uneavenly distributed resource because of uneaven population distributions, and hence, the uneaven allocation of resources within societies. Wisner et al. (2004, p. 5) described this as follows: 'People live in adverse economic situations that oblige them to inhabit regions and places that can be affected by natural hazards'. Wisner et al. (2004) stressed that different groups take risks for advantages voluntarily or involuntarily because of their economic needs. For example, they take risk of landslide to have a house on a slope for a better panorama or they live in a poorly built informal settlement on a slope in an urban area to access better job opportunities. The former highlights increases risks from voluntary actions, while the latter demonstrates the 'forced' acceptance of risk for (economic) survival.

Furthermore, according to Cutter (2005, p. 42), the global extent of risks is not equally distributed among all places or among all social groups. This spatial inequality approach is highlighted by Carson and his colleagues in Chap. 5 that the Great Deprivation (the Swedish famine of 1867–1868) '…is seen as the last of the European famines to result from natural events'. However, Carson and his colleagues also highlight that Northern Sweden was a territory for Swedish northern settlement advances encouraged by the vast mineral and forest resources there at that time. Crop failure triggered by the cold summer of 1867 hit especially those northern advance settlements and caused famine there. Wisner's interpretation would suggest the famine to be a result of northern advance into sparsely populated areas rather than the cold summer (the natural element in the disaster). But Carson also stressed that improving food supply chains and reducing reliance on local food production helped northern sparsely populated territories to avoid famines later on; hence, the famine was the result of unfavourable economic patterns at the time.

According to Lavell and Ginnetti (2013), economic development, better technologies and improving living standards are reducing vulnerability on a global basis. Alexander (2005, p. 32) stressed, however, there is an endless resurgence of vulnerability, because of growing socio-economic inequalities and polarisation throughout the world. Alexander's perspective was also echoed by Naik et al. (2007). Accordingly, disasters have a disproportionate effect on developing countries because of poor quality of construction and less compliance with building codes, and absence or non-application of land registration and other regulatory mechanisms (ibid. p. 19). Furthermore, according to Naik et al. (2007), there is a significant difference in the impact of disasters on developing and developed countries in terms of the type of loss: data show a higher death toll in developing countries compared to developed countries, but absolute economic losses are greater in developed countries because of higher concentration of economic assets in the area. This was also supported by Robinson (2003 p. 5), who stated, that between 1991 and 2000, 3 million people were killed by disasters, while only 2% of them were from highly developed countries while 60% were from Africa.

The examples on spatial-geographical context we highlighted in the previous paragraphs showed that space allocation has a slightly different meaning depending on its 'local' or 'global' connotation. But Cutter and Wisner have tended to use these interchangeably in their arguments when explaining the role of space allocation in vulnerabilities. But the scale is important, because the former brings to the fore local social inequalities, the latter features global geographical diversity and inequalities among nations. Hence, the size of areal unit under consideration can influence the phenomenon we are observing (see Chap. 5 and Koch and Carson 2012 on the modifiable areal unit problem).

On a global scale, spatial variety of the natural environment provides different types of environmental opportunities and risks; hence, global assessment on natural hazard hotspots has a high priority for development and aid agencies (Nadim et al. 2006; Strömberg 2007). For example, the World Bank (2005) conducted a study on natural hazard hotspots to estimate GDP losses and mortality as a consequence of disasters in various countries. The study distinguished and analysed eight types of disasters based on their 'natural' characteristics; drought, storm, flood, earthquake, volcano, heat wave, landslide and wildfire. Accordingly, the World Bank used the terms single hazard hotspots and multiple hazard hotspots and investigated the exposure to these risks and vulnerabilities for different countries. Another example is the work by Schultz and Elliott (2012) which used census and hazards database to estimate the demographic consequences of disasters at a county level in the USA. They found a positive correlation between cumulative hazard impact during the 1990s and changes in local population numbers. Further examples of hazard risk assessments are the estimates of global flood risk (Winsemius et al. 2013), global landslide and avalanche risk (Nadim et al. 2006), tropical cyclone risk (Peduzzi et al. 2012) and their effects on population and GDP.

Despite the demand for natural hazard hotspot assessments, there are a relatively small number of such studies and they are absent in sociology based disaster studies. This is likely because the 'social embeddedness' disaster school denies the concept of *geographic possibilism* discussed earlier. Despite this, the spatial-environmental diversity is important especially in the course of emerging climate change (or more explicitly the, climate emergency faced globally) which highlights the low resilience of our political, social and technical systems, constraints and limitations of human society when coping with 'nature'.

Indeed, the *impact of and adaptation to climate change* requires special attention, which forms the sixth approach of demography-disaster nexus. Climate change connects disasters, spatial-geographical diversity and migration. As an illustration, according to de Sherbinin and his co-authors (2011), climate change-induced mass relocation is a politically disputed and socially sensitive adaptation strategy but seems to be unavoidable in the very near future. Adding to this, Oliver-Smith (2013) summarised the potential forces which could lead to mass displacement during climate change impacts; such as evacuation because of rapid onset events (typhoons, floods), slow onset drivers for forced migration (drought, desertification), or displacement from climate change mitigation projects (resettlement from coastal areas, large constructions as reservoirs or coastal defence dams).

While climate change may have extensive impacts through forced migration, the international dimensions of this relationship have been neglected until recently (Hansen et al. 2012; Jankó et al. 2018). Hugo (2008) suggests this is because such events have affected mostly developing nations. Adding to this, Bouwer's (2011) literature review spotlighted that losses caused by climate change are not significant so far but will be significant in the near future. Extending this, Zander and her colleagues suggest climate change adaptation-induced migration (intention to migrate because of climate change related heatwaves) is also present within developed nations, such as Australia (see Chap. 7). However, others (McLeman and Hunter 2010; Carson et al. 2016) have stressed that weather-induced mass migration in developed countries is mostly seasonal-temporal. These include the *snowbirds* in North America and the *grey nomads* in Australia, who seasonally move between the tropical–subtropical and temperate zones of their respective continents.

Lavell and Ginnetti (2013) and Hugo (2011) draw attention to the fact that to date most climate change migration caused by environmental change has occurred within national boundaries. According to McLeman and Hunter (2010), these climate change-induced internal and intra-regional moves (within a region of one country) were up to now temporary. Hugo (2011) stressed that the occurrence of climate change-induced extreme weather events is just one factor among several others influencing migration decisions. Hence, a natural hazard alone does not lead automatically to displacement (Piguet et al. 2011 p. 23).

Furthermore, according to IOM (2012), climate change-induced migration can be both a challenge and a solution for the problems, as people move to less affected areas. According to Piguet et al. (2011), migration is an adaptation strategy which should not be considered as a negative outcome to be avoided. For example, according to Naik et al. (2007), environmental migration can affect development not only negatively (through the exodus of highly skilled people, loss of workforce, brain drain and so on), but emigration can ease pressure on the environment, while remittances and returning experienced people can also boost the economy and promote development goals. Additionally, Naik et al. (2007) focused on how migratory flows and migrant communities are impacted by disasters, and how kinship and support from diaspora affected migrant communities in the aftermath of disasters (through, for example, aid and technical assistance).

As an illustration of the potential scale of climate change-induced migration, Lavell and Ginnetti (2013) estimated the likelihood of disaster-induced displacement and quantified the number of people at displacement risk using a probabilistic risk model. Their estimate showed that almost 3 thousand per million people are displaced annually in Central America and the Caribbean as a result of climate change, equating to 300 thousand per year. In fact, the potential scale of future environmental-induced migration is the subject of debate and its impacts will be very different around the world (Piguet et al. 2011) because impacts are determined not just by the absolute exposure and the size of exposed population but by their conditions of resilience and vulnerability as well (Oliver-Smith 2012). Based on these complex interactions, Oliver-Smith (2012) questioned whether can we really speak about migration directly induced by climate change? Probably the direct connection is present in the case of small nations of the Pacific suffering from direct effects of sea level rise and considered as first victims to climate change (Farbotko and Lazrus 2012). But according to Hugo (2011), demographic hotspots (places, countries with population booms) and climate change hotspots overlap in space (these hotspots are in Africa, in South and Southeast Asia and in Central America and the Caribbean), which is generating a complex interactions with migration and will cause increased mobility in future affecting developed nations as destinations as well (Reuveny 2007).

The interaction of climate change, population booms and spatial inequalities has fuelled rural to urban migration and the urbanisation boom in developing countries (Hugo 2011). Hence, according to Gencer (2013), disaster studies should pay special attention to urban vulnerability and disaster risk reduction, linking disasters, the global trend of urbanisation (Clark 1996; Seto et al. 2011) and climate change. So, the seventh approach to the disaster-demography nexus is related to *urban vulnerability*. Donner and Rodríguez (2008) stressed that increasing urban vulnerabilities are particularly evident in rapidly growing coastal megacities of developing countries such as Jakarta, Dhaka and Lagos (see Tacoli et al. 2015; Di Roucco et al. 2015). More generally, rural to urban migration means that people arriving to high hazard risk urban areas from rural areas are generally characterised by low economic opportunities but also a lower probability of hazard impacts (Hugo 2011). For example, empirical evidence suggests per capita death tolls are higher from earthquakes in urban areas compared to rural regions (Donner and Rodríguez 2008). Wisner and colleagues (2004) suggest urbanisation as major factor in the growth of vulnerability, particularly for low-income families living within squatter settlements in developing countries. These informal settlements are exposed to physical vulnerabilities due to their construction practices or location in hazard risk areas. In these informal settlements, social vulnerability and exclusion are strongly related to hazard risk exposures such as floods (Amoako et al. 2018) or landslides (Chardon 1999; Alves and Ojima 2013). As a result, urbanisation and rapid population growth together have led to the concentration of population in hazard prone urban areas and hence put more people at risk.

While it seems obvious that the urban vulnerability context is related primarily to developing countries, that is not exclusively the case. For example, pandemics can spread rapidly across global cities of developed nations as well (Alirol et al. 2010; Grais et al. 2003) such as during the 2002–2003 SARS coronavirus (in East Asia) and during the 2009 H1N1 flu virus (in North America) epidemics (McLafferty 2010). Armenakis and Nirupama (2013) highlighted that risks of technological disasters related to certain industries (nuclear, chemical or biotech facilities, gas supply systems) are high in urban zones of developed countries as well. To cope with these hazard risks, properly designed rapid evacuation systems are needed based on geography, population sizes, distributions, compositions and vulnerabilities (Kendra et al. 2008). Adding to this, Singh in Chap. 10 has emphasised there are more complex and interlinked (and hence vulnerable) lifeline networks under risks with growing urbanisation (see also Tielidze et al. 2019). Furthermore Murao in Chap. 12 pointed out that cities can be considered as engineering ‘products’; however, unlike other products, they have never been tested before people start using them during their everyday life. In Chap. 9, Barnes argued that urban landscape, engineering, social and community aspects are linked together in urban disaster resilience and that has different outcome for females. Additionally, there is a need to develop 'age-friendly cities' (Buffel et al. 2012), because of the growing number of elderly in urban areas of developed countries representing a highly vulnerable group to disaster risk as well (Donner and Rodríguez 2008). Based on the various aspects, we featured here on urban vulnerability it is clear that this approach of disaster-demography nexus is strongly linked to other approaches such as demographic vulnerability and climate change as well.

13.3 Conclusions

In this chapter, we have laid-out the links between disasters and demography evident in the field of disaster studies and plotted major historical paradigm changes in the field. Of course, this classification is subjective and others may separate or merge some of these approaches in different ways. The collective case studies in this volume further expand the links by highlighting the complex and multifarious nexus between demography and disasters which may not simply be linked to vulnerability. While, for example, demographic conditions prior to a disaster may be the reason for high impacts (for example, loss of life); it may also reflect longer-term and more localised structural changes in the demography of towns or regions. In terms of demographic consequences for disasters, as some chapters in this book noted (for example, Chaps. 2 and 5), disasters may be an agent for speeding up pre-existing demographic trends, such as rural to urban migration. The demographic profile 'left behind' may consequently be quite different to pre-disaster but, without detailed examination of pre-disaster demography, it would be easy to suggest that the disaster fundamentally 'caused' a new demographic structure at the local level of impacts.

Taking a wider perspective of demography, which goes beyond the statistical analysis of populations, enables us to depart from the classical scope of demographically rooted disaster studies in which the disaster is traditionally singularised as a root cause for demographic shifts (the non-routine event approach). In this volume, we attempted a broader demographic purview in order to extend disaster science research.

References

Alexander D (2001) Natural disasters. Taylor & Francis Group, New York.

Alexander D (2005) An interpretation of disaster in terms of changes in culture, society and international relations. in: Perry RW, Quarantelli EL eds. What is a disaster? New answers to old questions. International Research Committee on Disasters.

Alirol E, Getaz L, Still B, ChappuisF, Loutan L (2010) Urbanisation and infectious diseases in a globalised world. *The Lancet Infectious Diseases*, 11(2):131–141.
Alves HPF, Ojima R (2013) Environmental inequality in São Paulo city: An analysis of differential exposure of social groups to situations of environmental risk. In: Boone C, Fragkias M eds. Urbanization and Sustainability. Human-Environment Interactions, vol 3. Springer, Dordrecht.
Amoako C, Inkoom DKB (2018) The production of flood vulnerability in Accra, Ghana: Re-thinking flooding and informal urbanisation. *Urban Studies*, 55(13):2903-2922.
Armenakis C, Nirupama N (2013) Estimating spatial disaster risk in urban environments. *Geomatics, Natural Hazards and Risk*, 4(4):289-298.
Barton AH (2005) Disasters and collective stress. in: Perry RW, Quarantelli EL eds. What is a disaster? New answers to old questions. International Research Committee on Disasters.
Becken S, Mahon R, Rennie HG, Shakeela A (2014) The tourism disaster vulnerability framework: an application to tourism in small island destinations. *Natural Hazards*, 71:955-972.
Bissel R (1983) Delayed-impact infectious disease after a natural disaster. *The Journal of Emergency Medicine*, 1(1):59-66.
Bolin B (2007) Race, class, ethnicity, and disaster vulnerability. in: Rodríguez H, Quarantelli EL, Dynes RR eds. Handbook of disaster research. Springer, pp. 113–129.
Bourque LB, Siegel JM, Kano M, Wood MM (2007) Morbidity and mortality associated with disasters.in: Rodríguez H, Quarantelli EL, Dynes RR eds. Handbook of disaster research. Springer, pp. 97–112.
Bouwer LM (2011) Have disaster losses increased due to anthropogenic climate change? *Bulletin of the American Meteorological Society*, 92(1):39-46.
Briand S, Bertherat E, Cox P, Formenty P, Kieny MP, Myhre JK, Roth C, Chir B, Shindo M, Dye C (2014) The international ebola emergency. *The New England Journal of Medicine*, 371:1180-1183.
Brown V, Jacquier G, Coulombier D, Balandine S, Belanger F, Legros D (2001) Rapid assessment of population size by area sampling in disaster situations. *Disasters*, 25(2):164–171. Buckle P (2005) Disaster: Mandated definitions, local knowledge and complexity. in: Perry RW, Quarantelli EL eds. What is a disaster? New answers to old questions. International Research Committee on Disasters.
Buffel T, Philipson C, Scharf T (2012) Ageing in urban environments: Developing 'age-friendly' cities. *Critical Social Policy*, 32(4):597-617.
Carson DA, Cleary J, de la Barre S, Eimermann M, Marjavaara R (2016) New mobilities – new economies? Temporary populations and local innovation capacity in sparsely populated areas. In: Taylor A, Carson DB, Ensign PC, Huskey L, Rasmussen RO, Saxinger G eds. Settlement at the edge. Edward Elgar Publishing, Cheltenham, Northampton.
Casey S (2015) Love in the age of measles: The anti-vaccination movement in America. PhD thesis. Arkansas State University.
Cernea M (2004) Impoverishment risks, risk management, and reconstruction: A model of population displacement and resettlement. Presented at UN Symposium on Hydropower and Sustainable Development, 27 to 29 October: Beijing, China.
Cernea M, Guggenheim S (1993) Anthropological approaches to involuntary resettlement: Policy, practice, and theory. inCernea M, Gugggenheim S eds. Anthropological approaches to resettlement: Policy, practice and theory. Westview Press, Boulder.
Chardon AC (1999) A geographic approach of the global vulnerability in urban area: case of Manizales, Colombian Andes. *GeoJournal*, 49(2):197-212.
Clark D (1996) Global world, global city. Routledge, London.
Crosby AW (1976) Virgin soil epidemics as factor in the aboriginal depopulation in America. *The William and Mary Quaterly*, 33(2):289-299.
Curthoys A (2005) Raphaël Lemkin's 'Tasmania': an introduction. *Patterns of Prejudice*, 39(2):162-169.
Cutter SL (2005) Are we asking the right question? In Perry RW, Quarantelli EL eds. What is a disaster? New answers to old questions. International Committee on Disasters. pp. 39-48

Di Roucco A, Gasparini P, Weets G (2015) Urbanisation and climate change in Africa: Setting the scene. In: Pauleit S et al. eds. Urban vulnerability and climate change in Africa. Future City, vol 4. Springer, Cham.
Donner W, Rodríguez H (2008) Population composition, migration and inequality: The influence of demographic changes on disaster risk and vulnerability. *Social Forces*, 87(2):1089-1114.
Dyson T, Ó Gráda C eds. (2002) Famine demography. Perspectives from the past and present. Oxford University Press.
Enarson E (2000) Gender and natural disasters. Working paper. Recovery and Reconstruction Department, Geneva. Accessed on the 6 December 2019: https://www.ilo.int/wcmsp5/groups/public/---ed_emp/---emp_ent/---ifp_crisis/documents/publication/wcms_116391.pdf.
Enarson E, Fothergill A, Pee L (2007) Gender and disaster: Foundations and directions. In: Rodríguez H, Quarantelli EL, Dynes RR (eds) Handbook of disaster Research. Springer, pp 130–146.
Farbotko C, Lazrus H (2012) The first climate refugees? Contesting global narratives of climate change in Tuvalu. *Global Environmental Change* 22(2):382-390.
Fernandez LS, Byard D, Lin CC, Benson S, Barbera JA (2002) Frail elderly as disaster victims: Emergency management strategies. *Prehospital and Disaster Medicine*, 17(2):67-74.
Flanagan BE, Gregory EW, Hallisey EJ, Heitgerd JL, Lewis B (2011) A social vulnerability index for disaster management. *Journal of Homeland Security and Emergency Management*, 8(1).
Fonseca VA, Smith H, Kuhadiya N, Leger SM, Yau LC, Reynolds K, Shi L, McDuffie RH, Thethi T, John-Kalarickal J (2009) Impact of a natural disaster on diabetes. Exacerbation of disparities and long-term consequences. *Diabetes Care*, 32(9):1632–1638.
Foote A (2015) Decomposing the effect of crime on population changes. *Demography*, 52:705–728.
Fothergill A, Maestas EGM, DeRouen Darlington JA (1999) Race, ethnicity and disasters in the United States: A review of the literature. *Disasters*, 23(2):156-173.
Fothergill, A. Peek LA (2004) Poverty and disasters in the United States: A Review of recent sociological findings. *Natural Hazards*,32:89–110.
Frankenberg E, Laurito M, Thomas D (2014) The demography of disasters.in: Smelser NJ, Baltes PB eds. International encyclopedia of the social and behavioral sciences, 2nd edition (Area 3). North Holland, Amsterdam.
Friedsam HJ (1960) Older persons as disaster casualties. *Journal of Health and Human Behavior*, 1(4):269–273.
Gencer EA (2013) Chapter 2: Natural disasters, urban vulnerability, and risk management: A theoretical overview. in: The Interplay between Urban Development, Vulnerability, and Risk Management. *SpringerBriefs in Environment, Security, Development and Peace*, 7:7–43.
Gould WTS (2005) Vulnerability and HIV/AIDS in Africa: From demography to development. *Population, Space and Place*, 11:473-484.
Grais RF, Ellis JH, Glass GE (2003) Assessing the impact of airline travel on the geographic spread of pandemic influenza. *European Journal of Epidemiology*, 18(11):1065-1072.
Gray CL, Mueller V (2012) Natural disasters and population mobility in Bangladesh. *PNAS*, 109(16):6000-6005.
Hall A (2013) The North Sea Flood of 1953. Environment & Society Portal, Arcadia, no. 5. Rachel Carson Center for Environment and Society. Accessed on the 15/12/2019 at https://www.environmentandsociety.org/arcadia/north-sea-flood-1953.
Hansen J, Sato M, Ruedy R (2012) Perception of climate change. *PNAS* 109(37):2415-2423.
Harvey, D (1996) Justice, nature and the geography of difference. Wiley-Blackwell.
Hugo G (2008) Migration, development and environment. IOM Migration Research Series, Geneva.
Hugo G (2011) Future demographic change and its interactions with migration and climate change. *Global Environmental Change*, 21:21-33.
IOM (2012) Climate change, environmental degradation and migration. *International Dialogue on Migration*, 18.
Isoda Y (2011) God's taste victims' attributes based on the list of victims. The 2011 East Japan Earthquake Bulletin of the Tohoku Geographical Association.

James H (2012) Social capital, resilience and transformation among vulnerable groups in the Burmese delta after Cyclone Nargis. Conference paper. ARC Discovery Project on 'Demographic Consequences of Asian Disasters: Family Dynamics, Social Capital and Migration Patterns', April 2012, ANU.

Jankó F, Bertalan L, Hoschek M, Komornoki K, Németh N, Papp-Vancsó J (2018) Perception, understanding, and action: attitudes of climate change in the Hungarian population. *Hungarian Geographical Bulletin*, 67(2):159-171.

Jia Z, Tian W, Liu W, Cao Y, Yan J, Shun Z (2010): Are the elderly more vulnerable to psychological impact of natural disaster? A population-based survey of adult survivors of the 2008 Sichuan earthquake. *BMC Public Health*, 10:172.

Johnson NPAS, Mueller J (2002) Updating the accounts: Global mortality of the 1918-1920 "Spanish" Influenza Pandemic. *Bulletin of the History of Medicine*, 76(1):105-115.

Jones DS (2003) Virgin soils revisited. *The William and Mary Quarterly*, 60(4):703-742.

Jones A (2017) Genocide. A comprehensive introduction. 3rd edition. Routledge, London, New York.

Kapuchu N, Özerdem A (2013) Managing emergencies and crises. Jones&Bartlett Learning, Burlington, USA.

Kata A (2010) A postmodern Pandora's box: Anti-vaccination misinformation on the Internet. *Vaccine*, 28(7):1709-1716.

Kendra J, Rozdilsky J, McEntire DA (2008) Evacuating large urban areas: Challenges for emergency management policies and concepts. Journal of Homeland Security and Emergency Management, 5(1) 22 p.

Koch A, Carson D (2012) Spatial, temporal and social scaling in sparsely populated areas—Geospatial mapping and simulation techniques to investigate social diversity. In: Jekel T, Car A, Strobl J, Griesebner G (eds.) GI Forum 2012: Geovizualisation, society and learning. Herbert Wichmann Verlag, Berlin, Offenbach.

Krug EG, Kresnow M, Peddicord JP, Dahlberg LL, Powell KE, Crosby AE, Annest JL (1999) Suicide after natural disasters. *The New England Journal of Medicine*, 338(6):373-378.

Kurosu S, Bengtsson T, Campbell C eds. (2010) Demographic responses to economic and environmental Crises. Proceedings of the IUSSP Seminar. Kashiwa, Japan, Reitaku University.

Lavell A, Oppenheimer M, Diop C, Hess J, Lempert R, Li J, Muir-Wood R, Myeong S (2012) Climate change: new dimensions in disaster risk, exposure, vulnerability, and resilience. in: Field CB, Barros V, Stocker TF, Qin D, Dokken DJ, Ebi KL, Mastrandrea MD, Mach KJ, Plattner GK, Allen SK, Tignor M, Midgley PM eds. Managing the Risks of Extreme Events and Disasters to Advance Climate Change Adaptation A Special Report of Working Groups I and II of the Intergovernmental Panel on Climate Change (IPCC). Cambridge University Press, Cambridge, UK, and New York, NY, USA, pp. 25–64.

Lavell C, Ginnetti J (2013) Technical Paper: The risk of disaster-inducted displacement. Central America and the Caribbean. Internal Displacement Monitoring Centre (IDMC). Geneva. Norwegian Refugee Council. 54 p.

Lechat MF (1979) Disasters and public health.*Bulletin of the World Health Organisation*, 57(1):11–17.

Lemkin R (2012) Lemkin on genocide. Lexington Books, Plymouth.

Levine JN, Esnard AM, Sapat A (2007) Population displacement and housing dilemmas due to catastrophic disasters. *Journal of Planning Literature*, 22(1):3-15.

Lindell MK (2013) Disaster studies. *Current Sociology*, 61(5-6):797-825.

Malone EL (2009)Vulnerability and resilience in the face of climate change: Current research and needs for population information. Prepared for Population Action International. Washington DC.

Martine G, Schensul D eds. (2013) The demography of adaptation to climate change. New York, London and Mexico City: UNFPA, IIED and El Colegio de México.

McLafferty S (2010) Placing pandemics: Geographical dimensions of vulnerability and spread. *Eurasian Geography and Economics*, 51(2):143-161.

McLeman RA, Hunter LM (2010) Migration in the context of vulnerability and adaptation to climate change: insights from analogues. *WIREs Climate Change*, 1(May-June):450–461.
McNeil WH (1976) Plagues and peoples. Anchor Press/Doubleday, Garden City.
Moore N (2001) Political underdevelopment. What causes 'bad governance'. Public Management Review, 3(3): 385–418.
Nadim F, Kjekstad O, Peduzzi P, Herold C, Jaedicke C (2006) Global landslide and avalanche hotspots. *Landslides*, 3:159-173.
Naik A, Stigter E, Laczko F (2007) Migration, development and natural disasters: Insights from the Indian Ocean Tsunami. IOM. Geneva.
Nicoll A, Timaeus I, Kigadye RM, Walraven G, Killewo J (1994) The impact of HIV-1 infection on mortality in children under 5 years of age in sub-Saharan Africa: a demographic and epidemiologic analysis. *Aids*, 8(7):995-1005.
Nobles J, Frankenberg E, Thomas D (2015) The effects of mortality on fertility: Population dynamics after a natural disaster. *Demography*,52(1):15-38.
Noji EK (1995) Disaster epidemiology and disease monitoring. *Journal of Medical Systems*, 19(2):171-174.
Oliver-Smith A (1996) Anthropological research on hazards and disasters. *Annual Review of Anthropology*, 25:303-328.
Oliver-Smith A (2012) Debating environmental migration: Society, nature and population displacement in climate change. *Journal of International Development*, 24:1058–1070.
Oliver-Smith A (2013) Catastrophes, mass displacement and population resettlement. in: Bissel R ed. Preparedness and response for catastrophic disasters. CRC Press, Taylor&Francis Group. Boca Raton, London, New York. pp. 185–224.
Orum P, Moore R, Roberts M, Sáchez J (2014) Who's in danger? Race, poverty and chemical disasters. A demographic analysis of chemical disaster vulnerability zones. Environmental Justice and Health Alliance for Chemical Policy Reform. Report.
Peacock WG, Morrow BH, Gladwin H (1997) Hurricane Andrew. Ethnicity, gender and the sociology of disasters. Routledge, London and New York.
Peduzzi P, Chatenoux B, Dao H, De Bono A, Herold C, Kossin J, Mouton F, Nordbeck O (2012) Global trends in tropical cyclone risk. *Nature Climate Change*, 2:289-294.
Piguet E, Pécoud A, de Guchteneire P (2011) Migration and climate change: an overview. *Refugee Survey Quarterly*, 30(3):1-23.
Plyer A, Bonaguro J, Hodges K (2010) Using administrative data to estimate population displacement and resettlement following a catastrophic U.S. disaster.*Population Environment*, 31:150–175.
Reuveny R (2007) Climate change-induced migration and violent conflict. *Political Geography*, 26:656-673.
Robinson C (2003) Overview of disaster. in: Robinson C, Hill K eds. Demographic methods in emergency assessment A Guide For Practitioners. Center for International Emergency, Disaster and Refugee Studies (CIEDRS) and the Hopkins Population Center; Johns Hopkins University Bloomberg School of Public Health, Baltimore, Maryland.
Schultz J, Elliott JR (2012) Natural disasters and local demographic change in the United States.*Population and Environment* 34(3):293-312.
Scudder T, Colson E (1982) From welfare to development: A conceptual framework for the analysis of dislocated people. in: Hansen A, Oliver-Smith A eds. Involuntary migration and resettlement. Westview Press, Boulder, CO.
Seto KC, Fragkias M, Güneralp B, Reilly MK (2011) A meta-analysis of global urban land expansion. *PLoS ONE*, 6(8):e23777.
de Sherbinin A, Castro M, Gemenne F, Cernea MM, Adamo S, Fearnside PM, Krieger G, Lahmani S, Oliver-Smith A, Pankhurst A, Scudder T, Singer B, Tan Y, Wannier G, Boncour P, Ehrhart C, Hugo G, Pandey B, Shi G (2011): Preparing for resettlement associated with climate change. *Science*, 334(6055):456-457.

Smith AB (1989) Khoikhoi susceptibility to virgin soil epidemics in the 18th century. *South African Medical Journal*, 75(1):25-26.

Smith SK (1996) Demography of disasters: Population estimates after Hurricane Andrew. *Population Research and Policy Review*, 15:459-477.

Smith RW (2014) Genocide denial and prevention. *Genocide Studies International*, 8(1):102-109.

Stough LM, Mayhorn CB (2013) Population segments with disabilities. *International Journal of Mass Emergencies and Disasters*, 31(3):384-402.

Strömberg D (2007) Natural disasters, economic development, and humanitarian aid. *Journal of Economic Perspectives*, 21(3):199-222.

Tacoli C, McGranahan G, Satterthwaite D (2015) Urbanisation, rural-urban migration and urban poverty. IIED-IOM working paper.

Tielidze LG, Kumladze RM, Wheathe RD, Gamkrelidze M (2019) The Devdoraki Glacier catastrophes, Georgian Caucasus. *Hungarian Geographical Bulletin*, 68(1):21-35.

Veenema TG, Thornton CP, Proffitt Lavin R, Bender AK (2017) Climate change-related water disasters' impact on population health. Journal of Nursing Scholarship, 49(6):625-634.

Vidal de la Blache P (1911) Les genres de vie dans la géographiehumaine (Ways of life in the human geography). Annales de Géographie, 111:193-212.

White GF (1945) Human adjustment to floods. Research paper No. 29. Department of Geography, The University of Chicago, Chicago, Illinois.

Wilson R, zuErbach-Schoenberg E, Albert M, Power D, Tudge S, Gonzalez M, Guthire S, Chamberlain H, Brooks C, Hughes C, Pitonakova L, Buckee C, Lu X, Wetter E, Tatem A, Bengtsson L (2016) Rapid and near real-time assessments of population displacement using mobile phone data following disasters: The 2015 Nepal earthquake. PLoS Currents, 8. Published online 2016 February 24. Accessed on 9 December 2019 at https://www.ncbi.nlm.nih.gov/pmc/articles/PMC4779046/?report=printable.

Winsemius HC, Van Beek LHP, Jongman B, Ward PJ, Bouwman A (2013) A framework for global river flood risk assessments. *Hydrology and Earth System Sciences*, 17(5):1871-1892.

Wisner B, Blaikie P, Cannon T, Davis I (2004) At Risk. Natural hazards, people's vulnerability and disasters. Routlege, London-New York.

WHO (2018) Antibiotic resistance. Accessed on 15/12/2019 at https://www.who.int/news-room/fact-sheets/detail/antibiotic-resistance.

World Bank (2005) Natural Disaster Hotspots. A Global Risk Analysis. Disaster Risk Management Series.

Zhou Y, Li N, Wu W, Wu J, Shi P (2014) Local spatial and temporal factors influencing population and societal vulnerability to natural disasters. *Risk Analysis*, 34(4):614-639.

The manufacturer's authorised representative in the EU is Springer Nature Customer Service Centre GmbH, Europaplatz 3, 69115 Heidelberg, Germany. If you have any concerns regarding our products, please contact ProductSafety@springernature.com

Printed and bound by CPI Group (UK) Ltd, Croydon, CR0 4YY
22/09/2026
02220946-0007